Love, Order, and Progress

Love, Order, & Progress

The Science, Philosophy, & Politics of Auguste Comte

EDITED BY
Michel Bourdeau,
Mary Pickering,
& Warren Schmaus

University of Pittsburgh Press

Published by the University of Pittsburgh Press, Pittsburgh, Pa., 15260
Copyright © 2018, University of Pittsburgh Press
All rights reserved
Manufactured in the United States of America
Printed on acid-free paper
10 9 8 7 6 5 4 3 2 1

Cataloging-in-Publication data is available from the Library of Congress

ISBN 13: 978-0-8229-4522-2
ISBN 10: 0-8229-4522-3

Cover art: (*back*) Postivist Temple planned for Rio de Janeiro. Courtesy of the John Hay Library, Brown University. (*fore*) Engraved portrait of Auguste Comte, ca. 1840.
Cover design: Joel W. Coggins

Contents

PREFACE		vii
METHOD OF CITATION FOR AUGUSTE COMTE'S WORKS		xiii
INTRODUCTION: The Significance of Auguste Comte *Warren Schmaus, Mary Pickering, and Michel Bourdeau*		3

I. Comte's Philosophy of Science

1. Comte's General Philosophy of Science — 27
   Warren Schmaus

2. The Analytical Construction of a Positive Science in Auguste Comte — 56
   Michel Blay

3. Astronomical Science and Its Significance for Humankind — 72
   Anastasios Brenner

4. Auguste Comte's Positive Biology — 93
   Laurent Clauzade

5. Comte and Social Science — 128
   Vincent Guillin

II. Comte's Social and Political Thought

6. Comte's Political Philosophy — 163
   Michel Bourdeau

7. Art, Affective Life, and the Role of Gender in
   Auguste Comte's Philosophy and Politics   190
   Jean Elisabeth Pedersen

8. The Religion of Humanity and Positive Morality   217
   Andrew Wernick

   CONCLUSION: The Legacy of Auguste Comte   250
   Mary Pickering

   NOTES   305
   BIBLIOGRAPHY   355
   LIST OF CONTRIBUTORS   389
   INDEX   393

Preface

The idea for this book was born in June 2008 at the biennial meeting of the International Society for the History of Philosophy of Science (HOPOS) in Vancouver. The editors of the present volume participated in a symposium on the philosophy of science in nineteenth-century France, focusing largely on positivism. We were struck at the time by the contrast between the obvious interest in the views of Auguste Comte, which was often expressed in a highly critical form, and the very few publications in English devoted to this prominent French philosopher. It was clear to us that there was a lacuna waiting to be filled, and we told ourselves that a general volume presenting Comte's work to an Anglophone public could help people understand the long-standing importance he has had in the philosophical world. The project took off and a few years later, during the 2012 HOPOS meeting in Halifax, there was a Comte symposium, where some of the presented papers were early forms of the chapters in this book.

Comte deserves attention for several reasons. The first has to do with the recent political turn in the philosophy of science. Philosophers of science today are considering such questions as how scientific funding decisions should be made in a democratic society and how more widely diverse points of view can be represented in setting research agendas and theory choice. Concurrently, scholars have been unveiling the political motivations of what has come to be called the Left Vienna Circle: Otto Neurath, Rudolf Carnap, Hans Hahn, and Philipp Frank. However, their brand of logical positivism left the status of political values unclear. For Comte, on the other hand, politics and philosophy of science were intimately connected, given that he sought a method to establish a scientific politics.

Since his early years, Comte displayed much interest in both science and politics. A gifted mathematician, he was well versed in the sciences, having attended the prestigious École Polytechnique in Paris. In 1817, a year after he left the school, he started to work for the reformer Henri Saint-Simon, who guided him in the direction of scientific politics. During the seven-year period of their collaboration, Comte concentrated on two phenomena. The first was social. The French Revolution of 1789 had shaken the foundations of society. With the end of feudalism, an entire system of social relations had ceased to exist, and with the development of industry, other, very different relationships were beginning to take their place. What was the nature of these relationships? The second phenomenon that preoccupied him was related to changes in industry and science. As the industrial revolution was the daughter, or sister, of the scientific revolution, it changed the status of scientists, who henceforth fulfilled an essential social function. Evaluating the role of science in society became crucial, and this task made it necessary to consider scientists themselves as an object of study. This field of inquiry quickly became incorporated into his philosophy of science. Comte was perfectly aware of the novelty of his approach to science and its practitioners.

Comte's work is thus unique because it stands at the intersection of two great axes that run through the history of philosophy. Ever since the ancient Greeks, philosophers have treated questions of knowledge separately from questions of political and social organization, even devoting different works to them, such as Aristotle's *Posterior Analytics* and *Politics*. This trend continues through works such as John Locke's *Essay Concerning Human Understanding* and *Two Treatises of Government*, which bear only the most tenuous relation to each other. Philosophical questions concerning knowledge and society come together in Comte, only to split again into separate traditions. To the first belong philosophers of science, such as Ernst Mach, Henri Poincaré, Pierre Duhem, Rudolf Carnap, and Thomas Kuhn. The second includes social thinkers, or sociologists, such as Alexis de Tocqueville, Karl Marx, and Émile Durkheim.

Comte may be viewed as one of the first philosophers of science in the modern sense because, like John Herschel, William Whewell, and John Stuart Mill in England, he made philosophy of science distinct from science. In earlier centuries, in the works of writers like René Descartes and Gottfried Wilhelm Leibniz, for instance, science and philosophy of science were combined even in the same books, such as the former's *Discours de la méthode* of 1637. Comte's work can also be compared with more recent trends in the philosophy of science. In spite of the fact that positivism has more recently come to be associated with reductionism,

Comte embraced the disunity of the sciences, having explicitly rejected the goal of subsuming all natural phenomena under a single law or set of laws as "chimerical." The only unity of science that mattered to Comte was the unity of method. However, in his characterizations of the methods of mathematics, astronomy, physics, chemistry, biology, and sociology, what they appear to share in common can be described only in the most general terms. It is the differences that stand out. Thus Ian Hacking claims Comte as one of his inspirations for the idea that there are various "styles" of scientific reasoning.[1]

In social thought, Comte was less successful in leaving his mark. One could argue that he pondered the sciences more profoundly than any other nineteenth-century thinker who argued for a scientific reorganization of society. Marx comes across as naïve about scientific methods. He says little or nothing about the natural sciences, leaving that task to Friedrich Engels. What Engels has to say about the sciences in works like *The Dialectics of Nature* is downright embarrassing. And yet his works and those of Marx have retained their strong place in the canon. In contrast, Comte's contributions to social science have been largely forgotten. For instance, there is a lot more of Comte in Durkheim than most people realize. Even the idea of tracing social institutions back to their origins in primitive societies—the method of "historical analysis"—is Comtean.[2] The notion that primitive societies have a different "logic" has its origins in Comte's account of fetishism. The concept was popularized by Lucien Lévy-Bruhl, an influential interpreter of Comte, in the early twentieth century, and it survived until at least the 1970s. But its roots in Comte's thought are rarely mentioned, if at all.[3]

As for his political philosophy, positive politics is without a doubt the part of his work that is most neglected today. It is also the part that has the worst reputation. If one were to believe Friedrich Hayek, Comte did everything wrong. Hayek suggested that Comte was a precursor of a reactionary totalitarianism, despite the fact that the positivists usually eschewed nostalgia for the past and worked diligently to promote progress.[4] For example, on the question of colonialism, Comte compares very favorably to Tocqueville.[5] He was not a strident nationalist or imperialist. With its dependence on sociology, positive politics aimed to bring people together in harmony, not to polarize the world.

The second reason for a revival of interest in Comte is historical. The word *positivism* is bandied about quite freely these days, but its meaning remains controversial. Today among philosophers, the word is usually associated with Carnap and other members of the Vienna Circle; hence, the term *postpositivism* should be used to refer to what would be better named *postneopositivism*. As a result, *positivism* is often employed to refer to ideas that are just the opposite of what Comte and other positivists have meant. For instance, in everyday language, the word *positivist* has become more or less synonymous with "blind admirer of science," which Comte, for sure, was not. Positivism is also often confused with scientific realism. At least a third meaning was introduced by Talcott Parsons in American sociology. He deemed Durkheim's *De la division du travail social* to be his most "positivist" work, considering that it treated social actors as thinking scientifically and making self-interested decisions in light of material conditions.[6] Parsons clearly departed from Durkheim's original intent as well as Comte's concept of positivism. This concept and its roots deserve renewed scrutiny.

Hayek is among those who regarded Comte as the incarnation of scientism, and yet, despite his antipathy, he aptly pointed out over a half century ago that for a long time Comte exerted an influence the importance of which we have a hard time measuring today. The same Hayek did not hesitate to compare him to Georg Hegel. If it is true that they are separated by the distance between the Stift (Lutheran seminary) of Tübingen and the École Polytechnique of Paris, the comparison is legitimate. These are two encyclopedic thinkers who constructed elaborate philosophies of history and made great appeals for a synthesis. In spite of his attacks against "polytechnician *hubris*," Hayek's preferences were clearly in favor of Comte.[7] From this point of view, the difference in the respect that one gives to Hegel and Comte today appears to be a deep injustice, which this work hopes to help correct.

As readers will quickly recognize, the essays collected in this volume rely on the prodigious efforts by Comte scholars during the past few decades. After diverse events organized in 1998, the year of the bicentennial of Comte's birth, there has been a true renaissance of Comtean studies, and although there is no one philosophical book presenting Comte to an English-speaking audience, there exist many outstanding works to which this volume owes a great deal. Besides the works of the contributors to this volume, we should cite a few titles of scholarly books published during the past fifteen years: Jean-François Braunstein, *La philosophie de la médecine d'Auguste Comte* (2009); Gregory Claeys, *Imperial Sceptics: British Critics of*

Empire, 1850–1920 (2010); Claudio De Boni, *Storia di un'utopia, la religion dell'Umanità di Comte e la sua circulazione nel mondo* (2013); Thomas Dixon, *The Invention of Altruism* (2008); Mike Gane, *Auguste Comte* (2006); and Annie Petit, *Le système d'Auguste Comte* (2016), as well as *Auguste Comte: Cours sur l'histoire de l'humanité (1849–1851), Manuscrit de César Lefort*, edited by Laurent Fedi (2017).

Apart from the work of Annie Petit, which appeared while this volume was in preparation, the above-mentioned books each treat only one aspect of Comte's thought. The titles alone reveal that they cover a wide range of disparate topics, including Comte's thoughts on medicine, imperialism, religion, and altruism. This is true of so much of the secondary literature on Comte that we thought it more appropriate to discuss it in each of the relevant chapters, rather than in the preface or introduction to this anthology.[8] It is necessary to give an in-depth overview of the entirety of Comte's thought, one destined for the English-speaking public at large. That is what we are trying to do. Our objective is to give a more faithful image of Comte, rather than a new one. However, prejudices against him are so strong, caricatures so ludicrous that the image we will fashion will inevitably be different from the current one and will hopefully be more accurate. Already the few examples that we have given in the preface show that it is crucial to revisit our idea of positivism.

We would like to end by acknowledging the help of various people who made this project possible. First we would like to thank Vincent Guillin for translating the contributions of Michel Blay, Michel Bourdeau, and Laurent Clauzade. This work required hours of labor for which we are most grateful. We would also like to thank those who attended our HOPOS sessions in 2008 and 2012 for their helpful comments, as well as those of the anonymous reviewers of earlier drafts of this work. Finally, we owe a special debt of gratitude to Abby Collier for her guidance and help in bringing this project to fruition.

Method of Citation for Auguste Comte's Works

The same method of citation for Comte's works will be used throughout all the chapters in this book. All translations from French to English will be our own, unless otherwise indicated.

There is no complete translation of Comte's first major work, the *Cours de Philosophie Positive*, which consists of sixty lessons published in six volumes from 1830 to 1842. All references will be to the two-volume edition edited by Michel Serres et al. and published in 1975. After the volume number, in order to help readers who may be using another edition the lesson number will be included following the capital L, followed by the page number. Thus, for instance, the citation "*Cours*, I, L 45, 856" refers to volume 1, lesson 45, page 856 of the *Cours*. *Introduction* refers to the English translation of the first two lessons by Frederick Ferré, which is cited only for the convenience of the reader. *Positive Philosophy* refers to Harriet Martineau's condensed and free translation, first published in 1853, which is cited where possible and appropriate.

Comte's second major work is the *Système de politique positive*, published in four volumes from 1851 to 1854. "Trans." refers to the English translation of the *Système* by John Henry Bridges et al., originally published in London in 1875–1877. This translation is divided into four volumes in the same way as the original. Thus the citation "*Système*, I, 58, trans. 45" refers to volume 1, page 58 of the French original and volume 1, page 45 of the Bridges et al. translation. However, translations of passages from the *Système* are our own, unless otherwise indicated, and references to this translation are cited only for the reader's convenience.

Comte included some of his earliest writings in the appendix to the fourth volume of the *Système* in order to persuade his readers of the continuity of his thought from the 1820s through the 1850s. *EPW* refers to H. S. Jones's more recent translations of Comte's early works in *Early Political*

xiii

Writings. Thus "*Système*, IV, appendix, 217, trans. 646; *EPW*, 229" refers to page 217 of the appendix to volume 4 of the original French; volume 4, page 646 of the Bridges et al. translation; and page 229 of Jones's translation. However, once again, all translations of quotations from Comte are our own.

Abbreviations Used in This Work

Correspondance	*Correspondance générale et confessions*, 8 volumes, 1973–1990
Cours	*Cours de philosophie positive* (1830–1842)
Écrits de jeunesse	*Écrits de jeunesse, 1816–1828* (1970)
EPW	*Early Political Writings* (1998)
L'esprit	*Discours sur l'esprit positif* (1844)
Évolution	*Auguste Comte. Évolution originale* (1813)
Introduction	*Introduction to Positive Philosophy* (1988)
Positive Philosophy	Martineau's *The Positive Philosophy of Auguste Comte* (1853)
PTS	*Plan des travaux scientifiques nécessaires pour réorganiser la société* (1824)
Synthèse	*Synthèse subjective* (1856)
Système	*Système de politique positive*
Traité	*Traité philosophique d'astronomie populaire* (1844)

Love, Order, and Progress

Introduction: The Significance of Auguste Comte

Warren Schmaus, Mary Pickering, and Michel Bourdeau

Despite Comte's remarkable influence, there has been very little published in recent years on his thought, especially in English. Perhaps the best general overview in English of Comte's entire philosophy is still the 1903 translation of Lucien Lévy-Bruhl's *The Philosophy of Auguste Comte*. The present anthology aims to help correct this oversight. Such a volume is particularly timely given the recent political turn in the philosophy of science.

Philosophers of science such as Philip Kitcher and especially feminist philosophers of science such as Helen Longino have recently turned to questions of how science should be organized to better serve human needs. Kitcher envisions an enlightened public playing a larger role in science policy making, while Longino would like to democratize not just science policy but scientific decision making itself, permitting more different points of view to be expressed.[1] At the same time, recent scholarship in the history of philosophy of science by Don Howard, George Reisch, and Thomas Uebel is revealing the political and social motivations of members of the Vienna Circle of logical positivists such as Otto Neurath, Rudolf Carnap, Hans Hahn, and Philipp Frank.[2] Sarah Richardson, on the other hand, questions whether the logical positivists can serve as a model for a political philosophy of science today.[3] The debate concerns whether the logical positivists just happened to be involved in the politics of their time or whether there was any connection between their politics and their philosophy of science.

The connection between political philosophy and philosophy of science was clearer at a much earlier stage in the history of positivism. Auguste Comte (1798–1857) conceived of philosophy of science as part of his political project. He saw the need for a new system of ideas to ensure peace, stability, and progress in the wake of the social upheaval resulting from the French Revolution and the Napoleonic Wars. Drawing on the empiricist tradition in philosophy, he articulated the positivist philosophy of science and used it to provide a methodology for a new empirical science of society that he called "sociology." This new science had two parts: social statics, which was concerned with social solidarity, and social dynamics, which was concerned with social development or progress. This division reflected Comte's motto, "Order and Progress." As he saw it, sociology would provide the intellectual foundation on which to build a new kind of society that would promote the general welfare and restore order in the wake of the collapse of the old monarchical and religious regime. Enlightened industrialists, including manufacturers, merchants, and financiers, would replace kings, aristocrats, lawyers, and the military in policy making and governing. Sociology would supersede religion and metaphysical philosophy as the basis for morality. Education would be removed from church control and placed under the direction of an elite class educated in all the sciences, including the science of society. These sociological philosophers would guide public opinion.

The study of Comte shows how not only intellectual developments but political and social motivations contributed to the emergence of the philosophy of science as a separate subdiscipline within philosophy. Philosophers and scientists had always discussed the sorts of issues that philosophy of science deals with, such as the roles of induction and deduction in the production of knowledge, the proper use of hypotheses in science, and the question of whether the theories and concepts of science should be regarded as describing reality or as merely useful tools for making predictions to guide practical applications. Two circumstances were responsible for bringing all these topics from metaphysics, logic, and what is now called epistemology together under the rubric of philosophy of science in the nineteenth century. For some English philosophers such as John Herschel and William Whewell, it was reflection on the surprising growth and development of the sciences themselves. This growth and development included a division of intellectual labor in which philosophy separated from science—it was Whewell, after all, who gave us the word *scientist*—and philosophy and science each divided into various specialties. For others, including Comte and John Stuart Mill, it was the wish to extend a secular, scientific way of thinking to social questions, an aspiration with sources in the Enlightenment. In Mill's case, Thomas Babington Macaulay's attacks on his father, James Mill's, deductive methods in his political writings led to the younger Mill's desire to develop empirical methods for the social sciences that could be used in political arguments, methods he articulated in *A System of Logic* (1843), drawing on his reading of Comte's *Cours de philosophie positive* (1830–1842) and Herschel's *A Preliminary Discourse on the Study of Natural Philosophy* (1830). Comte was responding at least in part to Saint-Simonian socialism, which sought a scientific organization of society, but with an insufficient sense of what science is, a problem subsequently shared by many Marxist scientific socialists.

Ultimately, Comte's philosophy is rooted in the French Revolution's attempt to replace a theologically grounded monarchy with a secular society serving the general interest. Many of the faculty and students at the École Polytechnique, France's elite engineering school in Paris, thought that the sciences could provide the intellectual basis for such a society. Comte was one of many students there who were attracted to Henri de Saint-Simon's politics. Other *polytechniciens* were attracted to Fourierism.[4] Comte's new science of sociology was to provide the intellectual basis for a new scientific organization of society, much as the political philosophies of thinkers like Thomas Hobbes, John Locke, and Jean-Jacques Rousseau had provided the basis for an earlier generation of political experimentation. The goal of Comte's positivist philosophy of science was to establish a method of empirical investigation appropriate to this new science.

In his popular introduction to his philosophy, the *Discours sur l'esprit positif*, Comte explained that he called his system the "positive" philosophy because of the positive role that it was to play in building a new society, in contrast to what he regarded as the negative philosophy that could only criticize and destroy the older regime of church and state.[5] The other meanings he applied to *positive* are: "useful," "certain," "precise," "relative," "organic," "sympathetic," and "real," as opposed to "chimerical," a pejorative term that he often applied to metaphysical and unverifiable hypotheses in science.[6]

These seven characteristics of positivism are conceptually linked with one another. When Comte characterized the positive philosophy as "relative," he meant that it seeks knowledge relative to the satisfaction of our needs rather than absolute knowledge for its own sake. Hence, it seeks "useful" knowledge, knowledge that has practical applications. Such knowledge must be "real," or verifiable, as well as "precise." Although our theories may become increasingly precise over time, they never achieve the exact truth but only approach this ideal limit to the degree required by the satisfaction of our needs.[7] Similarly, the degree of certainty that is called for is only one of practical certainty, and not the absolute certainty of unshakeable Cartesian foundations. Positive science seeks this certainty regarding the laws governing observable phenomena that have some practical bearing on our lives, rather than the ultimate causes of things.[8] The emphasis on practical knowledge is linked to the task of organizing or rebuilding society; hence the epithet "organic." The term *sympathetic* has to do with the moral sentiment of sociability or sympathy, which Comte believed would keep scientists focused on questions that concern human needs. The role of these social sentiments in modifying our egoistic instincts is finally understood when sociology becomes a positive science, making possible a new, positive morality.[9]

Educated in mathematics, the physical sciences, and engineering rather than in philosophy, Comte brought a fresh perspective to problems of knowledge, as Warren Schmaus explains in chapter 1. French academic philosophy at that time was dominated by the eclectic spiritualist tradition founded by Victor Cousin, drawing on the work of Pierre Royer-Collard and Pierre Maine de Biran. This eclectic philosophy blended Cartesian rationalism with the Scottish commonsense philosophy of Thomas Reid and was taught to students throughout France by professors who had pursued a purely humanistic, literary education, grounded in the classics and completely divorced from the sciences. Eclectic spiritualism sought a foundation for all of philosophy, including logic, metaphysics, and ethics, in an introspective study of the human mind. Comte, on the other hand, thought that philosophy should be based on a study of the best examples we have of claims to knowledge, mathematics, and the sciences, in their historical development. Comte's intellectual project in the first three volumes of his *Cours de philosophie positive* thus resembles Whewell's, who at about the same time based his *Philosophy of the Inductive Sciences* (1840) on his *History of the Inductive Sciences* (1837).

But perhaps more than his contemporaries in the philosophy of science, Comte recognized the social character of science. His turn from introspective methods to the study of the history of science as a basis for epistemology represents a rejection of individualistic approaches and reflects his belief that scientific knowledge is a social product rather than the work of an isolated Cartesian genius. He understood how scientists depend upon a community of researchers for input and criticism. He was also concerned with conditions in wider society that would allow for continued scientific progress and with the role that science should play in society. Scientists, with input from society at large, should work for the common good, but they should also direct the education of ordinary citizens in order that they have scientifically informed views of what is in fact good for them. Critics may disagree with Comte regarding the extent to which he thought the sciences should be directed at the solution of practical problems, but Comte was the first important thinker since Francis Bacon to direct philosophers' attention to the social role of science.

Although historical antecedents of Comte's positivism can be found in the works of the British empiricists George Berkeley and David Hume, Comte was the earliest philosopher to show how it worked out in detail in the various natural sciences. He understood that the development of new sciences involved the development of new methods of inquiry. According to his well-known law of three stages or states (*la loi des trois états*), as each of the sciences of mathematics, astronomy, physics, chemistry, biology, and sociology in succession passed through the theological and metaphysical stages and finally reached the positive state, it made its unique contribution to what he considered the positive method. Mathematics contributed the method of analysis; astronomy contributed observation and hypothesis; physics and chemistry, experimentation; biology, comparison; and finally sociology would develop the method of comparison into the historical method, as Schmaus will explain in greater detail in chapter 1.

Ian Hacking claims that Comte's views about how each science contributes to the positive method provide part of the inspiration for his notion of styles of scientific reasoning.[10] As Hacking sees it, it is not even possible for one to think certain ideas in science until the appropriate method or style of reasoning has developed. To take a Comtean example, in the absence of the method of hypothesis it would make no sense for a Copernican to talk about the earth orbiting the sun. Hypotheses are needed to construct the very phenomena under study in astronomy. Without hypotheses, there are no orbits; there is nothing more for astronomy than points of light in the sky.

Although critics such as Whewell, Charles Renouvier, and Antoine Augustin Cournot were able nearly right from the start to point out where Comte got his history of science wrong, thanks to Comte the history of science was now the field on which epistemological battles were to be fought, particularly in France, where his influence was especially pronounced. Comte's positivism provided the intellectual context for French philosophy of science in the second half of the nineteenth century, whether providing a starting point for proponents of positivism such as Émile Littré and Lévy-Bruhl or a target for criticism for philosophers such as Renouvier and Émile Boutroux in the development of their own philosophies of science.

Prior to Comte, there was in fact very little history of science written. Most of it was written by scientists themselves, either as part of a polemic aimed at persuading the public that scientific research was deserving of support by society because of the societal benefits it provides, or as the introductory chapter of a scientific work placing it within a tradition, serving a function something like that of the contemporary review of the literature. There were a few works devoted to the history of the mathematical sciences, such as Jean-Étienne Montucla's history of mathematics and Adam Smith's history of astronomy. But Comte's historiography of science was new in two ways. First, where Enlightenment thinkers emphasizing mathematics presented the history of science as reflecting the steady progress of the human mind, Comte, though also a firm believer in progress, showed more of a historical sense, attempting to understand the past on the basis of the standards of that time. Thus, for instance, he did not dismiss alchemy or astrology as mistaken, but regarded them as appropriate systems given what was known at the time. Similarly, he refused to regard the medieval period as a backward step in civilization. Second, Comte emphasized how the sciences interacted with each other as well as with the larger culture and society. By way of contrast, Whewell, writing at approximately the same time, did not attempt to understand the sciences in their historical context but was more interested in reviewing the history of science in order to glean methodological lessons from it for improving current science.[11]

Of course, one could argue that Comte's history of science does not meet contemporary standards of scholarship. He did not always consult the relevant primary sources in the history of science in their original language or visit archives. But to be fair, it must be pointed out that the standards of historical scholarship with which we are now familiar were only beginning to be developed in the nineteenth century. Regardless of Comte's unreliability as a historian of science, he nevertheless inspired a French tradition of historical epistemology. Later in the century, Comte's disciples Pierre Laffitte and Grégoire Wyrouboff were the first two occupants of the chair of history of science at the Collège de France. Even philosophers such as Gaston Bachelard and Georges Canguilhem, who turned against positivism, gave paramount importance to the history of science.

Through his influence on people like Lévy-Bruhl and Émile Durkheim, Comte has had at least as much impact on the social sciences as he has had on philosophy, especially in France. In sociology and anthropology, even those who would reject other aspects of the positive philosophy nevertheless maintain Comte's idea that there is a level of explanation in the social sciences that is distinct from and does not reduce to explanations of individual behavior. Also, the method of historical analysis he considered appropriate for sociology is reflected in the works of Durkheim, Lévy-Bruhl, Marcel Mauss, and Claude Lévi-Strauss, who thought that the way to understand contemporary culture, religion, and conceptual thought was to trace the origins of these complex forms back to their simple origins in so-called primitive societies. This social science methodology assumes that our present-day complex social phenomena have been compounded over time from simple elements, which can be revealed through historical study and through ethnographic studies of contemporary societies that are regarded as stand-ins for our earliest ancestors: hence books such as Durkheim's *The Elementary Forms of Religious Life* (1912) and Lévi-Strauss's *The Elementary Forms of Kinship* (1949). Michel Foucault's archaeology of knowledge represents a synthesis of this ethnological tradition with the historical epistemology also inspired by Comte.

The influence of Comte's political and social philosophy has been less on academic philosophy than on practical politics. As Mary Pickering explains in the conclusion, the secular, anticlerical aspects of Comte's doctrines appealed to the left, while the right liked his authoritarianism. In France, his influence on the liberal Third Republic as well as the far-right Action Française was particularly noteworthy. The Young Turks, such as Mustafa Kemal Atatürk, the first president of Turkey; Czechs, such as

Tomás Masaryk, the first president of Czechoslovakia; as well as Russians, Poles, and Indians embraced positivism as the key to the modernization of their respective societies. American progressives, such as Herbert Croly, used positivist ideas to make liberalism more favorable to a managerial elite and an interventionist state. Comte's political influence was perhaps greatest in Latin America. His philosophy offered the rising middle class a vision of orderly economic, social, and political modernization and a liberation from their Catholic and colonial past. Among Latin America nations, positivism had the most impact in Brazil. Comte's motto, "Order and Progress" (*Ordem e Progresso*), is still on its flag today, and there is a church of Comte's Religion of Humanity that is still active in Rio de Janeiro. Comte's doctrines also shaped politics and society in Mexico, Argentina, Uruguay, Paraguay, Peru, Bolivia, Venezuela, Guatemala, Costa Rica, El Salvador, Honduras, and Nicaragua.

We are very far from claiming that Comte's philosophy provides a blueprint for contemporary feminists and others who wish to pursue a political philosophy of science. For one, Comte was responding to very different social conditions from those we face today. In addition, Comte's political philosophy taken as whole may be too authoritarian for anyone to recommend. Nevertheless, there are some elements in it worth thinking about that are relevant to discussions of the role of science in society today. For instance, Comte saw the need for leaders to be educated in the history, philosophy, and sociology of science, given the important role of science and technology in contemporary society, which had already become apparent in the nineteenth century and is increasingly so today. However, in many countries today, including the United States, numerous political leaders who are wholly ignorant of science and how it works, both intellectually and socially, nevertheless make policy on issues related to science and technology, including energy, climate change, science education, stem cell research, the space program, genetically modified food crops, and funding for scientific research.

Since Comte also recognized that a government should rest on public opinion and not on force, he thought that it was important that the public be scientifically educated. Unfortunately, he never spelled out a political process by which this public could exercise control over its leaders in government. How to square the notion of a society guided by scientific methods with the idea of individual rights was left to later philosophers such as Renouvier, and remains an important philosophical question. Nevertheless, Comte took the need for a scientifically educated public very seriously, presenting for seventeen years free public lectures on the historical development of astronomy, hoping to introduce the methods of science to ordinary working people in this way. This project could still serve as a model today, when few college students, let alone the general public, have any understanding of the reasoning behind the adoption of many of the scientific theories they are asked to learn.

Comte's Life and Work

In the 1860s, two of Comte's disciples, Émile Littré and Jean-François Eugène Robinet, wrote key accounts of his life and works.[12] Later in the nineteenth century, prominent intellectuals such as John Stuart Mill, John Morley, Thomas Huxley, and Lévy-Bruhl published important analyses of his philosophy.[13] Much of the work on Comte in the twentieth century appeared as part of an overview of philosophy or sociology, in works by scholars such as Émile Bréhier, Raymond Aron, Friedrich Hayek, and Frank Manuel.[14] One notable exception was Henri Gouhier's three-volume study of Comte's early life, *La jeunesse d'Auguste Comte et la formation du positivisme*, which came out in 1931 to great acclaim. That same year he wrote *La vie d'Auguste Comte*.

The definitive intellectual biography of Comte is the three-volume work by Mary Pickering (published 1993–2009). Here we can provide but a brief overview of his life. Comte was born in 1798 in Montpellier, France. His father was a civil servant who worked in the tax collector's office. His mother was extremely religious. Deeply influenced by the French Revolution (1789–1799), Comte rejected at a young age the royalism and Catholicism of his traditional, bourgeois parents. He adopted the republicanism and social idealism of the revolutionaries and disliked the militarism of Napoleon. He became a rebel thanks in part to the influence of his republican teachers at his high school, who already noted his brilliance, especially in mathematics. In 1814 he entered the École Polytechnique. There he learned the role the sciences could play in improving social conditions, which became one of his main goals. He sought above all to reestablish the imagined harmony and stability that society had lost during the French Revolution. However, his rebel nature soon got the best of him—he was expelled in 1816 for insubordination.

Eager to join the elite of scientists and philosophers committed to helping the common people, in 1817 Comte began to work as a writer for the social reformer Henri de Saint-Simon. Years before, Saint-Simon had called for the creation of a new unified system of scientific knowledge that would include the study of society. This system, which he at times referred to as "positive philosophy," would lead to a new stage of history, where capable industrialists and scientists would replace do-nothing military leaders and priests. Comte devoted himself to realizing this goal, which by 1817 Saint-Simon had largely abandoned in favor of more practical projects. Unlike Saint-Simon, Comte had scientific training and a systematic, disciplined mind. He would develop Saint-Simon's ideas and achieve an originality of his own, blending arguments from left-leaning thinkers, such as Condorcet and the Idéologues, and conservative writers, such as Joseph de Maistre. Because of intellectual and generational differences, Comte broke with Saint-Simon in 1824, accusing his mentor of trying to take credit for his seminal essay, the *Plan des travaux scientifiques nécessaires pour réorganiser la société* (1824).

After this rupture, Comte gave private lessons in mathematics and wrote articles for various journals. In late 1825 and 1826 he wrote two series of articles for the Saint-Simonians' journal, *Le Producteur*, where he developed his idea of what he called "spiritual power." Asserting the importance of shaping opinions and ideas in modern societies, Comte argued that scientists with knowledge of all the sciences, including the science of society, should represent the new spiritual power, shaping the education of both children and adults and thus countering the bureaucratic despotism and materialism of temporal power, the men in charge of political life and practical activities. From this point on, he devoted himself to forming this new secular priesthood.

In 1826 Comte decided to give a lecture course synthesizing scientific knowledge—that is, the positive philosophy that this priesthood would espouse. The first lecture was attended by many prominent scientists, including Joseph Fourier, Alexander von Humboldt, Henri Ducrotay de Blainville, François Broussais, and François Arago. But after several lectures, Comte had a mental breakdown. Suffering from manic depression and paranoia, he spent eight months in Dr. Jean-Étienne Esquirol's asylum before being discharged as uncured. His wife, Caroline Massin, then helped him recover at home. The daughter of provincial actors, Massin had run a reading room in Paris before they married in 1824. Despite her best efforts, Comte battled mental illness on and off for the remainder of his life.

In 1838 he experienced another mental crisis and adopted a regime of "cerebral hygiene." Desirous to maintain his own sense of originality and to avoid learning of attacks from critics, he stopped reading contemporary books, journals, and newspapers. To relax, he developed an interest in music and poetry, which led to his "aesthetic revolution." He began to argue for the importance of the arts in cultivating the feelings and enhancing one's comprehension of society.

While struggling with keeping his sanity, finding a suitable professorial position, and maintaining his relationship with his wife (whom he found difficult), he published six volumes of his most famous work, the *Cours de philosophie positive*, from 1830 to 1842. The first three volumes promulgated his positivist philosophy of science, and the remaining volumes, beginning with the fourth volume published in 1839, introduced and named his new science of sociology. The law of three stages, which he had first mentioned in the *Plan des travaux*, and his sixfold classification of the sciences provided the organizing principles for his philosophy of science, which will be discussed in the next several chapters. Society, as a reflection of the reigning philosophy, also went through three stages, marked by changes in temporal and spiritual powers. In the theological stage, military leaders and priests ruled. In the metaphysical stage, lawyers and metaphysicians dominated. In the positive stage, industrialists and positive philosophers would be in control. Comte concluded that once the study of society became a science, the whole system of knowledge would be "positive," which meant certain, precise, real, constructive, useful, and relative. Knowledge would be homogeneous and unified in the sense that all branches would have the positive scientific method and a common object of study, that is, the betterment of society. People would agree on the most basic ideas. Intellectual harmony would help bring about social harmony. Society would at last regain the stability it had lost during the French Revolution.

The *Cours* attracted the attention of many thinkers throughout the West. However, partly because it was not a work of specialization, it could not help Comte obtain a job as a professor or gain entrance to the Académie des Sciences or the Collège de France, which he thought he deserved. Instead, he became a lowly teaching assistant in mathematics and an admissions officer at the École Polytechnique. He worked there from 1832 to 1851. To further develop his philosophy of mathematics from the *Cours* and show how it could be applied to the teaching of mathematics, he published the *Traité élémentaire de géométrie analytique* in 1843. This work was addressed to professors of mathematics, encouraging them to teach their subject in a historical manner. Rather than teach students only currently accepted methods, Comte wanted them to become familiar with older methods of geometric analysis so they could see their limitations and understand how newer methods were developed to solve more difficult problems.[15] However, since this work contained some material from his mathematics lectures at the *École Polytechnique*, it was perceived as a text that a student could use to prepare for the entrance examination, and thus its publication went against the rules of his employment in the admissions office. This breach as well as his uncongenial behavior and inattention to his teaching duties led to his dismissal from the *École Polytechnique* in 1851. Émile Littré, his principal French follower, organized the "Positivist Subsidy" to enable Comte to live on contributions from his disciples and admirers. But like Comte's main English follower, John Stuart Mill, who corresponded with him from 1841 to 1846, Littré eventually broke with the founder of positivism because of petty arguments over money. Doctrinal disagreements and personality clashes also contributed to this rupture, which hurt Comte's reputation because Littré was a notable intellectual.

Despite these setbacks, Comte commanded a great deal of respect. Large audiences of over four hundred people attended his free lectures on astronomy between 1831 and 1848. These lectures were so successful that he published them in 1844 as the *Traité philosophique d'astronomie populaire*. The opening lectures were published separately that same year as the *Discours sur l'esprit positif*, a short summary of his main principles that was also intended to spread positivism after he realized that the six volumes of the *Cours* were too daunting for most people. In addition, he met several times a week with disciples and admirers in his apartment at 10, rue Monsieur le Prince and carried on a voluminous correspondence with people throughout Europe and the Americas. Besides Littré and Mill, his prominent correspondents and friends included the Count of Limburg Stirum (Menno David), Charles Robin, Henri Ducrotay de Blainville, Pierre Proudhon, Armand Barbès, Antoine Étex, Pierre Laffitte, Julia Ward Howe, Fanny Wright, Alexander Wilkinson, George Henry Lewes, George Grote, John and Sarah Austin, and Harriet Martineau. The latter freely translated the *Cours* into English and condensed it into two volumes in 1853.

Martineau resisted his pleas to translate the *Système*, and their correspondence soon ceased. So did the exchange of letters with many of these other individuals. It is clear that Comte always had problems managing personal relationships. Such problems stemmed from his intense egoism, paranoia, and devotion to his work. His case is paradoxical, given that he founded sociology partly to create harmony in the social world—harmony that was lacking in his private life. His parents, sister, friends, colleagues, and followers found him difficult, as did his wife, who left him in 1842.

Three years after separating from his wife, Comte met Clotilde de Vaux, the sister of one of his favorite students. Seventeen years younger than Comte, she had been abandoned by her husband, who had fled France to escape gambling debts. Dependent on her parents, she sought to launch a career as a writer to liberate herself. Comte flattered her with his attention. She was intelligent, witty, and forceful, much like his wife. They grew close, and Comte enjoyed his new role of mentor. Indeed, his experience with her later inspired him to write in 1852 the *Catéchisme positiviste*, which consists of conversations between a positivist priest and a woman and seemed to target a female audience.

However, Comte's behavior soon went beyond what de Vaux wanted: discussions about intellectual matters. As Comte made unwanted sexual advances, de Vaux found herself in an awkward situation, as she needed his moral support for her career as well as his money. She was suffering from tuberculosis, and Comte's financial assistance enabled her to buy medicine. Nevertheless, she resisted his demands for sex, inadvertently contributing to her image as a model of purity in Comte's mind. After publishing one short story in a prominent journal and almost completing a short novel, she died of her illness in 1846, about a year after they had met. Stricken with grief and fearful of another mental breakdown, Comte made her into his muse and claimed she inspired his new religion, the Religion of Humanity. In this way, he sought to make her immortal, at least in the memory of posterity. At the same time, his increased hatred for his wife led him to accuse her in his will of having been a prostitute when they met. Massin has suffered from this allegation ever since. However, these characterizations reflect nineteenth-century moralizing tendencies and binary thinking: women were either good or bad. Just as de Vaux was Comte's chaste angel, Massin was the evil sexual temptress.

In addition to these difficulties in Comte's personal life, which were reflected in his writings about the role of women in society, another of his problems was that he kept fishing for supporters in opposite parts of the political spectrum, thus muddying his message. During the Revolution of 1848, he took up the workers' cause. He supported workers' rights to work and to form associations. Indeed, until the industrialists, who were greedy, selfish, and egoistic, were regenerated, he recommended the establishment of a positivist republic ruled by a dictatorship of three workers concerned with the interests of the entire society. To attract the support of workers, he created his own club, the Positivist Society, and wrote a manifesto of his philosophy, the *Discours sur l'ensemble du positivisme*. Yet after only a small group of workers rallied to his side, Comte put his faith in Napoleon III, hoping that he could convert him to positivism and use his support to win over the French people. One of Comte's last books was *Appel aux conservateurs* (1855), which targeted people on the right. His praise of Napoleon III's takeover disappointed many of his leftist followers; however, once Napoleon displayed the same imperial ambitions as his despised uncle, Comte became critical of his maneuvers and derisively called him a "Mamamouchi," that is, a good-for-nothing. Each time Comte switched his position, he seemed to lose supporters.

Comte's growing moralizing inclinations and his original desire to put society on a new footing led him to write his second major work: the *Système de politique positive*. It was published in four volumes between 1851 and 1854. The first volume incorporated his 1848 *Discours sur l'ensemble du positivism*, and the appendix of the last volume contained his most important pre-1830 essays in order to demonstrate the continuity of his positivist program. In the *Système*, Comte outlined his new science of morality, his new religion involving the worship of Humanity, and his proposals for political reconstruction. He argued that social harmony depends on not only intellectual consensus but emotional solidarity. To his thinking, people were growing not only more intelligent but more sociable and religious, that is, more interconnected. He coined the term *altruism* around 1850 to underscore the importance of human sociability in maintaining social solidarity and progress. To cultivate altruism, people should make humanity the center of their thoughts, actions, and emotions. The Cult of Humanity would be reinforced through positivist schools, a new commemorative calendar based on secular saints (such as Aristotle and Dante), public festivals, new sacraments (reflecting his Catholic background), rituals conducted in temples of Humanity, continual socialization by positive priests, and a rich culture created by artists and poets.

Comte's stress on this universal secular Religion of Humanity as the key to social unity reflected his lack of faith in big-government solutions to social fragmentation. Having lived through monarchies, republics, and empires, he criticized his contemporaries for focusing excessively on political experimentations, such as parliaments, which he viewed as unproductive. To him, there first had to be an intellectual revolution. It would lead to moral renewal and then a social and political restructuring—one that would lead to the emergence of small republics ruled in a just manner by regenerated industrialists (the temporal power) and positive philosophers or priests of the Religion of Humanity, who would be aided by women and workers. He outlined his political plans in the *Système*, hoping to temper class conflict and alleviate the problems of the common people.

In the last years of his life, Comte carefully composed his posthumously published *Testament*, which included the correspondence between him and Clotilde de Vaux. He also went back to his first academic love, mathematics. Insisting on the continuity in his intellectual interests, he wrote the *Synthèse subjective*, which was published in 1856. It was to be the first volume of a larger, four-volume work called the *Synthèse subjective ou Système universel des conceptions propres à l'état normal de l'Humanité*. It was supposed to show how love could benefit and more closely bring together all three aspects of human nature: the mind, morality (which involved our spiritual and emotional lives), and practical activities. In this first, nine-hundred-page volume devoted to the mind, he showed how feelings affected logic, especially mathematics. He sought to prove that the sciences, even mathematics, which was furthest from man, should be inherently religious, that is, should have some sort of social goal. This work, the *Synthèse subjective* of 1856, is the least read and most incomprehensible of all of Comte's books. He did not have a chance to complete the other volumes, which might have helped make it more understandable, because the "Great Priest of Humanity," as he called himself, died the next year, in September 1857, of stomach cancer. At the time of his death the Positivist Society had shrunk to thirty-seven members. Yet Comte's influence would prove to be more widespread than this small number would suggest.

The Problem of Comte's Intellectual Evolution

Many scholars, including John Stuart Mill, blamed Comte's love for de Vaux for having led him to abandon his scientific agenda and revert to theological and metaphysical thinking. In their view, Comte had a "second career," one of decline. Yet it is important to remember that in his so-called "first career," before his meeting with de Vaux, he had emphasized the need for a spiritual power to counter the temporal power and maintained that the emotions had an important role in shaping our ideas and actions.

Comte's system arose from a paradox deep within him. Disturbed by the growing skepticism of the postrevolutionary period, he experienced a religious calling but suffered from an inability to believe in God. This ordeal gave him a multidimensional picture of human nature, which was reinforced by the romanticism of the period. Comte knew that reason could not satisfy all human needs. In fact, in the *Cours*, he stated that philosophers erred when they portrayed man "against all evidence as an essentially reasoning being, executing continually, without his knowledge, a multitude of imperceptible calculations with almost no spontaneity of action, even from the most tender age of childhood."[16] Years before meeting de Vaux, Comte denounced the exclusive attention given to the mind and discussed the power of the emotions. As early as 1818, he wrote to a friend, "The gentle and tender affections are the happiest, the source of the only true happiness that one can get hold of on this miserable planet, and one could never have enough of them."[17] Reason could not possibly satisfy all human needs. "Daily experience" demonstrated that it was the "passions" that stimulated the intellectual faculties and constituted the "principal motives of human action."[18]

Thus early on Comte recognized that the needs of society were not only intellectual but emotional, and that its spiritual reorganization had to involve the heart at least as much as the mind. Even if a general doctrine were established, a solid social consensus could not exist without the growth of the sympathies; it was wrong to assume that "it is above all by intelligence that man can be changed and improved."[19] As he endeavored to respond to the demands of the heart and mind, Comte sought to create an intellectual system based on science that would appeal to all classes by satisfying the human need for faith. Dogmatism, in his eyes, was the natural mental state of humanity—the state that ensured the sanity of the individual and the community and allowed for the possibility of action. By creating a new set of respectable beliefs that would extend to all people and transform their feelings and values, positivism would establish the kind of intellectual, emotional, and moral consensus that was the salient characteristic of a religious system and a smoothly operating society. The creation of this belief system and a clergy to implement it remained Comte's spiritual mission throughout his life. He always considered his goal to be "spiritual" because his project involved the organization of people's ideas, sentiments, and values.

The first volumes of the *Cours* discuss the natural sciences in a style that Comte made deliberately dry and passionless. He avoided literary devices that would have made reading his works more pleasant in order to differentiate himself from other social thinkers, whom he considered dangerous rhetoricians. His difficult style was intended to make his study of society seem scientific and objective and thus worthier of respect.

Comte also initially focused on the sciences alone, for in his view social regeneration would fall into a "vague mysticism" if it treated the feelings without first systematizing ideas. He explained to Mill in 1845: "This is why my fundamental work [the *Cours*] had to address itself almost exclusively to the intellect: this had to be a work of research, and even incidentally, of discussion, destined to discover and constitute true universal principles by climbing by hierarchical degrees from the most simple scientific questions to the highest social speculations."[20] Only when social issues came up at the end of the *Cours* did Comte believe that he could logically develop his concepts of a spiritual power and a spiritual doctrine touching on the emotions. Broaching these subjects at the beginning would have ruined the antitheological and scientific impressions of his enterprise, which were initially most important to impart to his readers because they distinguished his philosophy from that of other social reformers.

At the end of the *Cours*, Comte therefore launched into the spiritual aspects of social reorganization. In one of the closing volumes, written years before he met de Vaux, he wrote that "universal love . . . is certainly far more important than the intellect itself in . . . our existence . . . because love spontaneously uses even the lowest mental faculties for everyone's profit, while egoism distorts or paralyzes the most eminent dispositions, which consequently are often far more disturbing than efficacious in regard to . . . happiness."[21] Moreover, he suggested that the belief system of positivism was religious because it would replace Catholicism and have its own Positive Church.[22] Armed with the rational, coherent system of positive philosophy, the positivist clergy would "finally seize the spiritual government of humanity" and ensure the triumph of a new, more effective morality, which in the closing pages of the *Cours* was already attracting Comte's attention more than sociology.[23]

Comte would further develop his ideas on these issues in the 1850s, when he established the Religion of Humanity in the *Système de politique positive* (1851–1854), which focused on the moral and emotional aspects of social regeneration. Because he had already established a system of "fundamental ideas," he maintained that he now had to describe their "social application," which would consist of the "systematization of human sentiments, which is the necessary consequence of that of ideas and the indispensable basis of that of institutions."[24] In the *Système*, he made morality a seventh science and explored ways of ensuring social consensus. Here he also emphasized the importance of collective memory for maintaining society, an idea that was developed further by Durkheim and his associates in sociology such as Maurice Halbwachs. Private and public acts of commemoration in the form of worshipping important figures from one's own past and that of Western civilization would provide a sense of continuity between past and present generations and help to unite society. Seeds of this concept can be found in the *Cours*, which states that individuals could best satisfy their natural "need for eternity" by contributing daily to the progress of humanity, especially through "benevolent actions" and "sympathetic emotions."[25] In the *Système*, he described the rituals and positivist culture that would rejuvenate people's emotional lives, bring them back into contact with the concrete, and stimulate the arts. The *Système* represented the understandable result of Comte's fervent desire to effect a return to the most emotionally intense form of religion found in fetishism, in a way suitable to modern society.

★ ★ ★

There was no sudden change of direction from Comte's "first" career to his "second," as most scholars have argued. The second part of his life, which involved setting up his religious and political system, flowed naturally from the first part of his career, which established the intellectual basis of that system. There was no break in Comte's development; from the start his solution to the malaise of his era was a new belief system that functioned as a religion. If anything, Comte's approach was indebted to Saint-Simon's conviction that the "positive doctrine" was a religion because all religions consisted of the reigning intellectual system—that is, the ideas and moral precepts held in common by the members of society.[26] Comte's interest in religion was a natural outgrowth of his concern with moral regeneration, an interest he had had since his earliest writings for Saint-Simon. In 1817, while working for Saint-Simon, he had proclaimed the need to organize "a system of terrestrial morality" that would replace

Christianity.²⁷ Later, he decided that the word *system* was too restrictive and intellectual. The Revolution of 1848 made him particularly eager to experience what he assumed would be the final, decisive clash between positivism and its main rival, Catholicism. To encourage this last battle, he began to call his system the "Religion of Humanity," a term he borrowed from the Saint-Simonians, who had invented it in the early 1830s.²⁸ Defending his terminology, he explained in 1849 that he had "dared to join . . . the name [religion] to the thing [positivism], in order to institute directly an open competition with all the other systems."²⁹ Though rational, this decision caused a rift among Comte's followers, even during his lifetime. Those disciples who believed his scientific program was of key importance fought against those who asserted that his religious ideas were most significant.

The Scope of This Book

There are two parts to this book. The first part focuses on various aspects of Comte's philosophy of science, while the second deals with his social and political thought. The volume begins with a chapter on Comte's general philosophy of science, followed by four chapters concerning each of the most important of the six sciences in Comte's encyclopedic classification scheme.

In the first chapter, Warren Schmaus explains how Comte introduced a historical and social approach to philosophical questions about the nature of scientific knowledge and broke with the eclectic spiritualist tradition of grounding philosophy in an introspective psychology. He also critically examines Comte's attempt to separate his philosophy from metaphysics. Finally, he asks whether Comte's philosophy of science provided any useful normative lessons for science.

The second chapter questions whether Comte gave good normative advice when he told us to separate science from metaphysics. Michel Blay argues that in attempting to do so, Comte obscured all the difficult conceptual work that went into the historical development of the science of mechanics. Similarly, even if it had been advisable in Comte's day to avoid speculation about the nature of light and to focus on geometric optics instead, this does not imply that it would continue to be sound methodological advice.

In chapter 3, Anastasios Brenner discusses two sorts of normative lessons Comte drew from astronomy. In Comte's public lectures on the history of astronomy, he sought to introduce laypeople to scientific reasoning rather than to present scientific knowledge as a finished product. For scientists, he maintained that astronomy provided a model for the proper use of hypotheses in science, which was to anticipate the laws governing the phenomena rather than to postulate causal entities to explain them. Brenner argues that Comte maintained a very liberal attitude toward the use of hypotheses, which derives from his antimetaphysical conception of the aim of science: it is to provide useful knowledge, rather than knowledge that corresponds to the real laws of nature.

In chapter 4, Laurent Clauzade explains how Comte sought a definition of life that avoided metaphysical vital principles and emphasized that biology was a theoretical science consisting of explanatory laws and not just descriptive natural histories. The chapter then describes the various methods of comparison that Comte thought biology should use. For Comte, comparison among different species is linked to their classification in a linear hierarchy. As Clauzade explains, this serial classification of species according to their increasing degree of complexity provided Comte with a model for the linear arrangement of societies according to their presumed degree of complexity, an issue that is taken up again in the following chapter on Comte's sociology.

The fifth chapter explains Comte's political and social purposes for founding the new science of sociology and his reasons for thinking that the time was propitious for just such a discipline. By historically situating Comte's sociology in both its intellectual and wider social context, Vincent Guillin makes what might otherwise be surprising aspects of Comte's new science appear reasonable, such as his exclusion of psychology and economics from his classification of the sciences. Guillin also criticizes some of Comte's other assumptions, such as his supposition that societies could be ranked in a linear classification scheme.

Part II of this book consists of three chapters devoted to Comte's social and political thought. In chapter 6, Michel Bourdeau provides an overview of Comte's positive politics, which Comte posited would be superior to the old regime because it would rely on the newly created science of society, which provided it with its two key principles. The first principle is that there is no society without government; there must be some kind of force to strengthen the fragile, social consensus that is weakened by the increasing division of labor. Yet since social phenomena are subject to natural laws, government power is necessarily limited, and a government can aspire only to modify the existing order. The second of Comte's principles is the separation of spiritual from temporal power. Spiritual power not only provides a moderating influence on temporal power but also helps to create social solidarity by shaping people's beliefs and feelings, especially through education. For Comte, scientists should represent this new spiritual power. The chapter presents some of the surprising, often very modern consequences of these two principles, touching on the relationship between local government and central government, colonialism, and our relations with animals.

In the next chapter, 7, Jean Elisabeth Pedersen explores the complex relationships between Comte's evolving views on art, emotional life, and the social significance of gender. Pedersen also defends the continuity of Comte's thought over the course of his life. She argues that Comte was never uninterested in the fine arts, contrary to received opinion. His early writings recognize their important social function as a way to reach the masses and to disseminate new ideas. Although his later thought may have placed more importance than the *Cours* did on the emotions rather than on intelligence, the break was not total; the heart always needed to be disciplined. As for "the woman question," despite exalting the "emotional sex" and establishing a cult of the Virgin Mother, the subordination remains: women are confined to domestic life and kept in a state of dependence and inferiority.

Andrew Wernick begins chapter 8 by noting how difficult it is to take seriously Comte's efforts to found a new religion. However, we should evaluate Comte's proposals in their historical context. Comte asked a question that we cannot avoid indefinitely: what religion can exist after the death of God? Providing an essential bond among the members of society, religion would only increase in importance as the coercive nature of the state would decrease in the future. Comte's positivist religion is actually not a religion of science, but a Religion of Humanity. Through a system of cult (worship), regime, and dogma that borrows heavily from Catholicism, it would strengthen the weaker instincts of altruism in their struggle against the stronger instincts of egoism. It would strengthen altruism by having us know, love, and serve Humanity (the Great Being), which, unlike God, is a temporal, human production, for it consists of all beings, past, present, and future. We become part of it in worshipping it.

In the concluding chapter, Mary Pickering traces Comte's considerable influence on both scholarly and political worlds and the modern mindset.

It covers the various thinkers throughout Europe, the Americas, and Asia who took up and developed various aspects of his thought or were even deeply influenced by him in a negative direction. The author points out the many difficulties in evaluating Comte's legacy, thanks in part to the ambiguity of his most important contribution to Western thought, positivism. Should positivism include just his philosophy of science, or his political philosophy and Religion of Humanity as well? Herbert Spencer, John Stuart Mill, Alexander Bain, and many others since have been considered positivists, but their views are very different from Comte's. As Comte himself was not above making metaphysical assumptions, can we consider even Comte himself a positivist? Defining positivism either too narrowly or too broadly may both present problems, in both philosophy and the political sphere, where it has taken on a very broad meaning. Auguste Comte produced a remarkable system of thought that besides attracting the attention of philosophers and academics held enormous appeal for a class of people who had little concern for traditional philosophical problems but a keen interest in the application of science to improving the human condition.

Part I

Comte's Philosophy of Science

1 Comte's General Philosophy of Science

Warren Schmaus

Philosophy of science and epistemology as specialized philosophical disciplines with their own names did not exist until the nineteenth century.[1] Prior to Auguste Comte (1798–1857), epistemology, philosophy of science, and philosophical theories of the mind were not distinct from one another. The possibility of obtaining knowledge, including mathematics and the sciences, was thought to depend on qualities of the individual mind. Thinkers otherwise as different as Bacon and Descartes nevertheless agreed that in order to do science, one had to properly prepare one's mind through philosophical reflection and the elimination of preconceptions and doubtful beliefs, and then rigorously follow a prescribed path. Even Kant blended the question of knowledge with philosophical inquiry about the mind, seeking a foundation for the sciences in the forms of intuition and the categories of the understanding.

Comte, on the other hand, considered scientific knowledge to be a social and historical product. He challenged the introspective, individualist approach to problems of knowledge by proposing to replace it with a systematic study of the sciences and their history. Thus Comte initiated philosophy of science as a distinct discipline in France much as John Herschel and William Whewell were doing at about the same time in the United Kingdom. According to Comte, "the true philosophy of every science is necessarily inseparable from its actual history."[2] Comte was one of the earliest thinkers to recognize that past works in science and philosophy may be the products of entirely different modes of thought than those we find at present. He also thought that there had been steady progress in these ways of thinking and that he could extract normative lessons from the history of the sciences.

Comte's historically informed philosophy of science established a precedent for a French intellectual tradition of historical epistemology. Cristina Chimisso traces the history of this intellectual tradition from Lucien Lévy-Bruhl—who wrote what is still one of the best works on Comte's philosophy of science—and Léon Brunschvicg through Gaston Bachelard, Georges Canguilhem, Hélène Metzger, and Alexandre Koyré.[3] Thomas Kuhn credits Koyré and Metzger among those historians who showed him "what it was like to think scientifically in a period when the canons of scientific thought were very different from those current today."[4] Kuhn's *The Structure of Scientific Revolutions* was a major contribution to the debates during the 1960s and 1970s over the relationship between the history of science and the philosophy of science, a question that is far from being settled.[5]

As we shall see in this chapter, Comte thought that the positive method grew over time by accretion, adding elements as it developed first from mathematics and astronomy and then through physics, chemistry, and biology. His project was to study these methods in order to establish a method for his new science of sociology. He faced a philosophical problem similar to that faced by those who would integrate the history and philosophy of science today: how is it possible to draw normative conclusions from a description of the past? It also confronts the ongoing problem of whether it is possible to extract methods for the social sciences from a study of the natural sciences.

But to appreciate the significance of Comte's contribution to philosophy, it would be more instructive to compare him to his contemporaries rather than to ours. Comte's contemporaries, who would base epistemology on a philosophical psychology, could equally be charged with attempting to derive prescriptions from descriptions, but in this case descriptions of what they found through the introspection of their minds. Comte's shifting the focus of epistemology from the individual mind to the history of science provided a common ground for philosophical debate. Because introspection is private, others cannot contest what an individual alleges to have found in his or her own mind. The history of science, on the other hand, can provide evidence to challenge another's normative and descriptive claims. As we shall see, Comte's nineteenth-century critics in fact drew on this history in their challenges to his philosophy of science. But these critics had to agree with Comte that science is the product not of some isolated Cartesian genius who has discovered the correct rules of method but of communities of interacting researchers who have experimented with different ideas, methods, and epistemic norms over the course of history.

Comte's Critique of the Philosophical Tradition

At the time that Comte was writing his major contribution to philosophy of science, the *Cours de philosophie positive* (1830–1842), academic philosophy in France was dominated by Victor Cousin (1792–1867) and his school of eclectic spiritualists.[6] Eclecticism, the original name that Cousin had given to his philosophy, encouraged the study of the history of philosophy and sought to synthesize what it considered the best ideas from systems of thought that might otherwise appear incompatible. Comte, on the other hand, thought that instead of attempting to synthesize incompatible philosophies, one should try to explain how each was compatible with its own historical situation.[7] Although Cousin continued to accept the name *eclecticism* for his philosophy, he came to prefer the term *spiritualism*, which he defined in terms of a belief in the spirituality of the soul, free will, morality, and God.[8] Hence, the terms are often combined to give the name "eclectic spiritualism" to this philosophy.

Spiritualism sought a foundation for philosophy, including logic, ethics, and metaphysics, in the introspection of the human mind. Cousin employed what he called the *méthode psychologique*, by which he meant two things. First, it referred to the grounding of all philosophy in an introspective psychology that inquired into the laws, scope, and limits of our cognitive faculties.[9] Second, it also referred to the very method of introspection this philosophical psychology used.[10] Cousin argued that we have a direct, unmediated internal perception of the self or spirit that reveals that we are not only substances but active causes.[11] His introspective method was in fact an eclectic synthesis—or confused jumble—of Descartes's *cogito*, Leibniz's apperception, Kant's transcendental apperception and empirical apperception, and Pierre Maine de Biran's (1766–1824) internal experience of willed effort. Cousin accepted Maine de Biran's argument that we can have direct perception of the activity of the will, unmediated by mental representations or ideas. This internal perception of the active self was supposed to provide an epistemological foundation for a universal and necessary principle of causality as well as for other categories such as substance, unity, and identity. Cousin argued that the universal and necessary principles associated with these categories are required for the sciences, mathematics, and morality. These universal and necessary principles must be distinguished from merely general principles based on experience, which can show us only how things are and not how they must be.[12] Comte, on the other hand, entirely opposed the principles and methods of the eclectic spiritualist philosophy, accusing the eclectics of promulgating "German metaphysics" under the guise of a science of psychology.[13] Comte's identification of the eclectics with German metaphysics is no doubt a reference to the influence of Friedrich Jacobi and Friedrich Schelling on Cousin, who had visited these philosophers in Germany.[14] Far from seeking a foundation for the categories of causality and substance in the causal powers of our will, Comte sought to dismiss these concepts from positive science.

Comte saw no place for that "illusory psychology, the last transformation of theology, which concerns itself with neither the physiological study of our intellectual organs nor the rational procedures that direct scientific research."[15] As Comte saw it, there are two ways to study the mind: the static, which considers the organic conditions on which it rests, and the dynamic, which looks to its products. Where the eclectic spiritualists were dualists, Comte identified the mind with the brain as early as 1828, in his "Examen du Traité de Broussais sur l'irritation."[16] In an obvious reference to Cousin, Comte said that Broussais's work had inspired a healthy reaction against "the deplorable psychological mania with which a famous sophist has momentarily succeeded in inspiring French youth."[17] In this essay, Comte directly confronted the eclectic spiritualist's claim that one can introspect the mind's activity, arguing that one cannot observe one's intellectual operations but only the organs responsible for them and the results they produce, and that these studies belong to physiology and the philosophy of the sciences, respectively.[18] Comte continued to maintain in the *Cours* that we can observe only the results of our mental activity, arguing that "the intellectual functions present, by their nature, . . . this particular character of not being able to be directly observed during their accomplishment itself, but only in their more or less proximate and more or less lasting results."[19]

There is a precedent for Comte's argument that we can observe only the results of our mental activity in Hume's *Enquiry Concerning Human Understanding* (1748).[20] Locke had argued that our minds obtain the idea of active power by reflecting on the power of our will over the parts of our bodies and the ideas in our minds.[21] Contra Locke, Hume argued that we experience no such power. Hume said that when we will the motion of a part of the body, the immediate effect is not on the limb but on the nerves, muscles, and "animal spirits" that convey the will's command through the nerves to the muscles. Thus, the mind wills an event different from the one immediately produced and the one immediately produced is unknown to us. If we could perceive the operation of the will on the body, we would be able to perceive this immediate effect, and vice versa. With regard to the power of the will over the mind's ideas, Hume objected that all we perceive are our ideas and the connections between them, and not the force or power by which ideas are called up. "We only feel the event, namely, the existence of an idea, consequent to a command of the will. But the manner in which this operation is performed, the power by which it is produced, is entirely beyond our comprehension."[22] Again he appealed to physiology, arguing that we are unable to explain the fact that our ability to command our ideas varies with health and sickness, the time of day, and the fullness of our stomachs. Maine de Biran, on the other hand, had rejected Hume's argument that we cannot introspect the power of the will and criticized him for assuming that interior observation, like external observation, is mediated by representative ideas.[23]

Comte maintained that, due to the growing influence of the positive method, the eclectics were only pretending to base their illusory science of psychology on the observation of facts by appealing to internal observation. He argued that much as one's eye cannot see its own retinal image, the brain cannot perceive its own inner workings. One cannot divide oneself in two, with one part reasoning and the other part observing the reasoning process. Since the organ observed and the observing organ would be one and the same, this would be impossible.[24] Jean-François Braunstein traces this argument about not splitting oneself in two back to Cicero and up through Pierre Gassendi's objections to Descartes's *Meditations*.[25] Ironically, Maine de Biran had also used it, although in a work that was still unpublished during Comte's lifetime.[26] However, where these earlier arguments concerned the mind-body connection, Comte's argument operates on a purely material level, concerning the impossibility of the brain observing itself.

John Stuart Mill attempted to defend introspection against Comte, maintaining that we can have direct knowledge of our observations and reasonings and that the mind can pay attention to more than one impression at a time, or at least hold one in memory and study it later. He said, "Whatever we are directly aware of, we can directly observe."[27] But the question is precisely one of what we are directly aware. Mill was talking about a kind of introspection of mental events linked by associations, while Comte was rejecting the metaphysical psychology of the French eclectic spiritualists, which involved the introspection of the activity of the will.[28] Cousin had insisted that we directly apperceive the power of the will and thus that the question of free will is not a matter for philosophical dispute,[29] whereas Comte maintained that the notion of free will belonged to the metaphysical stage of thought and that people were determined in their actions. Though no determinist, Mill similarly dismissed what he called "the figment of a direct consciousness of the freedom of the will," arguing that we are conscious of little more than that we could have done otherwise than we did if circumstances had been different.[30] However, Mill's attack on the notion of a direct awareness of free will was directed at British philosophers such as William Hamilton and Henry Longueville Mansel. He did not mention Maine de Biran, Cousin, or any other French philosopher in this context. Mill may not have even considered the possibility that Comte's target in attacking introspection was a philosophical position that he also disliked.

Robert Scharff correctly identifies the eclectic spiritualists as Comte's target, but maintains that their method of internal observation was simply an inwardly turned version of external observation, thus overlooking Maine de Biran's denial of the analogy between internal and external observation.[31] In a further criticism of *cette prétendu méthode psychologique*, Comte argued that it on the one hand recommends that one isolate oneself as much as possible from external sensations and refrain from all intellectual work, while on other hand, once one attains "this perfect state of intellectual slumber," one is supposed to observe the operations of one's mind while it is not doing anything.[32] Scharff finds it uncharacteristic of Comte to assume here either that observation is passive or that the object of observation must be active,[33] thus not grasping that it was not Comte who made these assumptions; rather, Comte found them in the position he was attacking. Scharff does not see that Cousin's whole purpose was to ground the category of causality in the observation of the activity of the will. The eclectics were pursuing quite different goals than were associationists such as Mill.

Comte then took on a much larger target when he argued that during the two thousand years that metaphysicians have cultivated psychology, they have produced nearly as many different opinions as there are individuals who practice the method of interior observation. Instead of advancing our knowledge, this method has yielded nothing but divergent, conflicting schools of thought that endlessly debate the "first elements" of their systems of thought. Any true notions they may have discovered, he contended, have actually been obtained not by introspection but by observing the development of the human mind as the sciences progress.[34]

As Comte never denied that we are able to introspect our sensations and feelings, Scharff finds it curious that Comte never considered the possibility that our consciousness of these things could yield facts or observations for a positive science of psychology.[35] After all, Hume had proposed three principles of the association of ideas—causality, resemblance, and contiguity—in spite of the fact that he also denied that we can introspect the manner in which these ideas are produced. Comte, however, borrowing another argument from Broussais, maintained that introspection at best provides knowledge of the workings of only the healthy, adult human mind, thus yielding an insufficient empirical basis for a positive science of the human mind.[36] Furthermore, the very idea of an introspective psychology runs counter to the positivist emphasis on publicly observable, repeatable phenomena. Comte may have been an even more thoroughgoing empiricist than Hume.[37]

Comte turned to the history of science as providing a better empirical basis for a theory of mind. Like Kant, Comte maintained that the best way to know the human mind is through an analysis of the knowledge it produces, and that mathematics and the sciences are the best examples of such knowledge. But where Kant investigated the logically necessary concepts that make the sciences possible, Comte was interested in how the sciences had developed over time. As the sciences are the product of many thinkers over many generations, the history of science provides a broader empirical basis for a theory of mind than the introspective reports of individuals, whose intellectual development he believed merely recapitulates that of their society in any event.[38] According to Comte, "it is only through the positive study of human evolution as a whole that one may discover the actual laws of the intellect" or the "logical laws of the human mind."[39]

Comte's use of the term "logical" suggests that these laws had normative content for him.

The Three-State Law

For Comte, the normative lessons drawn from the history of science were summed up in his three-state law of scientific progress. This law governs the historical development of scientific methods, especially methods of explanation. According to Comte, "the human mind, by its nature, successively uses in each one of its researches three methods of philosophizing of which the character is essentially different and even radically opposed: first the theological method, then the metaphysical method, and finally the positive method."[40] The theological method seeks absolute knowledge of essences and first and final causes, and offers explanations of natural phenomena that appeal to the activity of supernatural entities. Matter is conceived as utterly passive with active principles having their source in these supernatural beings. By habitually invoking only these active principles in explanations, people lose sight of the supernatural and the theological gives way to the metaphysical state of mind, in which explanations appeal simply to abstract causes and forces.[41]

The transition to the positive stage is more radical than that from the theological to the metaphysical stage. The positive method abandons the quest for absolute knowledge of causes and essences and seeks explanations that appeal only to general laws governing the phenomena, "which is to say their invariable relations of succession and similitude. The explanation of facts reduced in this case to its real terms, is henceforth nothing more than the linkage established between the various particular phenomena and some general facts of which the progress of science tends more and more to diminish the number."[42]

The sciences pass through the three stages in the order of Comte's sixfold hierarchy: first mathematics, then astronomy, physics, chemistry, biology, and finally sociology. His classification does not include psychology as a separate science between biology and sociology, since, as we have seen, he regarded it as an illusory science. For Comte, there are only two orders of phenomena that living beings present. Those that relate to the individual are the province of biology, while those that relate to the species, especially when it is a "sociable" species, belong to sociology.[43]

Although the positive method seeks to diminish the number of "general facts" or laws, Comte thought it would never be able to represent all the observable phenomena as particular cases of a single general law, such as the law of universal gravitation. He regarded this goal as "chimerical," comparing it to the way in which the theological method sought to attribute all the phenomena of nature to one god and the metaphysical method sought to attribute them to a single entity, nature. According to Comte, the human intellect is too weak and the universe too complex for the goal of reducing all the phenomena of nature to a single law ever to be within our reach, and people have an exaggerated idea of the advantages that would accrue from achieving it, even if it were possible.[44]

However, it would be wrong to conclude that theories were not important in Comte's positivism. For Comte, facts must be organized into laws and theories. "Every *science* consists in the coordination of facts; if the various observations were entirely isolated, there would not be any science."[45] As Larry Laudan argues, what distinguishes science from nonscience for Comte is testability and generality, not certainty or even high probability.[46] Isolated facts do not have the required generality, and for a scientific theory to be testable it must make predictions. As Comte put it, "prevision" is the criterion of positivity.[47]

Comte thought that ultimately the positive sciences will all be presented in what he called the "dogmatic" or theoretical mode of exposition, which he distinguished from the historic mode, the order in which they were developed. In the dogmatic mode, the sciences are "recast in a general system, in order to be presented following a more natural, logical order."[48] In his time, only highly developed sciences, such as geometry or mechanics, could be presented in this manner. However, once a science is presented in this form, its observational origins are obscured.[49] It then dispenses with "direct observation" as much as possible and allows for the derivation of the greatest possible number of consequences from the smallest possible number of givens.[50] Nevertheless, even in mathematics, Comte thought it highly instructive to introduce students to the historical development of the subject, as he attempted to do in the *Traité élémentaire de géométrie analytique* (1843).[51]

Presenting a science in the dogmatic mode does not change its epistemological status, for Comte. The truth of the science still depends on empirical verification. He made this somewhat clearer in a similar passage in the *Discours sur l'esprit positif* of 1844, where he used astronomy as his example, explaining that in a "true science" we substitute "rational prevision" for "direct observation."[52] That is, in astronomy, we substitute mathematically expressed laws for a catalogue of individual observations, which would be more of an almanac than a science. Nevertheless, the truth of these laws depends on the predictions we draw from them. Thus Laudan may have gone too far when he interpreted these passages as suggesting that Comte thought that in its advanced stages, "science can move from the laboratory to the armchair, from the tedious method of observation and fact-collecting to the more rapid methods of calculation and ratiocination."[53] Comte's account of the dogmatic mode has to do only with the mode of presentation of a science and not its methods of inquiry.

Defense of the Three-State Law

Comte claimed that his three-state law could be defended by both rational proofs and historical verifications. His rational proofs are premised on his view that observations in science must be guided by hypotheses. He then argued that, pressed between the need to have some guiding ideas and the lack of observations on which to ground them, people had to begin with theological hypotheses. Furthermore, only the hope of attaining knowledge of the intimate nature of things and of the origin and end of all phenomena could have provided sufficient stimulus for engaging in the long, arduous task of research. Thus astrology preceded astronomy and alchemy preceded chemistry. People also needed to find explanations of natural phenomena to make them less frightening. Finally, society needed some system of ideas to create order, and people could not wait the centuries needed for the development of the positive method, and so adopted the theological.[54] Comte subdivided the theological stage into the fetishistic, polytheistic, and monotheistic stages. The theological method began with fetishistic hypotheses, in which everything in nature is animated with spirits and people explain events in nature through analogies with their own actions.

Friedrich Hayek objected that the argument that early humans saw everything in nature as animated with wills analogous to their own is inconsistent with Comte's rejection of the introspection of the will.[55] However, for someone to believe *that* they have willed some action, it is not necessary that they are able to introspect *how* they have willed it, that is, that they can introspect the causal power, force, or energy of the will.

Humanity then progressed from the fetishistic through the polytheistic, monotheistic, and metaphysical stages through a process of abstraction.[56] However, the theological and metaphysical methods could not supply the sort of hypotheses humanity needed. One of the reasons we pursue knowledge of the natural world is to be able to make predictions to guide our actions.[57] But the theological method provides no guarantee of the regularity in nature that such predictions require since it explains natural phenomena as subject to arbitrary wills.[58] For this reason, the positive method has actually been used from the very earliest times in our day-to-day lives, while the theological method has dominated our "higher" speculations.[59] Thus, Comte's picture of early thought is one of an unstable mixture of positive and religious modes of thought.

In the *Considérations philosophiques sur les sciences et les savants* of 1825, Comte argued that the positive method is gradually supplanting earlier methods simply because it is more fruitful in practical applications:

> The vigor and the influence of a method are measured by the number and the importance of its applications: those which no longer produce anything soon absolutely cease to be used. Now, as since the last two centuries at least, the theological and metaphysical methods . . . have become entirely sterile; since the most extensive and important discoveries . . . have been, since that epoch, due uniquely to the use of the positive method, it is evident, by this fact alone, that it is to the latter that the exclusive direction of human thought henceforth belongs.[60]

In addition, Comte criticized metaphysical explanations as "empty," as they appealed to causal entities that were in fact nothing but the abstract names of classes of phenomena.[61] He compared their use in explaining social phenomena to the doctor in Molière's *Le Malade imaginaire* appealing to the dormitive powers of opium.[62]

In the *Cours*, Comte gave what at first appears to be a somewhat different account from the one he had provided in the earlier *Considérations* of the transition to the positive method. In the *Cours*, the decisive step toward the positive stage was what Comte considered the nominalists' victory over the realists in the interpretation of causal entities.[63] Indeed, he thought this was the most important development in philosophy until Hume's analysis of causation.[64] However, these two accounts are not necessarily inconsistent with each other. Perhaps what Comte meant is that people began with fetishistic and then polytheistic explanations, but since these methods could not guarantee regularity in nature, people abandoned them in favor of first monotheistic and then metaphysical explanations. As we shall see below, Comte spoke of a stage of "bastard positivism," intermediary between the metaphysical and positive stages, in which scientists invoked explanatory hypotheses involving such entities as invisible fluids and ethers. Then they came to appreciate that we do not need to think of these causal entities as real and so adopted a nominalist attitude toward them. Recognizing the vacuity of explanations that were nothing more than names given to the phenomena, they gave them up in favor of explanations in terms of Humean laws governing the phenomena of nature, which they realized had been yielding successful practical applications all along. Thus there is a sense in which the positive method both developed out of and in opposition to the theological and metaphysical methods.

Scharff interprets Comte as holding that the positive method simply grew out of the metaphysical and theological methods and not also in opposition to them. He sees Comte's positive method as "simply the last and finally successful expression of humanity's long struggle to explain and control nature."[65] However, it is not clear that, for Comte, the theological or metaphysical methods were ever directed at controlling nature. As Scharff concedes, prayers and rituals are not effective means for controlling nature.[66] That is why Comte held that the positive method was used from earliest times for practical purposes, while the religious was used for "higher" purposes. Religious thought provided those explanations of the world that were shared by the members of a society to make them feel at home in nature and to provide a kind of social glue.

In *The Poverty of Historicism*, Karl Popper charges that Comte's three-state law is no law at all, but at best a historical trend. Popper maintained that any sequence of three or more causally connected events does not proceed according to a single law, but must be explained in terms of several laws and initial conditions.[67] He thought that Comte had not done this, but had simply offered up an absolute historical trend grounded in human nature.[68] Although Comte's defense of the three-state law may be based on claims about human nature, he did not present it as an unconditioned or absolute trend. On the contrary, Comte recognized that progress in knowledge depends on social and historical conditions. As early as 1825, Comte argued that our intellectual development was not possible without the division of labor. The beginnings of this division were in the theocratic system, with the separation of a priestly caste. Subsequent advances in our knowledge required a much greater subdivision and one under a totally different sort of regime. It was necessary that the "culture of the mind become independent of the immediate direction of society, so that the division and perfection of our branches of knowledge could take place without compromising the existence of the political order."[69]

At the same time, scientists must be organized into communities of some sort, as they depend upon one another for input and feedback: "Either to observe or to meditate, every mind always depends on others, who prepare his materials and verify his results."[70] Comte added that a scientist needs to feel that his or her work is backed by that of his or her contemporaries or at least fits with the direction in which things are proceeding: "The boldest innovator rarely acquires a full confidence in his own discoveries, as long as he has not obtained some freely given approval. And he cannot even ever forgo such a sanction unless he feels sufficiently supported by the general march of humanity."[71] Indeed, Comte went on to say that the past can have an even greater constraining effect on a scientist than his community of contemporaries.[72]

Comte also recognized that scientists depend on society's language in order to fix their ideas. He distinguished two stages of concept formation in science, the first involving imagistic thinking in which our ideas are not yet communicable and the second in which they are. Putting ideas into words allows them to acquire sufficient precision and consistency, which is needed "to support any true discourse."[73] Communication "constitutes the sole decisive proof of the maturity of any of our ideas whatsoever."[74] Even when we are working out ideas only for our own use, we should regard them as not sufficiently thought out if we find ourselves unable to communicate them to others. The role of language is to "produce a continuous discourse, which develops and clarifies itself during meditation, of which the spontaneous development thus becomes difficult to distinguish from such assistance."[75] Although speech helps, the final stage of perfection is reached only with the written exposition of our thoughts.[76]

Popper had no quarrel with those who recognize that historical trends depend on initial conditions. However, he was still concerned about whether such trends are testable.[77] For Comte, what he called the historical verifications of his law are provided by his detailed accounts of mathematics and the natural sciences in the first three volumes of the *Cours*. In the fourth volume and then in the *Système*, he shifted the burden of proof to his critics, boasting that none of them had raised any serious objections to it, since they ignored the auxiliary role of his sixfold classification of the sciences.[78] According to Comte, mathematics and astronomy had already reached the positive state; physics, chemistry, biology, and sociology were in various stages of making the transition from the metaphysical to the positive. Thus, the coexistence of different modes of thought in different sciences is no objection to the law.

Later critics were not guilty of this error. Contra Comte, Whewell showed that in each of the sciences of celestial mechanics, optics, and chemistry, metaphysical discussion has been inextricably bound up with the discovery of phenomenal laws.[79] Herbert Spencer argued that Comte's distinction between astronomy and physics was arbitrary, as the same laws apply to both celestial and terrestrial matter.[80] Charles Renouvier pointed out that modern astronomy relies on methods of investigation drawn from chemistry and argued that biology in Aristotle's day had achieved a more positive state than physics.[81] Antoine Augustin Cournot maintained that the life sciences at that time contained more obscurities and mysteries than the social sciences as represented by political economy, which Comte considered part of the metaphysical philosophy. Cournot also questioned whether scientific progress consists in the gradual elimination of the metaphysical elements and, like Whewell, suggested that metaphysics and philosophy were inseparable from the theorizing that distinguishes a true science from a mere aggregate of empirical facts.[82] Many other such examples could be given. As Robert Flint pointed out, Comte himself conceded that some phenomena, such as those in astronomy, are so regular that we were never entirely in the theological state.[83] Comte also admitted that even today astronomy depends on physics, especially optics, to make to its observations corrections that are necessary because of things like parallax and atmospheric refraction.[84]

If Comte did not seem troubled by such exceptions to his three-state law, it is because he held that his more important contribution to sociology is its method: "If it were possible that I could be mistaken about the true law of the long human evolution, the only thing that would rationally result would be the necessity of establishing a better sociological doctrine, and I would have no less irrevocably constituted, in this subject, the sole method capable of leading to the positive knowledge of the human mind."[85] Comte regarded his methodological contribution to the study of the mind to be the replacement of introspectionist psychology with the historical and philosophical study of the sciences. He maintained that as each of the sciences became positive, they gave rise to new aspects of the positive method, including mathematical analysis, observation, experiment, comparison, and the historical method. As Ian Hacking puts it, as each science achieves the positive state, for Comte, it contributes a new "style of reasoning."[86] But for Comte, the development of these new methods at the same time also contributes to the Lamarckian evolution of the human mind.[87]

The Sixfold Hierarchy and the Unity of Science

The six sciences in Comte's hierarchy are what he called the abstract sciences, that is, those that aim at general laws, rather than the concrete sciences, such as natural histories and sciences like mineralogy and meteorology, which largely concern the application of general laws to particular cases.[88] In the *Cours*, he used chemistry and mineralogy as examples of the distinction between abstract and concrete sciences. Chemistry studies "all the possible combinations of molecules, and in all imaginable circumstances," while mineralogy is concerned with just those combinations found in nature.[89] The six abstract sciences are ordered according to what he saw as their degrees of simplicity, generality, dependence, and abstraction. The simpler sciences are those whose objects of study are composed of fewer parts. The simpler phenomena are also the most general, "because that which is observed in the greatest number of cases is, for that reason, disengaged as much as possible from the circumstances proper to each case."[90] Since the phenomena studied at any level differ from those at a more general level by the addition of complicating factors, they are subject to and thus dependent on the laws at the higher level of generality. The simplest and most general phenomena are also the most abstract and most distant from the interests of humanity.[91]

Comte's encyclopedic formula also orders the sciences according to the degree of "perfection" of their study.[92] For Comte, the degree of perfection has to do with the extent to which a science can be presented in the dogmatic mode, not the degree of certainty. The more complex sciences are no less certain than the simpler ones. In particular, Comte rejected the argument that sociology is less certain than the other sciences. He thought its critics confuse certainty with precision. For Comte, sociology may be less precise, but it is no less certain. He also rejected skeptical arguments that assume a distinction between testimonial and nontestimonial sciences, pointing out that all sciences rely on testimony to some degree, for example in accepting the testimony of scientists who have made observations that cannot be repeated. Thus the fact that sociology relies on historical testimony is no barrier to its being a positive science.[93]

That the sixfold hierarchy presented the sciences in their order of dependence does not imply that any science reduces to another on which it depends. For Comte, it is enough that the laws of science are "homogeneous,"[94] by which he seems to have meant that the laws at any level in his hierarchy cannot contradict laws at higher levels of generality. "In place of blindly searching for a sterile scientific unity, as oppressive as chimerical, in the faulty reduction of all phenomena whatsoever to a single order of laws, the human mind will finally regard the diverse classes of events as having their especial laws, in other respects inevitably convergent, and even, in some respects, analogous."[95] By a "single order of laws" it appears he meant the laws of one science, judging from the context in which this passage appears, in which he spoke of the legitimate independence (*la juste indépendance*) of each science.[96] It may appear that on the question of the dependence or independence of the sciences Comte contradicted himself between the first and the last volumes of the *Cours* or was careless with his choice of words. However, he consistently maintained that each science is independent of the others in the sense that its laws cannot be derived from the laws of some more general science, yet dependent in the sense that its laws may not contradict the laws of any other science.

Comte thought that biology especially would benefit from this positivist view of the relationships among the sciences, since it would no longer have to seek protection through the use of theological-metaphysical concepts from being encroached upon by the other sciences.[97] That is, biology would no longer have to appeal to things like vital spirits or entelechies in order to maintain its independence from physics and chemistry. For Comte, this strategy was both dangerous and insufficient. He saw the conflict over vitalism in biology as the sole remaining struggle in the sciences between materialism and spiritualism or dualism. He was equally opposed to both positions and thought they would disappear under positivism. Materialism tries to absorb what he regarded as the "superior" sciences into the inferior, while the spiritualist tendency in biology attempts to maintain the independence and dignity of this science in a way that is linked to the "dark" preservation of the ancient philosophy.[98] In other words, both materialism and spiritualism make the mistake of treating the relationships among the sciences as a metaphysical question: is there only one kind of stuff, or more? From the positive point of view, the question is not about how many kinds of entities there are or whether there are such things as souls or vital spirits, but simply about the logical relationships that obtain between the laws of one science and the laws of another.[99]

The only unity that is "indispensable" to the positive philosophy, Comte said, is unity of method.[100] Unlike the twentieth-century logical positivists, Comte did not try to argue that the positive sciences display unity of law or language. Nor did he place much importance on the reform of scientific language. He had no objection to the use of terms taken from older theories infected with metaphysics, as long as these terms were redefined in a positive way. For example, "gravity" and "attraction" are to be understood merely as general names denoting a class of phenomena and not as causes or modes of producing the phenomena.[101] Comte argued that the difficulty of separating the meanings of terms from theories has been greatly exaggerated. Using an example from optics, he spoke of how the word *ray* was taken from the emission theory and adopted by the undulatory theory and added that it was possible to attach a meaning to it that is independent of any hypothesis and simply relative to the phenomena.[102] Similarly, he had no objection to talking about a ray of heat, as long as it is separated from caloric theory.[103]

Comte's Positive Metaphysics

In spite of Comte's opposition to metaphysics, his philosophy of science assumes a metaphysical realism. Émile Boutroux and Lucien Lévy-Bruhl argued that Comte's sixfold classification of the sciences presupposes a kind of metaphysical realism just to the extent that he thought that natural phenomena form distinct kinds corresponding to his classification of the sciences and are subject to laws.[104]

However, one could argue that Comte's classification has no realist implications, since he did not regard the number of sciences as fixed. For Comte, the classes of phenomena particular to each science are only nominal and not real kinds: "One cannot objectively fix the number of sciences. . . . Basically, the name consecrated for each science merely designates the group of speculations for which the unity finds itself sufficiently recognized; which ought to vary in accordance with the time and the intellect."[105]

But even if one were to insist that Comte's classification of the sciences is purely nominal and has no realist implications, there are nevertheless realist implications of his conception of scientific laws, as Boutroux and Lévy-Bruhl suggested. For Comte, the laws of science "represent the universal order, to the extent that we have need of knowing it."[106] Comte's belief that there is a "universal order" exemplifies a philosophy for which Hacking has invented the term "inherent-structurism," since he finds that "realism" has taken on so many different meanings in philosophy.[107] But to the extent that Comte held that the laws of science represent this structure, the terms in these laws would have to refer to kinds of things in nature. Thus we arrive back at the assumption of natural kinds, although by a different route than before. Indeed, Comte regarded living species as constituting real kinds in nature, or "natural groups"—unlike, for instance, Charles Darwin, who thought of species as more or less artificial creations of the human mind.[108] Thus, although Comte may have maintained a nominalist or instrumentalist view of unobservable entities in science, he appears to have been a realist about kinds at least in the observable or phenomenal realm. He even appears to have been a realist about at least some concepts in mathematics. For instance, he regarded Gaspard Monge's classification of geometrical surfaces according to their mode of generation as forming "natural families," and saw this as somehow analogous to the formation of natural groups in biology.[109] There thus appear to be limits to Comte's relativism: for Comte, there is a real structure in nature that the laws of science must represent.

For Comte, the principle of the invariability of the laws of nature was not a metaphysical assumption but rather the result of an induction from the appearance of invariable laws in all "the essential orders of phenomena," including social phenomena.[110] However, he regarded the laws of nature as deterministic as well as invariable. For instance, in his social dynamics, he spoke of each social state as the "necessary result" of the preceding state and the "indispensable motor" of the following. According to Comte, the laws of social dynamics "determine the fundamental march of human development."[111] Given his rejection of the concept of causality as metaphysical, a serious problem arises: how can there be determinism without causality? Comte did not give any empirical or inductive arguments that the laws of nature are deterministic as well as invariable. Indeed, how could he have? Experience can show us only that one thing always follows another; it can never show us that these things *must* follow in this way. Comte appears to have perhaps unwittingly introduced a metaphysical assumption into his philosophy, that is, that the independently existing universal order in nature is a deterministic order.

There is nothing necessarily wrong with making metaphysical assumptions, as long as one is explicit about them. Comte's metaphysical assumptions about scientific laws, biological species, and a universal order of nature have implications for his methodology. For instance, although he defends an instrumentalist view about the use of hypotheses in science, his belief in a universal order of nature would appear to place some constraints on the kind of hypotheses scientists may adopt. The justification of these constraints would then entail the need to justify the metaphysical presuppositions behind that methodology.

Comte's Views on Method

What Comte considered the unity of method is perhaps more accurately characterized as the continuity and consistency of method. In particular, he saw no radical break between the methods of the natural and the social sciences. As I mentioned earlier, he held that as each science in the sixfold hierarchy achieves the positive state, the positive method develops through these corresponding stages: first mathematical analysis, then observation, experiment, comparison, and finally the historical method, all of which may be regarded as methods of analysis and induction, broadly construed. Each science supposedly builds on the one that went before, adding new aspects to the positive method, except for chemistry, which adds nothing new to the experimental method of physics: "Although the method may be essentially the same in all, each science develops in particular some one of its characteristic procedures."[112] However, what the *Cours* actually presents are similarities in the methods of sciences adjacent to each other in the encyclopedic scale; the resemblance is less the further apart the sciences. Thus although he may have spoken of the use of analysis in both mathematics and sociology, analysis looks quite different in these two sciences.

Mathematical Analysis

Comte meant more than one thing by "analysis" in his philosophy of mathematics alone. In the *Cours*, he distinguished the "abstract" analysis of arithmetic, algebra, and calculus from the "concrete" analysis of geometry and rational mechanics. Abstract analysis is only a set of deductive methods for transforming equations and cannot be used to establish the first principles of any science. Concrete analysis, for which there is no set of rules, is for discovering relations among quantities and expressing them as equations.[113] He attached special importance to analytic geometry, writing and publishing the *Traité élémentaire de géométrie analytique* in 1843 in addition to the lessons on mathematics in the *Cours*. Analytic geometry allows us to represent curves as equations by decomposing them into their perpendicular components. Comte's concrete mathematical analysis bears some resemblance to the positive methods in the other sciences, since these, too, make use of methods of decomposition. In a footnote to his translation of John Leslie's *Geometrical Analysis*, which Comte published in 1818, he discussed another sense of analysis, which has to do with resolving a relatively difficult geometrical problem, the construction of a regular pentagon, into successively simpler problems, and finally into dividing a line into its mean and extreme ratios.[114]

However, to maintain his unity of method thesis, Comte had to do more than argue that mathematics, like the empirical sciences, employs a method of decomposition. He also needed to show that mathematics is empirical. And in fact he argued that the basic phenomena of geometry are derived from observations. The reason this is not generally recognized, he thought, is the high degree of "perfection" this science has reached. That is, it can now be presented in the dogmatic mode. Nevertheless, he felt that at the basis of all our deductions in geometry there were certain primitive phenomena that could not be discovered through reason alone but must be founded on observation.[115] In his last work, the *Synthèse subjective*, he also argued that mathematics is inductive in the sense that it develops in the direction of seeking more general solutions to problems, such as finding the area under a curve.[116] Here he made explicit something he only suggested in the *Cours*, where he explained that the invention of analytic geometry allowed for a general solution to the problem of finding a tangent to a curve, instead of taking up the problem anew for each curve in the manner of ancient geometry.[117]

Observation and Hypothesis

The method of observation developed when astronomy reached the positive state. For Comte, observation in science must always be guided by a theory or hypothesis of some sort: "In order to apply itself to the making of observations, our mind has need of some sort of theory. If, in contemplating the phenomena, we do not immediately attach them to some principles, not only would it be impossible for us to combine these isolated observations, and, consequently, to draw any fruit from them, but we would even be entirely incapable of retaining them; and, most often, the facts would remain unperceived before our eyes."[118]

Comte made the role of hypotheses in observation clearest in his account of the methods of astronomy. Direct observation in astronomy is limited to measuring the angles of points of light in the sky and recording the times. Phenomena such as the shape of the earth, the curve described by the path of a planet, and the diurnal motions of the heavens "are for the most part essentially constructed by our intelligence."[119] According to Comte, "the mind must construct the form or the motion which the eye could not embrace.... The simple geometric sketch of the diurnal motion would remain impossible without an abstract hypothesis that one compares to the concrete spectacle, in order to connect the celestial positions."[120]

The need for hypotheses to guide observations is true for sciences other than astronomy as well: "In any possible order of phenomena, even with respect to the simplest, no true observation is possible except to the extent that it is first directed and finally interpreted by a theory of some sort."[121] Without having some hypothesis in mind, "the observer most often does not even know that at which he ought to be looking in the fact which is taking place before his eyes."[122] He cautioned that any attempt at pure empiricism risks resulting in our observations supporting unspoken metaphysical assumptions. It is better that our observations be guided by explicit positive hypotheses.[123] Comte maintained similar views as early as 1825: "Absolute empiricism is impossible, no matter what has been said about it. Man is incapable by his nature not only of combining facts and of deducing any consequences from them, but even simply of observing them with attention, and of retaining them with certainty, if he does not immediately attach them to some explanation."[124] According to Comte, the brain is not simply passive in perception, but "furnishes the hypothesis according to which each group of sensations determines the corresponding entity."[125] Comte in fact saw no real distinction between observation and reason. He understood Kant as having argued that our minds are simultaneously active and passive and that all of our opinions are objective and subjective at the same time. Comte claimed to be reinterpreting Kant from a positivist or naturalized point of view when he argued that our empirical knowledge represents a harmony between a sentient organism and the environment acting upon it, and is thus relative to both.[126]

Although Comte argued that hypotheses were necessary to direct observations in science, we ought not to read into this anything like the twentieth-century thesis of the theory-ladenness of data. As Laudan argues, given that Comte appears to have held that observations can be guided by theological theory and subsequently used by positive science, it seems he assumed that the observational part of some system of thought can be separated from the theoretical.[127]

For Comte, astronomy provides the best model for the proper use of hypotheses, because it seeks only the laws governing celestial motions and not their causes.[128] Positivism generally rules out hypotheses that postulate underlying causes or modes of production of phenomena. He questioned why physicists could not follow the model of astronomy and limit themselves to hypotheses concerning the laws governing phenomena, instead of speculating on their mode of production. He was particularly opposed to the hypotheses of ethers and fluids that were used to explain the phenomena of heat, light, electricity, and magnetism, which he regarded as untestable. For Comte, these hypotheses were holdovers from the metaphysical notion of a substance or substratum, belonging to a stage of what he called a "bastard positivism," intermediary between the metaphysical and positive stages. For instance, although Descartes's vortex hypothesis was at least an attempt at a mechanical explanation of the phenomena and represented progress over Johannes Kepler's mystical accounts of the solar system, it was only a transitional hypothesis to be replaced in turn by Isaac Newton's celestial mechanics.[129] The time for such hypotheses had passed.

Although Comte was opposed to hypotheses postulating ethers and fluids as causes underlying phenomena, he was not against all hypotheses concerning unobservable entities. For Comte, a hypothesis is an "anticipation of what reason and experiment would show immediately if conditions were more favorable."[130] Atoms and molecules, unlike fluids and ethers, are not defined as invisible in principle but are simply too small to see. Thus they are the sort of things that reason and experiment *could* show us if conditions were more favorable:

> The intimate structure of actual substances necessarily remains unknown to us. But in studying their properties, we are rationally permitted to introduce with respect to them all the hypotheses that may facilitate our thought, provided that these artifices always conform to the nature of the corresponding phenomena. Now, the molecular conception very well fulfills these two conditions in all inorganic speculations, above all in physics, where it is spontaneously linked to the progress of the inductive

spirit and the ascendance of experimentation. While studying the general properties of material existence, it is appropriate to attribute them to the least particles that we are able to conceive.[131]

However, Comte found the atomic and molecular hypotheses to be acceptable mainly in physics, and went on to warn against their use in biology and to caution that they are of little use even in chemistry, because the properties there are too complicated and too variable to be attributed to "inalterable atoms."[132]

Comte maintained a strictly instrumentalist attitude toward atomic and molecular hypotheses. He compared the atomic theory to the concept of inertia in mechanics, and warned that "our tendency to endow our subjective constructions with an objective existence still distorts either conception, by supposing they exactly represent external reality."[133] Similarly, in the *Cours*, he advised against assigning a "faulty reality" (*une réalité vicieuse*) to molecules.[134]

There thus appears to be a tension in Comte's thought between his antirealist instrumentalism and his metaphysical belief in a "universal order" that exists independently of human investigations into nature. But perhaps there is no contradiction here, since Comte also said that science should aim to represent this universal order only to the extent that we need to know it for our practical purposes. Approximate generalizations that are, strictly speaking, false may nevertheless be good enough for science, just as they continue to be used in engineering. Comte thought that hypotheses could be useful even when they are not strictly true. What is important is that they be able to direct research: "If the hypothesis is not at all exactly true, as it ought to happen the most often, it will for that reason no less always necessarily contribute to the actual progress of the science, in directing the whole of the active researches towards a clearly defined goal."[135] According to Laudan, Comte thought that a hypothesis that is only indirectly verifiable through its implications may be a fiction, but can still be scientific.[136] In sum, although Comte maintained a metaphysical belief in the universal order of nature, the aim of science was not to reveal that order so much as to provide useful generalizations to guide our practical affairs and improve human lives.

Comte did not live to see the development of the kinetic theory of heat and gases, which persuaded many to adopt a realist view toward molecules.

Nevertheless, his attitude toward atoms and molecules was at least more liberal than that of some later positivists such as Ernst Mach or Wilhelm Ostwald.

Experimentation

The method of experimentation, which developed when physics and chemistry became positive sciences, also makes use of hypotheses. Experimentation differs from ordinary observation for Comte insofar as it involves the comparison of cases. What is essential to the method is that we have two different cases that are exactly alike in all respects save one: the characteristic that is under study.[137] Whether these cases are artificially produced or occur spontaneously in nature is immaterial: "the principal philosophical characteristic of experimentation . . . consists above all in the freest possible choice of a case appropriate for best revealing the course of the phenomenon."[138] That is, the purpose of the experiment is "to discover which laws are being followed as each of the determining or modifying influences of a phenomenon participate in its accomplishment."[139] In other words, experiments are designed to choose among competing hypotheses concerning the succession of phenomena. In Comte's account, the logic of the method of experiment thus sounds like that of Mill's methods—in particular his method of difference. Mill characterized this method in *A System of Logic*, which came out in 1843, after the *Cours* was completed. Comte recognized the similarity and in the *Discours sur l'esprit positif* recommended Mill's work.[140]

The Comparative and Historical Methods

Comte held that it becomes progressively more difficult to find two cases that differ in only one respect as one passes from physics to chemistry, biology, and sociology.[141] By the time one gets to sociology, experimentation is of little value. Other methods are thus called for. Biology introduces the method of comparison, which goes beyond the method of experimentation in employing not just two but a whole series of progressively more complex cases.[142]

Finally, sociology introduces the historical method, the "last part" of the comparative method.[143] Comte thought that the comparative method of biology is of limited value for studying the intellectual and moral functions, as there are not a sufficient number of animal species that exhibit these phenomena.[144] Animal societies are useful for studying only the "first germs" of social relations, that is, the spontaneous formation of the first social institutions, the family and the tribe.[145] However, animal societies do not have histories. They do not manifest the phenomenon of the gradual and continuous accumulation of the influences of successive generations, and it is this phenomenon that distinguishes sociology from biology.[146]

The Subjective Method

Once sociology has attained the positive state, the "subjective" method can then come into play. The subjective method is one of theory construction or synthesis, combining the results obtained through "objective" methods of induction and analysis in a deductive system. Earlier attempts at a theoretical synthesis have resulted in theological or metaphysical systems such as Descartes's. According to Comte, the positive subjective synthesis had to wait until all the sciences had reached the positive state.[147] The subjective method unites deduction with induction, bringing both within the scope of the positive method.

Although the term "subjective method" was introduced only in the *Système*, some of the concepts that make it up appear already in the *Cours*. The subjective differs from the objective method in the following ways: First of all, it is synthetic, as opposed to analytic.[148] The second difference is that the subjective method takes the point of view of the whole, which is provided by sociology, while the objective focuses on details.[149] Sociology can provide this global view because it is the science that studies the historical development of all the other sciences, their methods, discoveries, theories, and concepts, all of which are social phenomena for Comte. The encyclopedic scale is now inverted and the other five sciences are regarded as parts of a single science, sociology.[150] A third difference has to do with the way in which our systems of knowledge seek an equilibrium between external sensations and internal inspirations. The objective method favors the external while the subjective method favors the internal.[151] The fourth and final but related difference is that while the objective method takes the external world as its point of departure, the subjective method takes humanity as its point of departure in the sense that it seeks knowledge concerning that part of nature in which it is possible for us to intervene in order to ameliorate our condition.[152] For Comte, improving our lot is a social task.

The Question of Normativity

Comte's philosophy raises several normative issues. First, one could argue that he has presented at best only a descriptive and not a normative philosophy of science in the *Cours de philosophie positive*. His philosophy of science is a naturalized philosophy of science, as it is a synthesis of the methods he found in all the sciences of his day. That is, he has simply characterized the methods of what he regarded as the most successful sciences, but has given us no argument that these are the most successful sciences, that their success is due to the methods that they used, or that these are the methods we should follow. The second problem concerns whether a study of the methods of mathematics and the natural sciences could tell us anything useful about the methods we ought to use in the social sciences. Indeed, it is not even clear that Comte's methods of mathematical analysis, observation and hypothesis formation, experimentation, comparison, and historical analysis all are part of one and the same "positive method." The third problem concerns how we are to draw any normative lessons for how we should live from a descriptive science such as sociology. Yet Comte would derive a seventh science, morality, from sociology.

The first problem is the most fundamental. Mill raised only a part of the first problem when he complained that the *Cours* provides only a logic of discovery and not a logic of justification or proof: "The philosophy of Science consists of two principal parts; the methods of investigation, and the requisites of proof. . . . The former if complete would be an Organon of Discovery, the latter of Proof. It is to the first of these that M. Comte principally confines himself. . . . We are taught the right way of searching for results, but when a result has been reached, how shall we know that it is true? . . . On this question M. Comte throws no light."[153] Scharff attempts to defend Comte by arguing that Mill was seeking rules of logic abstracted from actual practice, while Comte regarded abstract rules of reasoning as a holdover from the metaphysical and theological modes of thought and held that methods could not be separated from their applications.[154]

Although there is some truth to both Mill's criticism and Scharff's defense, neither of them gets to the heart of the problem of normativity. Statements concerning "the right way of searching for results" have just as much normative content as rules for proving the truth of these results. We expect a logic of discovery to provide guidance in making rational decisions about how to pursue our research. Renouvier, for instance, questioned whether Comte's positive philosophy was able to provide a useful guide for working scientists and argued that a normative philosophy of science requires the methods of conceptual analysis, argument, and criticism that Comte rejected along with the metaphysical mode of thought.[155] To be sure, Renouvier's criticisms are somewhat exaggerated. Comte did, in fact, give arguments, for instance in his critique of the eclectic spiritualists' method of introspection. Nevertheless, there is some justice in Renouvier's remarks, as Comte presented little if any analysis of concepts such as "progress" either in science or in society, in spite of the crucial role they played in his philosophy.

Comte simply assumed that the victory of the positive over the first two methods represented success, and maintained that this victory was due to the positive method's greater fecundity in providing practical applications. Only the positive method provides the general laws governing the phenomena that are needed for making the predictions that guide these applications. If practical success is what we value, then we should choose positive methods over metaphysical and theological methods. The normative lesson is that scientists should restrict themselves to testable general laws connecting observable facts and abstain from the realist tendency to attempt to explain them by postulating causal entities hidden behind the phenomena.

However, this can be only a partial answer to the question of normativity. We still need a definition of practical success and an argument as to why our choices should be made on this basis. Comte's position that science should be useful actually led to what many would consider very bad advice for scientists. For instance, Comte dismissed sidereal astronomy as useless, since he believed that the stars can have no effect on us. He advised astronomers to restrict their research to our own solar system.[156] Comte should have known that William and John Herschel had been studying binary star systems to see whether Newton's gravitational law was confirmed outside our solar system. Félix Savary, a faculty member of the *École Polytechnique*, had published calculations of the orbit of a binary star system in 1827.[157] The confirmation of Newton's law for binary star systems would seem to bear on the question of its reliability closer to home. Comte also disparaged the use of probability and statistics in the sciences, particularly the social sciences, which would appear to undermine his emphasis on the applicability of the sciences to practical problems.[158]

Whewell thought that Comte was wrong to condemn hypotheses

concerning the modes of production of phenomena, arguing that, for instance, Fourier could not have conceived his theory of heat, one of Comte's favorite examples of a positive theory, without the idea of a flux of caloric. In the cases of double refraction and polarization, there is no other way to explain the phenomena except through the wave theory of light.[159] In Comte's defense, one could argue that one should maintain an instrumentalist attitude toward such hypotheses. However, even if the sciences were to achieve a purely positive state in which all scientists were instrumentalists, there would remain the question of how to choose between competing hypotheses or theories when all are arrived at through positive methods. Comte's answer is that we should always choose the simplest hypothesis that is consistent with all the facts that are presented. Of course, he was hardly the first philosopher to recommend the principle of parsimony. Curiously, he did not draw this principle from his study of the history of science. Instead, his justification for it is that complexity is bad for the human mind.[160]

Comte was not the last philosopher to propose a naturalized philosophy of science or epistemology. More recent attempts have looked to studies of the human mind as well as the history of science to see what they can teach us about knowledge. In all fairness, although facts about nature and history alone are not sufficient to determine what sorts of things should count as knowledge and how we should pursue our researches, they are not irrelevant, either. However, exactly what role the history of science should play in a normative philosophy of science is not an easy question to answer. And even if Comte could answer it, two normative problems with the positive philosophy remain. Laurent Clauzade and Vincent Guillin will discuss the problem of the application of the methods of the natural sciences, especially biology, to sociology, in chapters 4 and 5, respectively. Michel Bourdeau, Jean Elisabeth Pedersen, and Andrew Wernick will take up Comte's positive morality and political philosophy in chapters 6, 7, and 8.

★   ★   ★

Comte's contribution to philosophy was to have helped to bring about the shift away from traditional epistemology grounded in philosophy of mind and toward a philosophy of science that answered to the history of science. That we may question some of the lessons Comte drew from history is less important than the fact that he turned the attention of even his critics to the history of science in the first place. In addition, his philosophy was an early source of the contemporary notion that scientific traditions should be evaluated in relation to the historical situation in which they were pursued. Comte was a historicist philosopher of science who understood that a method of explanation that allowed for such things as celestial vortices or electrical fluids may have been quite appropriate in the past even if that method is not to be countenanced today. In Comte's view, the theological and metaphysical methods are not so much wrong as obsolete. His view of scientific knowledge is thus relativist in an additional sense to that of being relative to our needs and abilities: it is historically relative. Even within the positive stage, knowledge is relative to the development of that aspect of the positive method appropriate to each science. For instance, what sense could we make of the claim that the Earth orbits the sun without the development of the method of hypothesis in astronomy? Finally, Comte broke with the philosophical tradition that took an individualist approach to problems of knowledge and recognized that science is a social endeavor that depends on critical interaction among a community of scientists.

2

The Analytical Construction of a Positive Science in Auguste Comte

Michel Blay

From the very outset of the third lesson of his *Cours de philosophie positive*, Auguste Comte was quick to emphasize "the importance of positive philosophy in perfecting the general character of each science in itself."[1] Moreover, "it is only since the beginning of the last century that the development of the various elementary conceptions of this great science [mathematics] became sufficient so that the true spirit of the whole it formed could manifest itself clearly."[2]

"The beginning of the last century": so, for Comte, it was during the first years of the eighteenth century that science seemed to actually have become science. What happened during that period? The answer is straightforward: that was the moment when differential and integral calculus achieved their first true successes, after years of difficult diffusion and fierce debates, but also the moment, correlative to the previous one, when analytical mechanics started developing and, more generally, when the science of motion underwent its first algorithmic transformations and reconstructions. In other words, for Comte, the science that was beginning was, properly speaking, neither that of Newton nor even that of Leibniz, but the science that came after their works and formed itself on the model of a nascent analytical mechanics.[3]

For Comte, science can only be analytical. There is more than meets the eye in such a proposition, for it assumes both that everything that is not subjected to analysis must be expelled from science and that analysis must be applied, in line with the latest developments in the field, as illustrated by the "work of the immortal author [Joseph-Louis Lagrange] of the *Théorie des fonctions* and of the *Mécanique analytique*,"[4] to anything aspiring to become an object of scientific interest.

In light of these introductory remarks, it indeed seems that, for Comte, the analytical model imposes itself on mathematical science. Accordingly, the object of the latter consists in the construction of relations: "The general method that is constantly resorted to, and the only one that can be conceived to ascertain magnitudes that cannot be measured directly, consists in connecting these magnitudes with some others that can be directly determined and through which the former can be discovered by way of their relations to the latter. Such is the definite object of mathematical science considered as a whole."[5]

On what objects do these relations bear? Which objects do they relate? Comte's answer to these questions lies in the presentation of a few examples: falling bodies, triangulation, ballistics. None of these examples is strictly mathematical. They are all drawn from mechanics, physics, etc. This means that for Comte, there is no mathematics as such, only mathematics for physics, which of course suggests an instrumental and linguistic conception of mathematics. Mathematics is used to translate relations between objects in the language of analysis. It is a mere instrument or tool in the service of science, or more simply of the set of "facts" or "observations" that might be put in relation. The adoption of such a perspective forces Comte to neglect, or even refuse a priori, the idea of a science of motion or, more broadly, of science *qua* mathematical physics, i.e., he fails to acknowledge the full mathematical import of mathematical physics in the process of coordination, coherence, and conceptual genesis. By giving precedence to post-Newtonian science, a science that was already analytical, Comte overlooks the properly mathematical conceptual work that allowed for the transition from Newtonian mechanics to analytical mechanics, that is, primarily, the construction of the concepts of speed at each instant and of accelerative force at each instant, around which has been built an algorithmic kinematics that differs greatly from the Newtonian procedures based on an intrinsically nonalgorithmic infinitesimal geometry. Given the properly mathematical nature of its contribution, it might be said that mathematics entertains with physics a relation of constitution or, rather, a constitutive relation, for it is indeed the dynamics of their relation—one might also talk of their coupling—and not the mere translation of one into the other, that produced new concepts, most notably those of the science of motion.

In that respect, it is worth noting that the first example given by Comte in the third lesson concerns falling bodies and does not mention any kind of conceptualization: "Observing this phenomenon, even the mind most unfamiliar with mathematical conceptions will immediately acknowledge that the two quantities involved, namely the height from which the body falls and the time occupied by its fall, are necessarily connected with one another, since they both vary together and remain fixed together; or, in the language of geometricians, they are functions of one another."[6]

Those are surprising claims. Indeed, although, according to Comte, "even the mind most unfamiliar with mathematical conceptions will immediately acknowledge" that the distinction between space, time, and their relation is obvious, it was not so for Galileo. In an undated *frammento* drawn up in Italian by Galileo, which probably dates back to 1604, one finds a demonstration relating to falling bodies that relies on the hypothesis according to which the degree of speed of a free-falling body is proportional to the space covered but not the time, as Comte obviously seems to think.[7]

The choice of time, instead of space, as the independent variable calls for some conceptual work that is not obvious for Galileo, a work that bears on both the choice of that variable and the assumption that time is discontinuous.[8] And it was indeed the case that the influence of discontinuist ideas, inspired by atomistic positions, remained very strong throughout the seventeenth century, as illustrated by the works of François Bernier or Edme Mariotte.[9]

This continuity of time and of the other variables (space, speed, etc.), which Comte seems to have readily accepted, fits nicely with his analytical conception of science. It is indeed only after Gottfried Wilhelm Leibniz had stated and defended, against the advocates of discontinuism, his principle of continuity, that analysis allowed for the transformation of physics according to the model of science endorsed by Comte.

★ ★ ★

In 1687, Leibniz wrote in a letter to Antoine Arnauld, "It is a defect in the reasoning of Descartes and his followers not to have considered that everything that is said of motion, of inequality, and of elasticity, should also be true if things are supposed to be infinitely small or infinite. In this case motion (infinitely small) becomes rest, inequality (infinitely small) becomes equality, and elasticity (infinitely prompt) is nothing else than extreme hardness."[10] Moreover, in 1704, in the *Nouveaux essais sur l'entendement humain*, Leibniz specified that "nothing takes place suddenly, and it is one of my best great and best confirmed maxims that *nature never makes leaps*. I called this the Law of Continuity. . . . There is much work for this law to do in natural science."[11] In other words—as phrases such as "will immediately acknowledge," "immediate data," or "precise observation suggests by itself the law" illustrate—Comte overlooks the basic conceptual work through which "facts" or "immediate data" are constructed with the help of concepts that can be handled mathematically and that can be expressed quantitatively.[12] Of course, these facts are never given; yet Comte holds them to be so and proceeds directly with the equations and mathematical resolution. But these two operations are indeed only possible provided that a fact be mathematically constructed and that the law of continuity allows for the use of differential and integral analysis.

Once the specifically conceptual work has been set aside, Comte is able to formulate an elegant definition of mathematical science that could please an *ingénieur polytechnicien*:

> We have finally managed to arrive at an exact definition of mathematical science, whose goal is the indirect measurement of magnitudes and which constantly aims at determining magnitudes by one another, according to the precise relations which exist between them. This definition, instead of merely evoking the idea of an art, as was the case with all ordinary definitions so far, immediately distinguishes it as a true science and instantaneously shows that it is composed of an immense sequence of intellectual operations that might of course become very complicated.[13]

Accordingly, for Comte, mathematical science boils down to the systematic elaboration of analytical algorithms and their development into increasingly richer systems of calculus by way of the establishment of relations between the various magnitudes of physics. "According to that definition, the mathematical spirit consists in always considering as mutually connected all the quantities that might be represented by any phenomenon whatsoever, so as to be able to deduce them from one another."[14] This is the general framework adopted by Comte. In order to grasp more fully how it was implemented in the lessons of the *Cours de philosophie positive*, we will focus on two examples, the former bearing on kinematics, the latter on optics.

The Construction of a Positive Kinematics

In the opening pages of the fifteenth lesson, Comte reminded his readers that, on the one hand, analysis cannot establish or ground a science but can only contribute to its development—mathematical analysis can only serve to relate or translate:

> One might even concede that the great improvement of rational mechanics in the last century, either with regard to the scope of its theories or their coordination, has caused some sort of regression in our philosophical appraisal of that science, whose contemporary presentations are quite less clear than Newton's. Because this improvement had been primarily the consequence of an increasingly exclusive use of mathematical analysis, the major importance of this wonderful tool has gradually accustomed us to conceive rational mechanics as consisting only of simple questions of analysis and, by way of an illegitimate—although natural—extension of such an approach, it prompted attempts to establish a priori, in accordance with purely analytical considerations, the elementary principles of that science, which Newton wisely contented himself with presenting as resulting from pure observation.[15]

On the other hand: "What establishes the reality of rational mechanics is indeed the fact that it is grounded in a few general facts, directly provided by observation and that any true positive philosopher must consider, it seems to me, as not amenable to any kind of explanation."[16]

The Comtean perspective remains the same: "general facts directly provided by observations" and some analytical procedures that link together these general facts, just as if these facts were "directly provided" by some sort of miraculous intuition that Comte calls "observation."

In accordance with this general framework, "causes" play no part in scientific inquiry and must be banned from science: "It is common to underline, and rightly for that matter, that mechanics neither considers the primary causes of motions, which exceed the scope of any positive philosophy, nor the circumstances of their production, which, although they constitute an interesting topic for positive investigations in the various parts of physics, are not in the purview of mechanics, which limits itself to the study of motion, with no interest whatsoever in the way it has been produced. Thus, in mechanics, forces are only motions produced or tending to be produced."[17] All this is fine, but a difficult problem remains, as noted in the seventeenth lesson. From the very first lines, Comte points out that the primary object of dynamics consists in the "study of the varied motions produced by *continuous* forces, the theory of uniform motions produced by *instantaneous* forces being only a simple direct consequence of the three elementary laws of motion."[18]

This definition assumes—Comte highlights the importance of terms such as *continuous* and *instantaneous* by using italics in the text—that a relation must be established between, on the one hand, instantaneous forces linked to the formulation of the principles or laws that produce uniform motions and, on the other hand, the continuous forces that result in varied motions. The question raised by Comte here reflects the difficulties associated with the use of fundamental laws or principles that have been built upon the model of collision in the analysis of problems involving a continuous action of forces (as, for instance, that of gravity). Comte is quite straightforward about these issues when he specifies, in the fifteenth lesson, that

> from this, it follows that all the mechanics of uniform motions or instantaneous forces can be entirely treated as a direct consequence of the combination of these three laws that, because they are extremely precise, can be of course expressed through easily obtained analytical equations. As for the most extended and important part of mechanics, that which is also the most difficult, namely the study of varied motions or continuous forces, one might conceive that, in general, it is possible to reduce it to the elementary mechanics we have just considered, which will enable us, through the application of the infinitesimal method, to substitute, for any infinitely small point of time, a uniform motion for a varied one, whence will directly result the differential equations relative to this latter kind of motion.[19]

This issue is important from the point of view of the elimination of causes. For, with respect to varied motion, variation is subject to an implicit model of collision. That is why Christiaan Huygens, in his *Discours de la cause de la pesanteur*, which was published in Leiden in 1690 as an appendix to the *Traité de la lumière* but which had been written, for the most part, twenty years before, as an offshoot of debates that had taken place within the Académie Royale des Sciences de Paris, argued: "Eventually, one can find here the reason for the principle Galileo used to demonstrate the proportion of the acceleration of falling bodies: which is that their speed increases equally in equal times. For bodies being moved successively by the parts of matter which try to take their place and which, as we have just seen, act constantly on them with the same force, at least for the falls that are within the limits of our experience, it necessarily results that the increase in speed is proportional to the increase of times."[20]

Huygens also emphasizes, in the first pages of the second part of the *Horologium Oscillatorium*, published in Paris in 1763, that the mathematization of the motion of falling bodies presupposes that the action of gravity, irrespective of any explicit hypothesis about its cause, must be continuous during all the equal time periods considered. For instance, it is stated in Proposition I: "And this latter [motion produced by gravity], obviously being the same in the first as in the second time, must have given the body during the second time [*ideo decursu temporis secundi*] a speed equal to that it had received at the end of the first." Similarly, in the next paragraph: "The force of gravity added some more during the third time."[21] Because he took for granted this continuist construal of the action of gravity and used the idea that the ratio of the spaces covered remains the same whatever the successive equal time periods considered,[22] Huygens obtained the Galilean classic law of odd numbers, which is characteristic of the accelerated motion of falling bodies.

Accordingly, one can distinguish between, on the one hand, the succession of impulsions, or collisions, and, on the other hand, the alleged continuous action of, say, gravity. How is one to pass from the conceptualization of action in terms of impulsions or collisions to the mathematization of an uninterrupted, continuous action—for example, that of gravity? Or, more precisely, how is one to confer mathematical rigor to the transition from discontinuity to continuity in the modalities of action?

I take it that the answers to these questions represent one of the central issues of Newton's *Philosophiae naturalis principia mathematica*, which was published in London in 1687. In that book, Newton faces problems involving the consideration of a continuous, uninterrupted action (central force, the motion of projectiles in resisting mediums). Now, what is stipulated by the second law or axiom of motion, which specifies the modalities of the action of force? "Law II. The change of motion is ever proportional to the motive force impressed; and is made in the direction of the straight line in which that force is impressed."[23] Of course, that law should not be conflated with the one, expressed in differential terms, that is known today as "Newton's Law" and is expressed as $F = ma$ or $F = m.d^2x/dt^2$. Most notably, Newton evokes here a "change of motion" without any precision regarding the time—not even a *dato tempore*—during which this change happens. If one were really willing to formulate this law in modern terms, the closest expression would probably be the following:

$F = \Delta (mv)$,

in which F is the motive force impressed, m the mass, and v the velocity, considering that $\Delta (mv)$ represents the "change of motion." In that perspective, one might say that the motive force that is impressed is not a force in the modern sense of the term, but an impulsion.

Accordingly, in the demonstration of the law of falling bodies that Newton offers as an example of a deduction from Law II, in the "Scholium" that follows the formulation of the laws and their six corollaries, the "motive force impressed" appears clearly, via the process of totalization resorted to, to be an impulsion: "When a body is falling, the uniform force of its gravity acting equally, impresses, in equal particles of time, equal forces upon that body, and therefore generates equal velocities; and in the whole time impresses a whole force, and generates a whole velocity proportional to the time. And the spaces described in proportional times are as the velocities and the times conjunctly; that is, in a duplicate ratio of the times."[24]

Moreover, two references to Law II given by Newton as part of the demonstrations elaborated in book 2 shed a similar light on the content of this law. Proposition III states that a series of rectangles "will be as the absolute forces with which the body is acted upon in the beginning of each of the times, and therefore (by Law II) as the increments of the velocities," and, a little bit later in Proposition VIII: "for the increment PQ of the velocity is (by Law II) proportional to the generating force KC."[25] In both cases, no precision regarding time is given. Accordingly, Law II indeed seems to be linked to a model of action based on impulsion, a model that presupposes a discontinuist conceptualization of that action.

But, if so, how is one to solve the problems that involve continuous action? Newton draws an implicit distinction between two situations, depending on whether he deals with first-order or second-order infinitesimals to overcome the problems generated by continuity.[26] In the former case, he resorts to transitions to the limit based on the manipulation of intervals or successive time particles (the laws of the areas, motion of projectiles in resisting mediums); in the latter case, he relies on Lemmas IX, X, and XI of the first section of book 1 of the *Principia*. That section is dedicated to the "method of first and last ratios" and provides the mathematics that are appropriate for addressing the second order of questions, such as that of the determination of the expression of force in the case of central force motions. In these analyses that bear on continuous action, the force of Law II is replaced either by reference to Lemma X, which introduces a quasi-acceleration ("The same thing is to be understood of any spaces whatsoever described by bodies urged with different forces; all which, in the very beginning of the motion, are as the forces and the squares of the times"[27]) or by an equivalent reference to an osculating circle (Lemma XI).

These problems faced by Newton reappear at the beginning of the eighteenth century in the work of Pierre Varignon on the algorithmization of the science of motion. Following Newton, Varignon does not avoid the problem of the transition from the continuous to the discontinuous because he inherits from him the same underlying modelization of collision.[28] That is the very problem Comte intends to solve, so that mechanics does not have to rely on any intuitive foundation.

Let us now return to the fifteenth lesson of the *Cours*. The Comtean formulation of the three fundamental laws of motion primarily rests on the idea that mechanics "limits itself to the study of motion, with no interest whatsoever in the way it has been produced. Thus, in mechanics, forces are only motions produced or tending to be produced."[29] In that respect, Comte agrees with Jean le Rond d'Alembert, for whom "motion and its general properties are the first and primary object of mechanics."[30]

The three Comtean principles or laws of motion are thus as follows: "the first law is the one that is inappropriately referred to as the law of inertia. It has been discovered by Kepler. It holds that all motions are naturally rectilinear and uniform, i.e., that any body impelled by any single force, acting instantaneously, will move in a straight line with a constant speed."[31] That first law of motion, which actually had not been discovered by Kepler[32] and which corresponds to Newton's first law, clearly stipulates that force, within this context, acts instantaneously. And here again, the implicit model of collision that is at work is the very same one that has presided over the conceptual elaboration of Huygens, Newton, and Varignon. "Newton provided us with the second elementary law of motion. It holds that there exists a constant and necessary equality between action and reaction; i.e., that whenever a body is moved by another body, the former exerts on the latter, but in the opposite direction, a reaction such that the latter loses, in proportion to the masses involved, as much motion as was received by the former."[33] One finds here the third law stated by Newton in the *Principia*. This third law, which does not appear in the 1685 version of the texts that preceded the writing of the *Principia*, enables Newton, in book 3, to formulate the law of universal gravitation in its full scope. "It seems to me that the third elementary law of motion maintains what I would suggest calling the principle of independence, or coexistence of motions, which directly leads to what is commonly called the composition of forces."[34] That law plays a central role in the approach adopted by Huygens in the *Horologium Oscillatorium*.

Finally, the formulation of these three fundamental laws, two of which, the first and the second, are present in Newton's work whereas the third is elaborated by Huygens in the three "hypotheses" of his *Horologium Oscillatorium*, ignores Newton's second law, which specifies the mode of action of force. For it is indeed the case that this mode of action is explicitly introduced by the adverb "instantaneously" in the first law. This absence of a law relative to force illustrates the fact that Comte clearly denies force any ontological or causal status.

Once these three laws have been stated, it is important for Comte, as it had been for his predecessors, to address the issue of instantaneous action, so that the science of motion can be provided with its analytic and continuist coherence. That is the central point of the seventeenth lesson, the opening lines of which, as noted previously, emphasize that dynamics primarily consists in the "study of the varied motions produced by *continuous* forces, the theory of uniform motions produced by *instantaneous* forces being only a simple direct consequence of the three elementary laws of motion."[35]

In the following pages, Comte takes up Varignon's algorithmic construction, which is also to be found in the first pages of d'Alembert's 1743 *Traité de dynamique*.[36] However, thanks to an ingenious choice of time intervals, Comte is able to eliminate the remaining ambiguities that still plagued Varignon's account:

> Finally, if one considers this order of questions in the most extended manner possible, one may claim that, in general, the definition of any varied motion can be given by an equation which includes these four variables at once, only one of which is independent, namely time, space,

speed, and force. The problem will amount to deducing from that equation the precise determination of the three characteristic laws relative to space, speed, and force, as a function of time, and then conjunctly. This general problem is always reducible to a purely analytical undertaking, carried out with the help of the two elementary dynamic formulas which express speed and force, as a function of time, when the law relative to space is assumed.

The infinitesimal method easily leads to these formulas. To get them, it suffices to consider, in accordance with the spirit of the method, motion as uniform for the duration of a single infinitely small time interval and as uniformly accelerated during two consecutive intervals. Thence speed, which has been assumed to be momentarily constant in accordance with the first consideration, will be naturally expressed by the differential of space divided by that of time; similarly, a continuous force, in accordance with the second consideration, will be measured by the ratio between the infinitely small increase of speed and the time needed to produce that increase. Accordingly, if t is the elapsed time, e the space covered, v the acquired speed and ϕ the intensity of the continuous force, the general and necessary correlation of these four variables will be expressed analytically by the two following elementary formulas:

$$v = de/dt$$
$$\phi = dv/dt = d^2e/dt^2$$

According to these formulas, all the equations relative to that preliminary theory of varied motion will immediately be reduced to mere analytical inquiries, which will consist in differentiations or, most often, integrations. When considering the most general case, in which the suggested definition of motion would be given by one single equation including the four variables, the analytical problem will amount to integrating a second-order differential equation, relative to the function of e, but which might often happen to be impossible to carry out, given the current extreme imperfection of integral calculus.[37]

Therefore, Comte elegantly succeeds in technically erasing the conceptual problems relative to the transition from the underlying colliding model of the action of force, which presided over the elaboration of the principles and the laws, to the analysis of that action in continuist terms.[38] With Comte, as was already the case with Lagrange, the science of motion gains its full coherence by freeing itself from the intuitive and spontaneous models of the action of force and by confirming the new functions assigned to algorithmic procedures.

The science of motion has become a "positive science." However, it still relies on a somewhat naïve relation to observation as well as on implicit conceptions of time, space, speed, etc., that remain tied to the principle of continuity, without which analysis, as practiced by Comte, cannot "work."

Accordingly, we have just shown that the positivist requirement, although it completely overlooks the whole process of conceptualization, transforms the science of motion through the elimination, most notably, of force understood, ontologically, as a cause. Let us now turn to optics to discover another way of constructing a "positive science."

The Construction of a Positive Optics

In the thirty-third lesson of the *Cours*, Comte, dealing with light, specifies that "the formation of this great science owes primarily to the philosophers who have contributed most efficiently, in many other crucial respects, to the foundations of positive philosophy, such as Descartes, Huygens and Newton; yet, the hidden influence of the old metaphysical and absolute spirit has prompted each one of them to create a hypothesis, which was necessarily chimerical, concerning the nature of light."[39] It was indeed the case that, quite rapidly during the seventeenth century, two trends emerged among scientists, depending on whether they considered light to be a body or the motion of a body without any displacement of matter. In the former case, one deals with so-called emission theories, whose main representative figure was Newton; in the latter case, one deals with so-called medium theories, which are best illustrated by Huygens's work. In such a context, how is one to build a positive science of light?

As noted previously, to become positive in the Comtean sense of the term, any study of the sciences must free itself from any relation that its results might have with "modes of production" associated with hypothetical or metaphysical entities. Comte specifies very clearly his position when he claims that "in physics, it will be admitted as a fundamental principle of the true theory of hypotheses that every scientific hypothesis, in order to be able to be genuinely appraised, must bear exclusively on the laws of phenomena, and never on their modes of production."[40]

Here it is also important to draw a distinction between two kinds of hypotheses, for, even if it is true that Comte considers hypotheses to be necessary instruments of scientific inquiry, one must be wary of not conflating hypotheses bearing on entities such as waves or corpuscles, and hypotheses that serve to introduce the "laws of phenomena":

> A careful distinction must be drawn between the various hypotheses resorted to by physicists today: the former, which so far remain few, are merely relative to the laws of phenomena; the latter, which are quite more common, concern the determination of the general agents which are to account for the various kinds of natural effects. According to the fundamental rule laid down previously, only the former are admissible; the latter, which are chimerical by nature, do have an antiscientific character and can do nothing but radically hinder the real progress of physics, instead of furthering it.[41]

From these general remarks, it obviously follows that waves and corpuscles must be banned from the field of positive physics; accordingly,

> after the general discussion of the twenty-eighth lesson on the fundamental theory of hypotheses in natural philosophy, it would be entirely superfluous to consider here, in a specific manner, either Newton's fiction about light, or that, which is necessarily as pointless as its predecessor, by which it has been replaced, following Descartes, Huygens and Euler: anybody could apply to both, with due respect for their differences, all the essential principles of the new philosophical doctrine. The radical worthlessness of these antiscientific conceptions, when considered in relation with their direct purpose, does not need to be formally acknowledged; it is enough to wonder, independently of any ordinary scholastic prejudice, whether the luminous faculty of bodies is really explained in any way by the very fact that it has been transformed into the property of emitting, at an incomprehensible speed, chimerical molecules, or of causing the vibration of motionless particles in an imaginary fluid, the elasticity of which cannot be appraised.[42]

The research program of optics is indeed straightforward, although it seriously compromises future investigations since, if one follows Comte, "light will remain eternally heterogeneous to motion or sound":[43] the field of physics, or at least most of it, must be freed from the hold of metaphysics or, more precisely, from the use of antiscientific hypotheses on the inner nature of phenomena.

Finally, a few lines below, Comte argues as if he were in search of the "physicist who, imbued with the philosophical doctrine established in this book, would undertake the writing of a special treatise so as to carry out properly that final purification." By doing so, by getting rid of waves and corpuscles, he "would, if I dare speak so, do science a crucial favor."[44]

The name of this distinguished physicist might have been Pierre Duhem, who put all his efforts into the completion of the Comtean endeavor by writing his "Fragments d'un cours d'optique." From the outset of his lectures, Duhem clearly stipulates the perspective he will adopt: "It seems that the time has arrived for optics to steer clear of these hypotheses on the nature of light and to imitate thermodynamics which has, little by little, given up the mechanical theory of heat."[45]

Accordingly, a reorganization of optics becomes possible on the basis of a few principles stipulated more or less arbitrarily or freely, and freed from any consideration regarding the nature of light or transparent mediums: "We have endeavored, in the last few years, in our teaching, to show that the most well-established discoveries of optics can be presented logically, without any reference being made to any hypothesis regarding the nature of light; this attempt has led us to reorganize quite completely the form of some of the chapters of this part of physics."[46] Similarly, in the first pages of the second fragment, Duhem specifies that "we merely conceive light as a *quality* and physical optics as a *system of symbolic equations* whose goal is to *represent* and not to *explain* the features experimental physics analysis identifies in that quality."[47]

It is not possible here to delve into the very subtle and intricate mathematical construction that Duhem elaborates, with analysis in the background, in the "Fragments." But it is worth noting that the first fragment deals with Huygens's principle of the envelopes of the waves and that Duhem intends to reformulate it independently of any hypothesis concerning the existence of metaphysical entities:

> The principle we are about to deal with has been discovered by Huygens and introduced by him in the *Traité de la Lumière* he published in the Hague in 1690: he discovered it by making *the hypothesis, which never was obvious nor certain*, that light essentially consists in a very fast motion of the parts of the lit body and by assimilating that motion to that which, *not hypothetically but as a truth of experience*, constitutes sound. We will try to give Huygens's principle a definite form, whilst freeing it from any hypothesis on the nature of light.

> We will take for granted the general definitions of *homogeneous* or *heterogeneous*, *isotropic* or *anisotropic* medium.[48]

That being said, Duhem successively introduces the hypotheses and definitions, just like in an account of thermodynamics, he needs for the mathematical reconstruction, understood as a deductive sequence of differential equations, of Huygens's principle.

The second fragment is dedicated to Thomas Young's and Augustin-Jean Fresnel's optics, which Duhem considers mere developments of Huygens's and which he thinks are able to account for the phenomena of interference and diffraction because of the renewal of the field of mathematics:

> In a previous fragment, we have introduced, whilst freeing it from any hypothesis concerning the nature of light, Huygens's optics. Today... we tackle Young's optics. Once again, we set aside the hypothesis that light is motion, whatever its historical role in the development of optical theories. Resolutely breaking... with the law Descartes wanted to impose on physics, which, for two centuries, it had tried to shake through muted and unconscious efforts, we only want to conceive light as a *quality* and physical optics as a *system of symbolic equations* whose goal is to *represent* and not to *explain* the features experimental physics analysis identifies in that quality.[49]

Here, light is treated, just like heat, as a quality; and in that respect, Duhem totally agrees with Comte.[50] The goal of scientific inquiry consists in the construction of differential equations that might "represent" or describe the processes through which light phenomena actually happen, with no reference being made to any explanatory hypothesis concerning the nature of light. Comte could only be pleased with that.

The third fragment is a followup to the second and deals more specifically with Fresnel's optics. In the closing lines of this fragment, Duhem forcefully emphasizes again the theoretical and epistemological significance of such a reconstruction, its truly positivist and antimetaphysical aspect: "We will end here these fragments, in which we have presented the most developed part of Optics; we have said enough so as to be able to prove that a rational optics can be elaborated without any reference to any hypothesis on the inner constitution of transparent mediums or the nature of light."[51]

Comte called for the construction of a "rational optics" that would not venture hypotheses on the nature of underlying objects. In his mathematical work, Duhem has shown that it might perhaps be possible, but, in the light of the development of physics, that strictly positivist optics came to naught. In 1905 Albert Einstein (1879–1955) simultaneously published his paper "On the Electrodynamics of Moving Bodies," which establishes the foundations of relativity theory, and his paper on the photoelectric effect, which actually introduces the hypothesis of quanta of light. And this later paper indeed questions the strictly wavelike and continuous nature of light advocated throughout the nineteenth century, and which also inspired Duhem in his attempt to ground his construction in analysis and in the implicit reference of the Leibnizian continuity principle previously evoked.

It is that "dual nature" of light that Louis de Broglie (1892–1987) tries to understand and interpret in his 1924 *Recherches sur la théorie des quanta*. In his PhD thesis, de Broglie assumes that the wave-corpuscle duality is a general feature of microscopic objects and that matter, just like light, manifests that dual aspect. That hypothesis will be rapidly confirmed by the observation of diffraction phenomena with electrons (Clinton Davisson and Lester Germer's experiments in 1927, G. P. Thomson's and Emil Rupp's in 1928).[52]

Does this mean that hypothetical entities defeated both Comte and Duhem? In any case, the mathematical sciences as Comte understood them reflected a certain state of physics that was characteristic of the methods favored by the engineers trained at the École Polytechnique during the nineteenth century. The glorious justification of that science in the *Cours de philosophie positive* relies both on a naïve appraisal of the status of mathematics, which overlooks the significance of concepts that can be mathematically handled and expressed quantitatively in the construction of "facts" that are never directly given by observation, and on a somewhat superficial understanding of the status of hypothetical entities and of the role of a priori assumptions such as, for instance, that of continuity.

Comte's positive science is indeed an artifact.

3   Astronomical Science and Its Significance for Humankind

Anastasios Brenner

Astronomy was of primary importance for Auguste Comte. Historically, it was one of the first branches of science to take shape; following right on mathematics in the encyclopedic order, astronomy opens the series of fundamental disciplines dealing with the world. And Comte's only research paper, devoted to the formation of the solar system or cosmogony, comes under this field. Furthermore, one of his most successful occupations was the public course he delivered on astronomy. Pursued for many years, it was well received and provided him with evident satisfaction. Comte came to publish this course under the title *Traité philosophique d'astronomie populaire*.[1] This major work, to which was appended the methodological tract *Discours sur l'esprit positif*, appeared in 1844, a time that could be taken to signal the middle of the author's career. Comte would allude frequently in his later writings to his description of astronomical science. He had furnished thereby an illustration of scientific method, but he had also provided a key to the elaboration of the science of man. What is characteristic of Comte's approach is that he did not separate astronomy from the other sciences. He was intent on bringing out its significance for humankind. Astronomy was even to serve as a preparation for his political philosophy.

Comte is recognized as one of the founding fathers of philosophy of science. Among the themes he contributed to the field, one can mention the role of hypotheses, the nature of scientific progress, and the value of experimental testing. And astronomy provided him in this respect with his main examples. The logical positivists—or logical empiricists, as they are also named—acknowledged some debt to Comte.[2] Like the Vienna Circle positivists, Comte formulated an empirical criterion of meaning in order to exclude metaphysics. He also adopted, with respect to the development of science, a resolutely sociological approach. One has the impression of reading Rudolf Carnap or Otto Neurath. Yet Comte did not reduce philosophy of science to a logical analysis of scientific language. The question then arises of the meaning of Comtean positivism and its relation to the various doctrines going under the label of positivism.

Comte gave prominence to astronomy. Such is not the case of all forms of positivism. In particular, the logical positivists do not have much to say about this science. Comte thus offers an instance of an antimetaphysical doctrine that allows for astronomy, a science seemingly far removed from our everyday concerns.

Comte's public course on astronomy provided him with an opportunity to come into contact with workers. In this instance his ambitions coincided with the workers' expectations. Of the diverse attendants at the course, they proved to be the most receptive and appreciative. Scientists, on the other hand, were in the main hostile to positivism. This was a strong incentive for Comte to refashion his philosophy. He was to reflect in greater depth on the relation between abstract study and concrete proposals. We have a means here of reassessing what a disciple such as Émile Littré perceived as two distinct doctrines: the sober, scientific positivism of the *Cours de philosophie positive*, and the extravagant, religious positivism of the *Système de politique positive*.

My aim in this chapter is then to reexamine Comte's astronomy and to make explicit its role within his philosophy. We have at our disposal several texts devoted to this science. Comparisons among the passages of the *Cours* and the *Système* will allow us to confirm, to complement, and to reveal some significant evolutions of his thought.

Comte's Account of Astronomy

Comte gave at least three significant presentations of astronomy. The first is included in the lessons devoted to this science in the *Cours de philosophie positive*, which were published in 1835.[3] The second is that of the *Traité philosophique d'astronomie populaire*, printed some ten years later. Comte returned again to astronomy in 1851 in chapter 2 of the introductory part of the *Système de politique positive*, under the title "Indirect introduction, essentially analytic or cosmology." We thus have three versions of Comte's astronomy, one from each decade of his career. The presentations of the *Cours* and of the *Traité* follow basically the same order.[4] A major difference, to which we shall return, is the omission in the second work of the substantial development on cosmogony. There are others, for example, the introduction of the concept of objectivity. The *Traité* follows the same series of problems as the *Cours*; but the treatment is more detailed. The *Système* provides only "a very summary appreciation," in Comte's own words.[5] We can then concentrate on the picture of astronomy delivered in the *Traité*, while calling on the *Cours* in order to bring out the relation to the other sciences. Recourse to the *Système* will help us in turn to define the place of astronomy with respect to the social and political theory as well as to locate changes in Comte's outlook.

One way of bringing out Comte's originality is to examine how he presented his work with respect to that of his predecessors. He duly mentioned two earlier endeavors of a similar nature: Bernard le Bovier de Fontenelle's 1686 *Entretiens sur la pluralité des mondes* and Pierre-Simon Laplace's 1796 *Exposition du système du monde*. The former was praised by Comte and obviously served as a model. Although inspired by Cartesian physics, its author was considered successful in his attempt at popularization. Fontenelle brought the layperson—in this case a marquise rather than a worker—to an understanding of the "new science" of the seventeenth century. One reads easily between the lines a typical Enlightenment lesson on emancipation.[6]

In contrast, Comte was dismissive of Laplace. He was of course obliged to situate himself with respect to Laplace's book, owing to its huge success. He objected to it for several reasons: rather than initiate his reader to reasoning in astronomy, Laplace gave a simplified account of his discoveries. He tended to promote his own research, and his book, directed rather to the upper classes, was an attempt to enhance the author's status. Comte's account is the very opposite. One should place the disparagement of Laplace's popularization in the setting of Comte's general attitude toward him. He disagreed with the latter's method and conception. Laplace had a penchant for theoretical speculation. In his physics, he introduced hypothetical fluids and imaginary entities, which was contrary to the positive spirit.[7]

Let us note that the *Traité* was published some forty years after Laplace's *Exposition*. We may establish another reference point, for our own needs: in 1880, some forty years after Comte, Camille Flammarion would publish his *Astronomie populaire*, which offers a speculative, optimistic outlook and signals the appearance of a new attitude. Flammarion takes up with enthusiasm all those topics that Comte had forbidden: the composition of the sun, other planetary systems, extraterrestrial life, etc. He was truly a contemporary of Jules Verne. So we have three characteristic examples of the popularization of astronomy: Enlightenment, Romantic, and Belle Époque.

In several passages of the *Traité*, Comte asserts the originality of his approach: several things have not been well understood until now, namely, the significance of astronomy, the basic nature of its method, and what humanity has learned from contemplating the stars. This is true not only of philosophers or journalists but even of professional astronomers and physicists. What of course comes as a surprise to the modern reader are the limits that Comte is constantly setting on research. We have noted that he no longer finds it necessary to present the hypothesis on the formation of the solar system. He mentions it only to dispense with it more easily.[8] He wants to impose numerous restrictions on speculation; he perceives barriers everywhere: we cannot go any further; we shall never know more; it is beyond our means. There is a restraint, a holding back, a pessimism that comes as a surprise in the field of astronomy.

Comte's account of astronomy formulates a strong contrast between ancient astronomy and modern astronomy. The latter is marked by the adoption of the heliocentric system and its consequences: an uncentered Earth and new methods of investigation. Comte provides his audience with a general knowledge of astronomy: the phenomena involved, the mathematical language employed, and the instruments used. He gives a series of tables summarizing the particularities of our solar system: the distances of the planets to the sun, their diameters, the periods of their rotations, the periods of their revolutions, and their masses.[9] Comte views science as a cumulative process: approximations of first, second, and third order are considered. Thus planetary motions are conceived first as circular, then as elliptic, and finally as the complex result of gravitational forces.[10] They represent so many stages in the pedagogic exposition. Comte emphasizes this gradual and continuous process in what resembles the science of textbooks. Kuhn has taught us to beware of such presentations, which correspond to normal science and leave out extraordinary science.

A characteristic feature of Comte's account is that he incorporates instruments. He speaks at length of the various instruments employed in an observatory: telescopes, clocks, and gnomonic projections. He takes pains to explain their functioning. He is careful to offer some elements pertaining to the history of technology, such as the invention of the reticle, the vernier, and the achromatic lens. This is in tune with his general outlook: scientific development is to be located within general progress. He thereby makes explicit the "extreme precision" of modern astronomy; it is the result of theoretical corrections and what have come to be called "precision instruments."[11] In this respect, Comte provides precious insight. Later thinkers will develop the thesis that the employment of an instrument in science presupposes theoretical ideas concerning its manner of operating—a theory of the instrument.

Comte's philosophical treatment does not offer a formal analysis of the structure of astronomical theories, following the method of the logical positivists. It incorporates much history. It remains, however, to understand how the past is put to use. Comte avoids erudition; he does not call on philology, paleography, and archeology. His history is cursory. He is intent on transmitting an idea of the overall evolution: the three stages of the development of humanity, the fact that the positive mind was present from the very beginning and served as the motive force behind progress. His study is supposed to be based on a "true philosophy of history."[12] Comte gathers his historical information from the classic works of Jean-Baptiste Delambre, whose judgment he praises.[13]

These historical studies were becoming obsolete, soon to be surpassed by new research. Paul Tannery would go back to the original sources, submitting them to critical scrutiny and reconstructing on this foundation a new interpretation, thus in his 1893 *Recherches sur l'histoire de l'astronomie ancienne*. Our vision of the past was profoundly changed. In retrospect, Comte's historical narrative is now only of antiquary value. We can spell out some of its shortcomings: Comte shows no understanding of pre-Socratic thinkers. He merely repeats the unlikely claims of tradition with respect to the likes of Thales and Pythagoras.[14] Likewise, his attributions of discoveries to the Middle Ages appear to be based merely on the conviction that change must occur gradually, rather than on any precise study of medieval astronomical texts. Tannery would nevertheless call on Comte in order to debunk Enlightenment claims to a highly developed science kept secret by a sacerdotal class: scientific progress must be on a par with the general development of a society.

Comte was much more interested in synthesizing the results of astronomy, especially those established over the past hundred years. He perceived a breakthrough toward the middle of the eighteenth century. Theoretical knowledge and empirical discoveries had led to a new order of accuracy, extreme precision, hence heralding in the positive spirit. Comte sought to draw a lesson with respect to method. Astronomy must be conceived correctly. It should be directed toward attainable goals, efficient solutions. Comte is careful to exclude exuberant speculations, unsolvable problems. Thus, he condemns inquiries concerning the chemical composition of the sun, which spectroscopy would soon make possible. Comte noted cases in which we come up against the complexity of phenomena, for example, the theory of atmospheric refraction. He could not imagine a complete solution to this problem and was willing to remain content with the explanation at hand. Modern theories of complexity were not on his mind. Obviously, they would carry us away from the style of research Comte was advocating.

The Political and Social Implications

After examining Comte's account of astronomy, we now turn to his audience. He gave his public course, which became the *Traité*, for seventeen years, requesting no salary in return. Comte was obviously pleased to point out his disinterestedness.[15] It was directed primarily to workers. According to him, the proletariat was to play a crucial role in the reorganization of society. The upper classes were indoctrinated by the educational system, one that maintained the old order and the reproduction of the elites. The workers would be more receptive to the positivist doctrine.

Comte framed his series of lectures with such an aim in view. The reasoning was to be comprehensible to those who had not received more than elementary schooling. The author of the *Traité* thus avoids any mention of integral calculus. Mathematical language is limited to geometry, complemented by a number of diagrams, which are collected at the end of the book. The teacher nevertheless sets high standards. He develops his reasoning at length, only omitting fastidious calculations. For example, Comte expounds in full detail the theory of aberration, that is, the apparent displacement of a heavenly body from its true position, owing to the motion of the Earth and the time required by light to reach us. As Comte phrases it, this is due to the "continual agitation of our terrestrial observatory."[16] The development in question takes up a good part of the chapter concerned, which amounts to one full lecture, and Comte returns at intervals to the diagrams. What the author is trying to get across is the kind of reasoning taking place in astronomy. Such reasoning involves the application of mathematics to empirical problems. Comte does not fail to address the philosophical problem of the relation of mathematics to the world. We learn how mathematical elaborations help to guide us in the exploration of complex phenomena—for example, how Kepler's laws lead, by means of an inductive method, to Newton's theory of gravitation.

It should be noted that the *Traité* was published in response to the request of the audience.[17] The book obviously benefited from feedback: Comte was able to put to the test his pedagogic and philosophical ideas. This process of elaboration explains the nature of the final result, a clear, precise, well-mastered treatment of the subject matter. We could take Comte to task with respect to scientific research, but this would be to miss his point: a highly technical, formal philosophy of science loses sight of the essentials, that is, the connections with the other sciences, the concrete practice of science, the relation with social issues, etc.

From the very outset, Comte expressed the wish that such a course would be useful for the workers, by offering them clear and accurate notions on fundamental issues. He also thought that they would be more receptive than the upper classes. Here is his proposal for the course in 1830: "To present to the workers an elementary course on general astronomy, whose main aim is the reasoned exposition of the essential phenomena of the system of the world, followed by an indication of its most important applications."[18] The formulation employed in this passage, "exposition du système du monde," obviously echoes the title of Laplace's corresponding work, which Comte is seeking to replace. As Comte went on to prepare his lectures for publication, some years later, he disclosed his intentions to John Stuart Mill: to provide a genuine popular teaching that would serve to promote positivism, that is, a system comprising both a theory of science and a theory of politics. He mentioned the clash with the religious authorities that had taken place in 1842 on account of his pronouncements in the introductory lectures. In the *Cours* he had already noted the difference between astronomy and religion: "For those minds that are unfamiliar with the study of celestial bodies . . . astronomy still has the reputation of being an eminently religious science, as if the following verse '*Caeli enarrant gloriam Dei*' had retained its force. It is nevertheless evident . . . that any real science is in radical and necessary opposition to any theology;

and this feature is more pronounced in astronomy than in any other place, namely because astronomy is as it were more of a science than any other."[19] His account in his public course on astronomy was to combat any proofs of God drawn from the physical world. And Comte would stress more and more that positivism aims to offer a critique of traditional religions and a substitute in the form of a Religion of Humanity.

Mill reacted favorably to the *Traité philosophique d'astronomie populaire*. He shared with Comte several convictions concerning education. He agreed that one could not count on those who had already been educated in science. As he states in a letter written in French to the author: "It is in every respect absolutely indispensable to provide as soon as possible those minds that are suitably inclined but lack specific scientific training with the positive knowledge necessary to assimilate the fundamental ideas of scientific method."[20] He went so far as to suggest that Comte underestimated the importance of this new formulation of his philosophy and to predict that it would hasten the establishment of the positivist school.

In favor of Comte's claim that his auditors had incited him to publish his public course, one could call on the testimony of Auguste Francelle. The latter, a clockmaker by profession, had expressed his enthusiasm for Comte's "sublime philosophy."[21] Having lost his faith as a Catholic, Francelle acknowledged that positivism provided him with a sense of purpose in life. Eventually, workers found out about Comte's astronomy course. Among them was Fabien Magnin, who became a dedicated disciple and an active member of the Positivist Society when it was established in 1848. He was entrusted with important missions within the society; he thus drafted a report on communism in order to counteract its attraction on the masses. In his obituary of Comte, Magnin would recount in detail how he and his fellow workers became part of the positivist movement. He explains why Comte chose astronomy as a vehicle for transmitting his doctrine:

> [Auguste Comte] knew very well that astronomy was the grand motor that transformed the spirit of human populations, leading them to the rational stage witnessed today; and that the main advances which we enjoy have for basis the propagation of the general principles of astronomy among the populations, in which it stimulates the germs of theoretical aptitudes present in the minds of each person. Finally, Comte knew very well that astronomy is the best means to recognize, among the minds disposed for abstraction, those that are capable of dealing readily with the difficult problems of sociology and morals, and to distinguish those

that are both modest and energetic enough to discipline themselves so as never to assert more than what is exactly known.[22]

Comte was attentive to the reaction of his audience. In the published version of his course he even offers a precise example of a remark from a worker. The context is that of explaining why the sun and moon appear larger on the horizon. He recalls Euler's explanation, which draws on the artifice by which painters tarnish the objects that they intend to appear distant. And he proceeds to discard this solution: in the first case, the sensations are simultaneous—we take in at a glance the whole picture, whereas in the astronomical case, we compare observations made at different times, the sun or moon on the horizon as opposed to these bodies directly overhead. Euler, who had contributed so remarkably to astronomy, could at times prove to be very superficial. On this occasion Comte acknowledges his debt in a note: "This decisive objection was brought to my attention, with as much clarity as sagacity, in a noteworthy letter that I received some ten years ago from a clearjudging worker in the printing industry, a regular attendant of my public course on astronomy."[23]

This would take us roughly back to 1834, in the early years of the course. The attendant in question was to remain anonymous, but it is telling that Comte kept this instance in mind when he came to write up his course. The objection mentioned above was congenial to Comte's general outlook. It attracted attention to our means of observation by way of our senses. Comte criticizes metaphysical theories, be they rationalist or empiricist, and calls for a scientific study, based on biology. He suggests that a comparison with animal behavior would probably show that the optical illusion in question is not due to an unconscious judgment but rather to the physical and physiological conditions imposed on observation. This goes to show the uneven development of the sciences; the so-called exact sciences are much more advanced than the others. Positivism is then an effort to give a balanced treatment of knowledge and to allow for feedback among the sciences, what Comte calls the subjective viewpoint, to which we shall return later. Comte would thus have been receptive to the future efforts of Hermann von Helmholtz and Mach to bridge the gap between physics and physiology. One may infer that Comte's critical evaluation of the hypotheses proposed by professional astronomers was sharpened by the interaction with a lay public. He encouraged the development of a critical mind; his pedagogy is not dogmatic.

The public course continued for four more years after the published version, and Comte continued to enrich his lectures, emphasizing the social and political consequences. He benefited from exchanges with the workers in his audience, associating them in his efforts to publicize the goals of his course. As he wrote to Magnin in 1846: "My intention is to characterize Positivism as representing the genuine philosophy of the people, in a more straightforward and insistent way than I have done so far."[24]

Over the years Comte became more and more concerned with addressing the working class. He usually employs the terms *ouvriers* and *proletariat*. But he also speaks of *travailleurs*, or *masses actives*. In fact, most of the workers with whom Comte came into contact were artisans, often highly skilled workers. Several became members of the Positivist Society. Although there are some similarities with Marx, Comte's contemporary, the differences between them run deeper: they diverge in their analyses of society and in their plans for its reorganization. Comte was not seeking to overturn industrial means of production, but to regulate the relations between industrialists, workers, and positive philosophers.

The *Traité philosophique d'astronomie populaire* as published is merely a freeze-frame of Comte's extended course on astronomy, from his first lectures of 1831 to the final lessons of 1848. This is apparent in the introductory part, which was expanded from two lectures to twelve and ultimately resulted in the *Discours sur l'ensemble du positivisme*. Comte then appended this text to the *Système de politique positive*, which formulates an ambitious program of social reform. He thereby suggested a connection between the two periods of his thought.

★ ★ ★

Comte's strictures on speculation aim at bringing out the tendency of scientists to promote what they are doing at the expense of any kind of usefulness. In speaking of geometers or mathematicians, he writes: "Succumbing to a blind analytic ardor, they have complicated and made obscure the true mathematical theory of tides, by an immature affectation to attain a degree of precision and specialization that it does not possess."[25] And in a note he lets escape his deeper motivation: "As if the mathematical education of humanity should last indefinitely, absorbing the best efforts to the detriment of contemplations at once more elevated and more useful."[26] There is an urgent need to develop a science of society.

What lesson then did Comte want to get across to his audience? Astronomy is to initiate us to scientific method and progress. As stated in the concluding sections of the book: "It is also here that the notion of progress appears spontaneously in its clearest purity, with the twofold character of filiation and continuity that is essential to its complete rationality as well as its profitable application."[27] The terms *filiation* and *continuity* define Comte's view of historical development, the gradual unfolding of our ideas and their uninterrupted transmission from one generation to the next. Astronomy, the earliest science to have taken shape, offers a model of scientific change. These indications lead to the general theory of progress that is expounded in the preliminary *Discours sur l'esprit positif*.[28]

Progress not only concerns scientific development; it also pertains to social and political advancement. Thus, Comte attempts this rather audacious parallel in speaking of the three major laws of physics—the law of inertia, the law of the composition of motions, the law of action and reaction: "Considered in particular with respect to political bodies, the first law pertains to their strong instinctive tendency to conserve any acquired disposition, the second characterizes the elementary harmony between order and progress, and the third defines the necessary solidarity of all the various agents."[29]

The reestablishment of a spiritual authority and a speculative class was not going to be an easy task. Comte must have been conscious of this, hence the usefulness of being able to call on the long record of astronomy in order to show the relevance of the effort required.

Comte's Philosophy of Science

How does Comte's description of astronomy contribute to our understanding of scientific method? One of the main problems that Comte encountered in the science of his time was the role of hypotheses. He singled out this problem for particular attention. Such is the purpose of the development he labels "the fundamental theory of hypotheses," which appears in lesson 28 of the *Cours*. It was in the field of physics that difficulties cropped up: the recourse to various fictitious entities that have no direct counterpart in observation, such as caloric, ether, electric fluids, magnetic fluids, atoms, etc. Because of his antipathy to fruitless speculation, Comte formulated strict rules concerning the role of hypotheses in science. A hypothesis should be merely an "anticipation" of future observation. It is only an "artifice" conceived in order to overcome the difficulties that a direct analysis of a particular phenomenon may present. Once the theory is constituted, the hypothesis can be dispensed with. Thus, the role of the hypothesis should be linked to the central intuition of positivism: according to Comte, science consists of the search for laws, and not the quest for causes. The theory of hypotheses is aimed in particular at Laplace's realism. Comte rejected any recourse to fluids and other fictitious entities, in which he perceived a lapse into metaphysics.

In fact, Comte was basing his conception on astronomy. He makes this clear: "If celestial science was at first unique in teaching us what constitutes the positive explanation of a phenomenon, excluding any inaccessible inquiry into its cause—be it first cause or final cause—nor its mode of production, at what clearer source could we draw such a lesson today? Physics, more than any other natural science, should mainly aim to imitate such a model."[30]

It is in the area of astronomy that Comte would receive the sharpest criticism. Thus, for example, Henri Poincaré writes:

> Auguste Comte has said somewhere, that it would be idle to seek to know the composition of the sun, since this knowledge would be of no use to sociology. How could he be so shortsighted? Have we not just seen that it is by astronomy that, to speak his language, humanity has passed from the theological to the positivist state? He found an explanation for that, because it had happened. But how has he not understood that what remained to do was not less considerable and would be not less profitable? Physical astronomy, which he seems to condemn, has already begun to bear fruit, and it will give us much more, for it only dates from yesterday.[31]

Poincaré justifies the cost entailed by astronomical research in view of the immense value of its results. He voices the optimism of the turn of the twentieth century. Poincaré is not alone in targeting Comte's astronomy. We can mention Joseph Bertrand, Édouard Le Roy, and Gaston Milhaud. The latter is particularly interesting, as he published one of the first general studies of Comte's philosophy. To follow Comte, theoretical concepts would be reduced to a mere "residue of experience." Milhaud completely refuses this conception: humankind invents notions that make it possible to organize reality. Comte's conception must be reformulated, and Milhaud adds to his third stage of human development a fourth stage, which corresponds to the innovative trends of the science of the turn of the twentieth century: "If Comte had paid more attention to some of his own reflections, he would have been ready to give up his demands for a rigorous positivism, which made him perceive constantly in the state of contemporary science the ultimate term reached by the activity of the mind and led him to circumscribe future progress within such narrow limits."[32]

Yet Milhaud can single out passages in which Comte went beyond the frame of his doctrine. Let us cite one such passage that concerns astronomy. When Comte described the activity of the astronomer, who determines planetary orbits, proceeding by way of successive approximations, he proffered the following reflection: "Modern astronomy by overthrowing definitively the initial hypotheses, considered as the real laws of the world, has maintained carefully the permanent and positive value of these hypotheses, the property to represent *conveniently* [*commodément*] the phenomena as a rough draft. Our own resources in this respect are even more extensive, precisely because we entertain no illusions concerning the reality of hypotheses, which allows us to use without any scruples, in each case, the one that we judge the most *advantageous* [*avantageuse*]."[33]

According to Milhaud, this remark escaped Comte; it is not in harmony with his other assertions. But it shows that he had an inkling of the difficulties raised by his efforts to regiment science. Milhaud notes that such passages anticipate what Poincaré will come to say. Milhaud provides some other similar passages from the *Cours*, and we can find remarks along the same lines in the *Traité*. How is it that Comte came to suggest ideas that were to be developed later? Comte's assertions about an "advantageous hypothesis" leads one to expect him to bring out the consequences more clearly. The passage quoted by Milhaud is borrowed from the *Cours*. Turning now to the *Traité*, we do not note any change of mind: a hypothesis can be, strictly speaking, false and yet help us make progress. Comte does not move toward anything like Poincaré's conventionalism.

It seems that Milhaud, like the majority of philosophers of science, was reading into Comte his own views. Comte was not seeking to provide a rigorous analysis of theory structure. He was looking for well-established facts, sure and solid knowledge that could serve as a basis for human action. It was not a matter of possible hypotheses nor even probable hypo-theses, but rather plausible hypotheses—that is, reasonable conjectures upon which we may construct something real. Once again what Comte was interested in was accumulated knowledge of our world, knowledge that would help us act for the better, improve our social and political conditions. He shunned vain curiosity. The scholar is not sustained by society merely so that he may pursue his own favorite hobby. Comte's solutions may raise problems of their own, but his remarks, the difficulty he perceives in the scientific organization of his time, carry weight. We cannot act simply on possible hypotheses; we cannot even regulate action on the most probable hypotheses, for a true determination of probability is not to be achieved by probability calculus in the case of theories or applied research. There are prudential considerations to take into account. Some external social control of science is necessary, for science is not self-regulating. Scientists have allied themselves with all kinds of political regimes or metaphysical systems.

Obviously, Comte rejected d'Alembert's history of science and his theory of knowledge. He was generally dissatisfied with what the *Encyclopédie* has to offer on such issues. He criticized Euler and Laplace for their sketchy philosophical reconstructions. And if he frequently praised Lagrange, it was for having generally abstained from drawing unfounded conclusions.

Comte's attempts to go beyond empiricism and rationalism are recognizably original. He insisted once and again on the need for both observation and reasoning. He furnished some fine examples of scientific reasoning and offered some acute arguments. Hypotheses are underdetermined by observation, and we are left with several options. Although he went somewhat further than his predecessors, his general conception of scientific method does not depart entirely from that of d'Alembert, Euler, and Laplace. We find the same procedures: inductive schemas, crucial experiments, general facts, etc. But in concluding thus, we are perhaps missing his point. The description of scientific method in the *Cours* and in the *Traité* was only a means to another end. The real lesson he was trying to get across was implicit in his addressing the Parisian workers attending his course: to build a new society.

Rational Values and Comte's Notion of Scientificity

Drawing on recent studies of constitutive notions, I shall now seek to fathom Comte's conception of scientificity. What model of scientific reasoning is he trying to get across? We have seen that Comte characterizes astronomy, as it had evolved in the previous hundred years, by its "extreme precision." This new degree of approximation was made possible by the combination of different strands of scientific practice: the systematic use of instruments and their integration within theory. Comte was taking up a concept that had played a crucial role in the discourse on science since the seventeenth century.

Let us recall that the term *precision* as well as its near synonym *exactness* acquired new meanings at this time. Deriving from Latin, *exact* in its first sense designates the quality of what has been made complete; *precise*, what has been cut short. We have in mind something completely different when we use these words today. It is surprising that these terms, which carry a lot of weight in our discourse about science, have received but scant attention. Some years ago, Heinz Weinmann sought to remedy this omission in a short study of Galileo's early works. He thus writes: "It is not the least paradox of the socalled exact sciences that they have 'forgotten' to define the very terms used to describe the rigor of their new methods: precision and exactness."[34] Indeed, these terms come up prominently in Galileo. They are used to characterize the rigor of his science, notably with respect to Aristotle. Galileo's use is still somewhat fluctuating and vague—the concepts in question were new and not yet sanctioned by usage. He resorted to a number of other terms as well to describe his method, notably "accuracy," which the English language was to adopt, unlike French. Galileo thus describes his telescope as "exact," where modern usage would have "precise." The received expressions we use today—*exact sciences*, *precision instruments*, *degree of accuracy*—would come later. Galileo's expressions caught on like wildfire. They came to designate the "new science," that is, an investigation of nature based on mathematics and observation. The major scientists of the seventeenth century took them up: Descartes, Boyle, Newton, etc. In the century following, a series of fundamental debates hinged on these notions: what falls under the exact sciences? How can the strict, absolute notions of mathematics be applied to the world? What constitutes a precision instrument? Comte encountered such questions in the works of his predecessors.[35]

Although Comte was no more attentive to the historical evolution of the concepts he was using than most philosophers of science until recently, his observations on the nature of scientific precision are retrospectively instructive. That this concept was central for him is made clear by its occurrence in his definition of the positive mind given in the *Discours sur l'esprit positif*. He thus delineated the positive or scientific spirit by means of the following series of terms: *reality*, *usefulness*, *certainty*, and, especially, *precision*. These traits were supposed to lead to a new attitude; they express

Comte's desire to get rid of the great metaphysical systems of the past and elaborate a philosophy inspired by science.

Interestingly enough, Gaston Bachelard still found it necessary to refer to Comte in his dissertation of 1928, *Essai sur la connaissance approchée*. He singles out the passage of the *Discours* we have been referring to for commentary:

> It is perhaps the fourth characteristic [precision] that, as far as measured phenomena are concerned, entails the others. Indeed, the results of a measurement can sometimes be so *precise* that no account is taken of the very small errors that they still retain. These measurements, free from sensitive discrepancies, give rise unquestionably to a general consensus. It is by means of *precise* measurement that an object can reveal itself as permanent and fixed, in other words as a duly recognized object.... *Precision* takes precedence; it gives to *certainty* so solid a character that knowledge seems to us truly concrete and *useful*; it gives us the illusion of touching the real.[36]

Bachelard is careful to leave aside positivity, a term of which Comte made immoderate use. He emphasizes precision and shows how the other attributes derive from it. To be sure, Comte had some interesting remarks to make: precision differs from certainty; it admits of degrees. This comes out in his presentation of the various meanings of the term *positive*. Let us now give the text commented on: "A fourth commonly received meaning . . . consists in contrasting the *precise* with the vague: this sense evokes the constant tendency of genuine philosophical spirit to obtain in all areas *the degree of precision* compatible with the nature of the phenomena and in accordance with our true needs; whereas the old manner of philosophizing invariably led to vague opinions, which only achieved the necessary discipline on account of the continual pressure based on supernatural authority."[37]

Comte brought out the specificity of modern science and outlined the cognitive evolution of humanity. But his main concern was to avoid all metaphysics, going so far as to impose limits on science. As for Bachelard, he rejected positivism. In the *Essai sur la connaissance approchée*, he was still influenced by Poincaré as well as his teachers, Léon Brunschvicg and Abel Rey, but he was quickly moving toward his own distinctive brand of rationalism and realism. Science overturns obstacles; it renews the conditions of cognition and transforms rational values. In consequence, he could concentrate on the procedures whereby we achieve precision.

In the introductory lessons of the *Cours*, Comte was careful to distinguish between the degree of precision and the degree of certainty: a very precise statement can be uncertain, and a very imprecise assertion can be certain. As he put it: "The various sciences must necessarily present a very unequal precision, but this is not at all true of their certainty. Each science can yield results as certain as those of another, as long as it knows how to circumscribe its conclusions within the degree of precision that the corresponding phenomena possess."[38]

Let us turn now to objectivity. This notion has come to be used as the hallmark of science, and expresses something different from accuracy or exactness. On this issue we can find ample material in the study *Objectivity*, by Lorraine Daston and Peter Galison. The authors bring out the multiple and profound transformations that our understanding of science has undergone. Before the modern notion of objectivity emerged, the main imperative for scientists was truth to nature—I add that this is something that the notions of exactness, precision, and accuracy suggest. Objectivity then started to gain currency during the nineteenth century. It was an effort to repress the subjective tendencies of the observer, by developing new techniques of automated registration of data.

Comte is part of this history. We have noted earlier that he employs the term *objectivity*, which, over the years, takes on importance. It is time now to explore this issue. *Objective* as opposed to *subjective* was originally a scholastic concept. Descartes continued to use the term in this sense, that is, as a representation of the mind. Then along came Kant, who turned the meaning around, and *objective* came to designate a reality in itself, independent of knowledge.[39] Daston and Galison mention Comte only in passing. Their aim is to locate this concept within a long-term history, from pre-Kantian uses to current ones. They are after the ordinary, dominant sense of the word in relation to scientific practice. The subtleties of Kant's original definition, bound up with his complex philosophical system, were soon forgotten by all but historians of philosophy. Scientists retained what was serviceable for them: a new way of characterizing their activity. Daston and Galison refer to Helmholtz, Claude Bernard, and Thomas Huxley. It should be noted that the occurrences they give are later than those we encounter in Comte. It is then a useful task to complete the picture. What we may gather is that Comte's use of the concept occurs at an early stage in its history; it is also somewhat peculiar, which may explain why Daston and Galison chose to overlook it. Perhaps he contributed in part to the fashioning of the modern sense, but later scientists and philosophers left behind in turn the particularities of Comte's formulation. Furthermore, Daston and Galison are somewhat wary of positivism, confounding it with later versions that depreciate history.

The main passage on the distinction of objectivity and subjectivity in the *Traité* is the following: "The theory of aberration, as it is used continuously in astronomy, completes . . . the highly complex system of multiple corrections that all exact celestial observations must undergo, in order to be stripped of any *subjective* illusion and acquire a purely *objective* character, making them worthy, as far as possible, to concur to the precise presentation of corresponding laws of inorganic nature."[40]

Let us underline the expression "stripped of any subjective element"; this could be taken as a characterization of objectivity: the result of an effort to go beyond the subjective conditions of observation, toward a rigorous, sustained, collective effort. One should hasten to add, however, that Comte does not go on to separate the subjective from the objective as two spheres of activity, the sciences and the arts. Science comprises necessarily some degree of subjectivity, and a complete view, a philosophical view, must take into account this aspect.[41] Objectivity corresponds to the observed thing; subjectivity, to the observer.[42] In a remark that could be taken as a critique of Kant, Comte notes that external harmonies do not always correspond to our internal expectations.[43] In this passage Comte is speaking of the absence of relation between the order of sizes of the planets and that of their distances. This is a contingent fact. We could perhaps outline the process as follows: one seeks to establish facts; one recognizes the input of the observer; and, finally, one seeks to locate these facts within the encyclopedic order. One could call on an expression that Bachelard coined for his own endeavor: "objective subjectivism."[44]

Comte was careful to relate his conception to Kant. Thus, in speaking of the substitution of the relative for the absolute: "The greatest of modern metaphysicians, the illustrious Kant, truly deserves eternal admiration, as he was among the first to escape straightforwardly the philosophical absolute, by his famous conception of twofold reality, both subjective and objective, which shows a so appropriate sense of sound philosophy [*saine philosophie*]."[45]

Comte, at pains to distinguish his philosophy from that of Kant, went on to explain that this enterprise ultimately came to fail: Kant remained a metaphysician. Mary Pickering has shown that Comte knew a good deal more about Kantian philosophy than he was prepared to admit. His disciples, in particular Gustave d'Eichthal, even provided him with translations from the German as early as the 1820s.[46] Thanks to Kant, he could call on "objectivity" alongside "precision" in order to better characterize his method. Science was to be seen not only as precision, or exactness, but also as objectivity. The latter makes it possible to address more directly such important issues as the relation of our theoretical constructs to reality.

Comte's use of *objectivity* does not correspond to the received meaning; it is more complex and does not imply a clear-cut division between science and other forms of human activity. Unlike precision, objectivity is not a category bound up with mathematical physics, but a broader category allowing for a sociological approach.

Emphasis on objectivity characterizes the account given in the *Traité*. Comte would continue to use this concept in the *Système*. Let us quote one passage: "We owe to astronomy the first systematic development of the art of observing, and, in consequence, that of genuine induction . . . . Not only does the need for material observation become here undeniable, but one may distinguish also the intellectual elaboration that always accompanies it, and which does not come out as clearly elsewhere. There is no absolute separation between observing and reasoning. No observation can or should be purely *objective*. As a human phenomena, this first mental operation is likewise *subjective*, in each case, to a degree proportional to its complication."[47] Comte invites us here to think twice before separating observing from reasoning. This is stronger than just asserting the theory-ladenness of facts.

★ ★ ★

What can we gather from this study of Comte's astronomy? Comte showed that a science dealing with remote objects can play a fundamental social role. Yet he was not able to foresee the later development of the field. His account circumscribes too narrowly the domain of phenomena; it leaves out promising directions of research. Perhaps predicting the future of science is a hopeless enterprise, as Poincaré would put it later. But Comte's conception of science is obviously not open enough. He was too preoccupied with establishing a sound, solid basis.

One could turn Comte's case around: his failed attempt to outline the program of future research and to implement his politics of science may offer a negative lesson. Such a lesson was drawn already by the thinkers of the next generation. Scientific research cannot be subordinated to social usefulness; it cannot be submitted entirely to economic planning. Some degree of freedom must be allowed the researcher. Furthermore, the hope to provide foundations—ultimate knowledge, absolute facts—is in vain. Science must continually question its hypotheses, its justifications, and its facts. Research is essentially "tentative" and "relative," as Comte himself admits. But one must go further in this direction. Here we may call on critics of positivism, whose arguments bring out the shortcomings of Comte's philosophy. For Karl Popper, there is no bedrock on which we can build once and for all. We are tempted then to turn around Comte's conception: for all its rigor, care, precision, thoroughness, science has a ludic dimension. In other words, intuition, invention, and speculation are essential. In the process of creating new theories a scientist may break former rules, go against common sense or ordinary experience; he or she may even refashion rational criteria.

In his effort to give a sound basis for science, Comte was led to synthesize various trends of research, to promote certain theories and to demote others. He sought to impose on research a particular orientation, on the basis of a carefully thought out program, bringing science under a collective decision issuing from the political authority. Comte imposed undeniably excessive limitations on freedom of research. He subordinated science to social order.

According to Milhaud, by the end of the nineteenth century a new stage had emerged in the development of science, a fourth stage, and there is no reason to believe that this process of elaborating new forms of rationality should ever cease. Bachelard, who was reacting to the pervasive positivism of the early years of the French Third Republic, would insist on the pluralism of science: science draws on different philosophical doctrines; it combines them in novel ways. Science does not ally itself to one particular coherent and homogeneous philosophical system. The new scientific mind is open, constantly evolving. Science is not merely the accumulation of discoveries, the consolidation of a paradigm. Following Kuhn, normal science ends up giving way to extraordinary science; new paradigms replace earlier ones. Many thinkers have gone so far as to question even the stability of normal science. Several research traditions can be in competition at one and the same time. Science is a multiplicity of options. Current theories of scientific change and historical studies encourage us to view science as a complex, pluralistic, and ever-evolving activity.

The lesson Comte makes is that science and its development are bound up with general history; they are part of the history of humankind. By bringing out the larger implications, he suggests that knowledge is power to transform, that is, to shape society, to mold people. One can then understand how Michel Foucault, though a critic of the positivist ideas of his time, could gather insights from Comte. Such modern perspectives can in a sense be traced back to him, and our exploration of these origins can help us to become clearer about the conditions of possibility of contemporary thought.

Historians generally have the contemporary era commencing during the second and the third decades of the nineteenth century: more decisive than the political revolutions was industrialization and its repercussions. This corresponds to Comte's formative years. He is thus a key witness to this momentous transformation. But his attitude is ambivalent: he sought both to hasten the demise of the old order and to guard against some deeper changes. The transformations were not only political and economic; they were scientific and philosophical. Mentalities were in flux. Comte noted some of the changes, such as Antoine Lavoisier's new chemistry, the constitution of biology, and he called for the founding of a science of man, sociology. He also noted the appearance of new branches of physics, such as the theory of heat—what would become thermodynamics—and emphasized the novelty of its method, conforming to the positive spirit. However, Comte did not perceive those other emerging theories: non-Euclidian geometries, probability calculus, mathematical logic, nor interdisciplinary fields such as astrophysics, physical chemistry, and biochemistry. The participant in an event is liable not to grasp it fully, and this applies to Comte. But even in this negative sense, he may still help us to grasp the complex and multiple processes leading to the advent of the contemporary mindset.

4

Auguste Comte's Positive Biology

Laurent Clauzade

Auguste Comte was not a biologist, but a philosopher of biology, and undoubtedly one of the first, if by "philosophy of biology" one understands an epistemological reflection on a unified field (or at least a field in the process of being so) bearing on the phenomena of life. At the turn of the nineteenth century, biology was indeed undergoing a process of unification and institutionalization, and Comte's views greatly contributed to the reflection on the epistemological and philosophical foundations of this process of constitution. Therefore, it was no accident that, in France, the founders of the Société de Biologie, through Charles Robin (1821–1885), inherited the most distinctive categories of Comtean epistemology.[1]

In this chapter I will try to offer an overview of Comte's philosophy of biology. I will first describe in detail the extremely consistent and insightful architectonics of biology that Comte developed in the *Cours de philosophie positive*, for it was indeed that work, which aimed at providing biological studies with a proper conceptual structure, that most impressed Comte's contemporaries. I will then turn to the evolution of positive biology in the *Système de politique positive* and address the issue of the relations between biology and sociology.

A New Science

Let us start with specifying the origin and meaning of the word *biology* in Comte's epistemology. Comte did not invent the word, but he contributed more than anyone to its popularization and diffusion in France and Europe. More precisely, the term first appeared in the *Cours* in 1835, during the writing of lesson 36. It was officially introduced in a footnote: no philosopher could study living beings "without, in some sense, being naturally forced to use the fortunate expression of *biology*, which has been so judiciously coined by M. de Blainville, and for which the name of physiology, even purified, would only offer a feeble and ambiguous analogue."[2]

A New Word

Biology replaced the older term of *physiology*. The new term was quite convenient because it cleared up an ambiguity: before the introduction of *biology*, the same word was used to refer both to biological science as a whole ("physiology") and to one of its parts ("physiology strictly speaking"). Quite surprisingly, neither Blainville (1777–1850) nor Comte had borrowed this word from Jean-Baptiste Lamarck (1744–1829), who was the first to use the term in the French tradition. Blainville, in the extract quoted by Comte,[3] had only mentioned some "German" biologists; one might think that he was there alluding to Gottfried Reinhold Treviranus (1776–1837), whose work Blainville was familiar with and who happened to be one of those who had introduced the term. In fact, Blainville had borrowed another word from Treviranus: *zootomy*, from which Comte himself derived the term *biotomy* so as to substitute it for *anatomy*. This reference to the Germans should not be taken as proof of a decisive influence over Blainville (and Comte). In this great era of terminological innovation, conceptual appropriations often were eclectic: Blainville also took up from Erasmus Darwin (1731–1802) the word *zoonomia*,[4] which Comte once again recycled as *bionomy*.

However, there is not much use trying to make sense of these terminological lineages: Comte never mentioned Treviranus, and the only German biologist he quoted significantly, Franz-Joseph Gall excepted, was Lorenz Oken. On the other hand, the direct relation to Blainville was decisive and, as we shall see, it was not merely a word that Comte borrowed from him, but his idea of the architectonics of biological science.

The Architectonics of Positive Biology

Biology primarily obeyed the standards of positive philosophy. As a "fundamental" science, it had to be abstract and theoretical. This dual characterization excluded, on the one hand, concrete biological studies (such as the natural history of specific organisms and pathology) and, on the other hand, practical studies (such as education and "medical art").[5]

Comte introduced biology properly speaking—the theoretical and abstract study of vital phenomena—as the reunion and articulation of three different kinds of investigations that had thus far been separated: anatomy, physiology, and the "natural method." The latter was both a discipline—the study of the hierarchy of living beings—and the basis of the primary method of biology, i.e., the comparative method. The articulation of these three kinds of studies was carried out through the distinction between the static and dynamic points of view.

Indisputably, it was Blainville who provided Comte with this crucial distinction.[6] This legacy was acknowledged at the very outset of the first lesson of the *Cours*: "To explain what I mean on this point I must first recall a philosophical conception . . . set forth by Blainville in the fine introduction to his *Principles of Comparative Anatomy*. According to him, every active being, and especially every living being, may be studied in all its manifestations under two fundamental relations—the static and the dynamic; that is, as fitted to act and as actually acting."[7] Toward the end of the *Cours*, Comte would develop the theory of "general laws," according to which some laws can apply to all orders of phenomena. Most notably, he would show that mechanics offered a first instance of the static/dynamic distinction. However, it would be erroneous to think that the biological distinction was merely derived from mechanical concepts: Blainville's pair was clearly an original instance of that distinction, which specifically applied to the field of the organic sciences. It was indeed to biology that Comte referred when he introduced the distinction in sociology.[8]

Anatomy and the study of the "great biological hierarchy" corresponded to the static part of biology, whereas physiology, properly speaking, corresponded to its dynamic part. To refer to these different parts, Comte introduced neologisms that he seldom used, but whose formation indicated his willingness to demonstrate the consistency of the new science. These terms, derived from those coined by Blainville, were the following: *biotomy* stood for anatomy, *biotaxy* for the study of the "natural method," and *bionomy* for physiology. Of these three terms, *biotaxy* was the only one that Comte really used, certainly because it designated the original association of two heterogeneous kinds of studies: "the logical theory of rational classifications" and "the hierarchical coordination of all known, or even possible organisms."[9]

Three principles governed the way in which these three kinds of studies had to follow one another. Static biology had to precede dynamic biology, "because the structure must be known before its action." Although the order within the static part of biology was quite arbitrary, biotaxy had nonetheless to follow biotomy, for "none but known organisms can be classified."[10] Finally, the arrangement of these three parts was organized according to the encyclopedic principle of classification laid out in the second lesson: one had to proceed from the most general and most elementary studies to the most specific and complex. Accordingly, one had to start with the phenomena of organic life, then turn to those of animal life, and finish with those relative to intellectual and moral life. This latter principle, although it was respected in all parts of biology, manifested itself most clearly in the dynamic part, which was divided into three successive lessons.

The following chart (table 4.1) sums up the overall architectonics Comte proposed for biology.

Such an architectonics, as previously pointed out, was inspired by Blainville's theses. Its main concepts and its global structure came from the latter's *Plan du cours de physiologie générale et comparée*, with only slight differences. The major change Comte brought to this plan was the insertion of the lesson on biotaxy in between that of biotomy and bionomy. But positive biotaxy also had its origin in Blainville's theory.

However, one would be wrong to think that Comte showed no originality in his appropriation of Blainville's scheme. On the contrary, through the formal categories of positive philosophy, such as the distinction between the concrete and the abstract, or the principle of classification laid out in the second lesson, but also and most notably through the philosophical analysis of life Comte produced, he imposed a new coherence to this *Plan*.

The Philosophical Basis of the Architectonics: From the Definition of Life to the Definition of Biology

Comte's definition of life was twofold: "I know no other successful attempt to define life than that of M. de Blainville, proposed in the introduction

Table 4.1

Static		Dynamic		
Anatomy (biotomy)	Theory of classification (biotaxy)	Physiology (bionomy)		
		Organic life	Animal life	Cerebral functions
Lesson 41	Lesson 42	Lesson 43	Lesson 44	Lesson 45

to his treatise on comparative anatomy. He characterizes life as the double interior motion, general and continuous, of composition and decomposition, which in fact constitutes its true universal nature. I do not see that this leaves anything to be desired, unless it be a more direct and explicit indication of the two correlative condition of a determinate organism and a suitable medium" ("milieu" in the original).[11] The first part of this definition was borrowed from Blainville and characterized life as the minimal feature shared by all living beings, namely organic life, which was a process of composition (or absorption) and decomposition (or exhalation).[12] This organic or "fundamental" life, as the *Système* would put it, was a feature common to the whole biological hierarchy, from plants to man. The first part of this definition would play an important role in ethics, for it implied that all animals, even those highest in the hierarchy, such as man, would have first to satisfy the needs of this fundamental life. On it rested the claim of the primacy of selfishness, which could not disappear but could be moderated by social instincts.

The Idea of *Milieu*

It was the second part of the previous definition that gave the architectonics I have just described its coherence. The elementary condition for life was the harmony between the living being and its "milieu." Let us first specify what Comte meant by *milieu*. The use of the term *milieu* in the singular was a Comtean innovation: until then, the term was used in the plural to refer to the fluids "in which the organism is immersed"—more exactly, as emphasized in the *Système*, "our twofold fluid envelope," gaseous and liquid.[13] In the singular, *milieu* now meant "the overall set of external circumstances, whatever their kind, necessary for the existence of every organism."[14] The forty-third lesson specified the nature of these circumstances: they were in fact physical and chemical parameters. Some of them were mechanical, such as the action of weight, of pressure (liquid or gaseous), the physiological influence of motion and rest; others were purely physical, such as the thermological action of the milieu (as opposed to the organic production of vital heat), light and electricity. The chemical parameters mainly had to do with the classical notion of *milieu*: the influence of air and water. But the more restricted general action of some specifics could also be included in the study of organic milieu.

Comte's theory thus considered the idea of *milieu* narrowly, since it only included the abiotic elements of the natural environment. Accordingly, the interaction existing between one organism and others did not play in it the crucial role it would have in Darwinism. Several reasons might be invoked to explain such a limitation: the first is that Comte partly inherited the conceptions current at the time, most notably Blainville's theory of "external modifiers" elaborated in the latter's *Cours de physiologie*.[15] This theory of external modifiers was indeed quite more restrictive than Lamarck's, which included among the circumstances the habits and influences derived from "the way of living, of defending oneself, of reproducing."[16] Another crucial reason for this conception of *milieu* derives from the fact that Comte modeled the relation between organism and milieu according to the encyclopedic relation existing between inorganic and organic sciences. For it was indeed the case that the pages of the forty-third lesson dedicated to the *milieu* systematically followed the order of presentation of the physical and chemical phenomena. From that perspective, the environment can be constituted only by inorganic parameters.

The definition of life as "the harmony between the living being and the corresponding *medium* [*milieu*]" allowed both for a rejection of Marie François Xavier Bichat's vitalism (1771–1802) and the precise definition of the object of biology.[17]

An Antivitalistic Definition of Life

According to Bichat's famous definition, "life is the totality of those functions which resist death":[18] "Such is in fact the mode of existence of living bodies that everything which surrounds them tends to their destruction. . . . [They] would necessarily soon be destroyed, did they not possess a permanent principle of reaction. This principle is life; not understood in its nature, it can be known only by its phenomena."[19] This definition is the core of Bichat's vitalism: a living body, isolated from its environment, can rest only on its own vital principle. Comte, on the contrary, considered such a conception not only false but also metaphysical, since it conveyed the "idea of an absolute antagonism between dead nature and living nature": "The state of life would thus be very viciously characterized as an imaginary independence from the general laws of the nature that surrounds us, as a fantastic opposition to the whole of external actions."[20] Comte, because he emphasized harmony and correlation, conceived of life as a relative idea. And relativity, as opposed to what is absolute, was a positive property. "Nothing is absolute in this world, everything is relative"[21] was indeed one of the first and earliest principles of positive philosophy.

In that respect, Comte's harmony differed radically from Bichat's antagonism. To a certain extent, it was true that both conceptions presupposed that there existed a relation between the living body and the milieu that surrounded it. But in the case of Bichat's antagonism, that relation took place between external bodies and a living body that is supposed to possess life within itself, whereas, with respect to harmony, it was the relation between the organism and the milieu itself that "constitutes the general idea of *life.*"

However, one would be mistaken in ascribing to Comte a materialistic position. On the one hand, Comte developed some sort of encyclopedic antimaterialism. On the other hand, positive philosophy indisputably manifested a vitalistic tendency that expressed itself in two different ways: through a latent anti-Lamarckism, which claimed that organic spontaneity was not reducible to the influence of the environment, and through the major themes of Comte's later philosophy relative to the struggle between living nature and the inorganic world.[22]

The Definition of Biology

But the definition of life as a correlation between the living organism and its environment primarily served to provide the architecture previously described with its epistemological legitimacy. The definition of the scientific goal of biology indeed directly followed from the definition of life: "We have seen that the idea of life supposes the mutual relation of two indispensable elements,—an organism, and a suitable medium [*milieu*]. . . . It immediately follows that the great problem of positive biology consists in establishing, in the most general and simple manner, a scientific harmony between these two inseparable powers of the vital conflict, and the act which constitutes that conflict: in a word, in connecting, in both a general and special manner, the double idea of organ and medium [*milieu*] with that of function."[23] The correlation between organ and milieu led to the idea of function understood as "the act that constitutes the conflict." Within this framework, the act or function became quite abstract, since it was described as some sort of hypostasis of the conflict between organism and milieu. As such, this act/function was more substantial than the classical notion of function, which was defined as the act of an organ. For the conflict could indeed result in two different kinds of consequences, relative either to the organism or to the milieu: if the milieu acted upon the organism, the organism in turn acted on the milieu.

Accordingly, the definition of biology produced three terms biology would have to account for: *organ*, *milieu*, and *function*. However, the "general purpose"[24] of biology would be formally defined only after two qualifications had been made. The former amounted to the decision of focusing exclusively on the organic outcomes of the conflict. As for the consequences of the act on the milieu, when important enough—which only happened in the case of the human species—it was for natural history to study them. Therefore, in this first step, only the physiological meaning of the term *function* was retained. The second reduction logically followed from the first: if the milieu remained stable, it could be eliminated from the fundamental equation of biology: "Biology, then, may be regarded as having for its object the connection, in each determinate case, of the anatomical and the physiological point of view. . . . The surrounding system being always supposed to be known, according to the other fundamental sciences, the double biological problem may be laid down thus, in the most mathematical form, and in general terms: *Given, the organ or organic modification, to find the function or the act, and reciprocally*."[25] One finds here, grounded epistemologically and resting on a definition of life, the major structure of biology: the division between statics and dynamics. It was because of this conceptual work that studies that had been so far undertaken separately could appear now as the complementary parts of one single science.

The Scientific Basis of the Biological Architectonics: Bichat's Tissular Anatomy

The scientific goal of biological science was thus to establish a correspondence between anatomical analysis and physiological analysis. The fundamental theory that would allow such a goal to be achieved was Bichat's theory of tissular anatomy. Bichat's conceptions, both in anatomy (the anatomical decomposition of organs into elementary tissues) and in physiology (the distinction between organic life and animal life) indeed provided Comte's biology with its foundations. One might confidently assert that tissular anatomy was the theory that, in Comte's time, made the unification of the biological field possible. In certain respects, cell theory would play the same role at the end of the nineteenth century.

Tissular anatomy, as introduced in Bichat's *Anatomie générale*, assumed that tissues were the ultimate elements of which the organs were made.

> All animals are an assemblage of different organs, which, executing each a function, concur in their own manner, to the preservation of the whole. . . . Now, these separate machines are themselves formed by many textures of a very different nature, and which really compose the element of these organs. Chemistry has its simple bodies, which form, by the combinations of which they are susceptible, the compound bodies. . . . In the same way anatomy has its simple textures, which, by their combinations . . . make the organs.[26]

Bichat also underlined that these elements themselves were organized: "these are the true organized elements of our bodies."[27] That precision mattered, for it echoed a positivist principle that also governed sociology: "every system must be composed of elements of the same nature with itself."[28] Accordingly, the basic unit of sociology was not the individual but the family; similarly, the elements of organized bodies had to be organized structures themselves, which was indeed the case for tissues. That was the same principle that explained why, in the *Cours*, Comte had rejected cell theory.[29] The alleged simplicity of "organic monads" was incompatible with the very idea of organization.

Granted this scientific basis, Comte's work, in accordance with biology's goal, would mostly consist in ensuring that anatomical and physiological categories rigorously corresponded to one another. To achieve such a correspondence, Comte first took advantage of the modifications Blainville brought to Bichat's study of anatomy. These modifications aimed at simplification and lent themselves more readily to such a process of rationalization. Bichat listed twenty-one unhierarchical tissues, whereas Blainville hypothesized the existence of a primitive tissue, the cellular tissue (whose name derived from the fact that it displayed some sort of sockets). Its cellular structure made it suitable for organic life, which merely amounted to absorptions and exhalations. This primitive tissue could undergo two kinds of modifications: on the one hand, mere structural modifications (through condensation) allowed for the production of most of the organism's structural tissues (dermous, fibrous, cartilaginous, bony, sclerous, and kystous tissues). On the other hand, modifications relative to the very composition of the primitive tissue gave rise to the two tissues characteristic of animal life: muscular and nervous tissues.[30]

This new tissular theory was an indisputable step toward the harmonization of statics and dynamics. But for the correspondence to be properly established, an in-depth criticism of Bichat's system was needed, most notably with respect to the relation between vital properties and the properties of tissues. This criticism was carried out formally through the regularization of the nomenclature that enabled one to order the correspondence for each level. The following chart summarizes these organizational distinctions, which Comte borrowed from Blainville.[31]

The criticism considered the most elementary level of the correlation, that of tissues and vital properties. What was questioned was not the idea of vital properties in itself, for Comte readily acknowledged that sensibility and irritability, the properties of animal life, were truly original.[32] The disagreement concerned whether the properties of tissues and vital properties had to be distinguished or assimilated. Bichat indeed drew a distinction between the properties of tissues and vital properties. According to him, the former "arise from their texture, from the arrangement of their particles, but not from the life that animates them," that is, not from their vital properties.[33] Yet to Comte, vital properties in some way intersected with the twenty-one tissues listed by Bichat: several tissues could display varying degrees of the same vital property, such as sensibility. Comte's system precluded a complete overlap of anatomy, which studies tissues, and physiology, which studies vital properties. That was why Comte introduced the following principle: "For no property can be admitted in physiology without its being at once vital and belonging to tissue."[34] The independent vital property of a given tissue was a metaphysical entity, which could not be included in a positive biological theory. Because of this principle, Comte objected to the idea that all tissues were necessarily sensible and irritable, merely differing by the "simple differences in degrees" with which these properties manifested themselves: such a stance resulted in his dismissal of Bichat's categories of organic sensibility and contractibility. Sensibility and contractibility could emanate only from nervous and muscular tissues respectively.

Table 4.2

Statics (anatomy)	Dynamics (physiology)
Tissues	Properties
Organs	Functions
Apparatus of organs	Functions
Organisms	Results

The rigorous coordination of tissues and properties was crucial for the harmonization of structural and functional hierarchies. From there, physiology as a whole could be ordered according to the various degrees of anatomical analysis. The structure of the two lessons dedicated to physiology (the forty-third and the forty-fourth, which dealt successively with the phenomena of organic life and animal life) clearly illustrated that point: Comte adopted an order of presentation that strictly followed a hierarchy that now was both anatomical and functional. He first considered the elementary tissues characteristic of both forms of life (cellular tissue for organic life, muscular and nervous tissue for animal life), then turned to the properties with which they were associated, reviewed the main functions, and finally concluded by taking stock of the results at the level of the organism as a whole.

The Limitations of the Architectonics

This architectonic construction, though coherent with respect to both its philosophical and its scientific tenets, nonetheless faced several problems. I will mainly focus on two of those, namely, the issue of approaching life analytically and the issue of the milieu, before I elaborate on a few distinctive features of Comtean biology.

The Analytic and the Synthetic Perspectives

Comtean biology was an abstract science; moreover, its scientific foundations—Bichat's histology—seemed to turn it into an analytical science, which broke down living beings into their basic organized elements. The overall distribution of the lessons, divided between statics and dynamics, also seemed to follow that analytical orientation, although the connection between the two aspects was the ultimate goal of biological science. This analytical character was indeed a problem Comte was aware of, for he kept on emphasizing the unity of the phenomena of life. That was why he tried, through different means, to draw attention to the importance of synthesis throughout the biological lessons of the *Cours*.

The architectonics I have just described, even though it was not totally artificial, owed part of its legitimacy to the fact that it was "philosophically indispensable to our intelligence." That was why the various divisions it was built on had to be conceived synthetically: "However, it must be always borne in mind that any system will have to undergo a general revision, with a view to bringing out the essential relations of its parts: the relations, not only of the two sections of static biology, but of both to the dynamic . . . . and the only reason why such a revision appears more necessary in biology than in the other sciences is, that there is a profounder accordance [*consensus*] between its departments than we find in theirs."[35] Similarly, in anatomy, we "artificially" break down organisms that "necessarily [are] individual wholes."[36]

The emphasis on the "artificial" character of this analytical process, which concerned both science and the organisms it studied, led to a new understanding of the architectonics of biology. For an analytical reading of the lessons, divided into two complementary parts whose reunion was the proper goal of biology, Comte progressively substituted a more finalistic reading, which construed statics as a preamble and dynamics as the true study of life. This interpretation was fully embraced in the structure of the biological lessons Comte proposed in the *Système*, but it had already been suggested in the *Cours*.[37] Comte would be increasingly vocal in his claim that specialized studies must not remain isolated: "their study will be begun and carried on with a distinctly synthetic purpose, that of forming a clearer conception of the general relation between the organism and the environment."[38]

The Problem of the Milieu

The architectonics Comte proposed faced another difficulty: the lack of clarity regarding the exact situation of the milieu. Although Comte stressed that the milieu was the "third elementary aspect" of the science of organized bodies, in fact, it had been eliminated from the final formulation of the biological problem. This absence resulted in the underdetermination of the exact place of the study of the milieu within Comtean biology.

In the *Cours* this theory was introduced at the beginning of the forty-third lesson, that is, at the juncture of statics and dynamics. In the *Système* it had to be studied after statics and dynamics. In both cases, Comte invoked the influence of other thinkers to justify his choices—something he rarely did. In the *Cours*, he followed Blainville[39] and in the *Système* he ascribed the rectification to Louis Auguste Segond (1819–1908).[40]

Comte's relatively complex attitude toward the problem of the environment might account for these hesitations. On the one hand, from an antivitalist perspective he had to define life as a relation between the organism and the milieu. But, on the other hand, he was wary of explanations, both in biology and in the theory of knowledge, that turned the milieu into the crucial factor that shaped minds or organisms. Claude Adrien Helvétius, for theory of knowledge, and Lamarck, for biology, were the best examples of an attitude which was characteristic of the eighteenth century and which therefore belonged to the past, according to Comte. This was indeed the basis of one of his disagreements with Mill, who, in order to build his ethology, needed to give a more important role to "circumstances."

Comte's stance toward biological evolutionism might be described as some sort of Lamarckism in a fixist context. His arguments were premised on the assumption that the specific type could not be altered: organisms could evolve only within fixed species boundaries. However, within this framework, Lamarck's claim of the fixation of acquired traits by inheritance was fully accepted. Accordingly, the modifications due to evolution could only be, to use a word Comte favored, "secondary": the use and disuse of particular functions resulted in the growth or decline of the corresponding organs. But use or disuse could neither create a new organ nor lead to the final extinction of an existing organ. That was what Comte highlighted in the *Système*: "The new biology will be in a position to determine the true influence, special or general, of environment on organisms; including the extreme cases where it acts as a disturbing force. While retaining the great principle of the substantial permanence of species, we shall thus discover the natural limits within which they may vary."[41] Comte evoked two paradigmatic cases. In the human species, under the influence of society, certain cerebral functions could shrink, such as those that favored the warlike instinct, to the benefit of other faculties that favored altruism. The second case, which was one of the first "positive utopias," consisted in the hypothesis that herbivores would become carnivorous animals by a gradual diminution of their digestive apparatus.[42] These two cases strictly obeyed the restricted Lamarckism endorsed by Comte, which claimed that evolution could influence only the use of existing organs.

This paradoxical Lamarckism was part of Comte's more general theory of the limits of variation. Due to the fixity of the species, the organism had only a restricted power of adaptation. Beyond certain modifications of the environment (the importance of which was directly proportional to the complexity of the organism considered), life was no longer possible: the influence of the milieu finally became "disruptive." The adaptation of the organism could take place only within these limits. It is perhaps in the context of this limited adjustment of the organism to the environment that Comte's interpretation of the tricky "principle of the conditions of existence" is best understood. Quite surprisingly, Comte did not evoke explicitly the issue of the environment, whereas Georges Cuvier (1769–1832), from whom Comte borrowed this principle, had: "As nothing can exist without the reunion of those conditions which render its existence possible, the component parts of each being must be so arranged as to render possible the whole being, not only with regard to itself but to its surrounding relations."[43]

According to Comte, this principle merely stated "that there is no organ without a function, and no function without an organ,"[44] and it ran counter to the "dogma of final causes." For if correlation between function and organ was indeed a biological principle, the adjustment between organ and function was never perfect: "We must entirely rid ourselves of the superstitious notion that the action of every organ is always perfectly healthful, or that every function has always the structure best adapted to its performance."[45] This imperfect adaptation explained why for Comte the principle of the conditions of existence did not result in the "antiscientific admiration" characteristic of theological optimism.[46] In other words, the positive conception of the relations between the environment and the organism enabled one to make sense of the antitheological interpretation of the principle of the conditions of existence. However, once again the environment was absent from the explanation.

The Major Biological Chapters of the *Cours*

The strength of Comte's biological philosophy did not only derive from the highly consistent architectonics whose main philosophical and scientific foundations I have previously described. Its value also lay in the manner in which Comte incorporated and, in some sense, summarized, within this framework, the major achievements of the biological sciences of his time. I have already showed how Bichat's tissular anatomy provided the scientific basis of Comte's biological architectonics. I will keep on recovering the richness of that philosophy by reviewing, in the order of presentation adopted by Comte, the lessons dedicated on the one hand to methodological questions and on the other hand to vegetal and animal physiology.

Biotaxy and the Problem of Methods

As previously pointed out, biotaxy represented both one of the important achievements of positive biology (the ordering of the hierarchy of organized beings) and the principal support of the main biological method, the comparative method.

The Comparative Method

The complexity and holistic nature of organized phenomena led Comte to reject both quantitative and experimental methods. Like Bichat, he opposed, from his very early writings, "any attempt at actually applying . . . mathematical theories to physiological questions."[47] Comte argued that the complexity of vital phenomena prevented the use of quantification and calculation. Another similar argument justified, if not the total exclusion, at least quite stringent restrictions on the use of experimentation, because the consensus characteristic of vital phenomena did not allow for the control of the perturbations that we artificially introduced within living bodies. The only mode of experimentation Comte really accepted was "pathological investigation,"[48] which was conceived of as a spontaneous kind of experimentation whose relevance was grounded in the "eminently philosophical" principle borrowed from François Broussais: this principle, to use the terms of the *Système*, stated "that the phenomena of Disease are essentially homogeneous with those of Health, the only difference being one of comparative intensity."[49] Pathological analysis would play an increasingly important role in Comte's thought, not only in biology but also in sociology.[50]

Comte's distrust of experimentation alienated him—and that was one of the limitations of his biological philosophy—from an important trend in French biology at the time: the experimentalist trend, whose most notable figures were François Magendie and, later, Claude Bernard, and which also originated in Bichat's works.[51] Comte represented another branch of the Bichatian lineage, which was characterized, in a positive way, by the endorsement of tissular anatomy, and, in a negative manner, by the rejection of experimentation.

The true method of biology was comparison, "which is so specially adapted to the study of living bodies, and by which, above all others, that study must be advanced."[52] What made such a method perfectly adapted to its object had to do with the very nature of the "system of biological science":

> The fundamental condition of its use is the unity of the principal subject, in combination with a great diversity of actual modifications. According to the definition of life, this combination is eminently realized in the study of biological phenomena, however regarded. The whole system of biological science is derived, as we have seen, from one great philosophical conception; the necessary correspondence between the ideas of organization and those of life. There cannot be a more perfect fundamental unity of subject than this; and it is unnecessary to insist upon the almost indefinite variety of its modifications,—static and dynamic.[53]

In the *Cours*, Comte drew a distinction between five modes of comparison: (1) comparison between the different parts of the same organism; (2) comparison between the sexes; (3) comparison between the various phases presented by the development of the same organism; (4) comparison between the different races or varieties of each species; (5) comparison between all the organisms of the biological hierarchy. However, he elaborated on only three of these modes, to which he would return in the *Système*: the first, third, and fifth modes.[54] The first mode—comparison between the different parts of the same organism—was the one by means of which Bichat had achieved most of his discoveries in tissular anatomy.[55] The third mode, which consisted in comparing the different ages of an organism, found its legitimacy in the idea that the early steps of development of a higher organism enabled one to represent the characters of lower organisms, a principle of recapitulation (although it did not refer to any kind of phylogenesis) that Comte used in fields other than biology: for instance, in the idea that the education of the individual had to go through the different states humanity had gone through. It must be pointed out that, even in biology, that principle did not really play an important role and that it is difficult to find instances of the application of that second kind of comparison.

The fifth mode was the crucial one for the logic of comparison: the comparison of all the organisms of the biological hierarchy. The comparative method is usually grasped through that mode, for it was indeed only with the support of the biological hierarchy as a whole that the logic of comparison could fully deliver what it promised, considering that it presupposes that the object studied displays both unity and the greatest diversity of modifications: "The peculiarity of this largest application consists in its being founded on a very protracted comparison of a very extensive series of analogous cases, in which the modification proceeds by almost insensible graduated declension."[56]

The Biological Hierarchy

Just as the primary mode of the comparative method was dependent on the constitution of the biological hierarchy, the latter was in turn dependent on the method of classifications. The two philosophical principles, which constituted the "so-called natural method," of the formation of natural groups and of their hierarchical ordering,[57] entrusted biological classification with the task of ordering organisms in a single linear hierarchy (a "scale"). The three main principles that governed that hierarchy were the following:

> First, that the animal species present a perpetually increasing complexity, both as to the diversity, the multiplicity, and the speciality of their organic elements, and as to the composition and augmenting variety of their organs and systems of organs. Secondly: that this order corresponds precisely, in a dynamic view, to a more complex and active life, one composed of functions that are more numerous, more varied, and better defined. Thirdly: that the living being thus becomes, as a necessary consequence, more and more susceptible of modification, at the same time that it exercises an action on the external world that is continually more profound and more extensive. . . . Hence results the possibility of conceiving of a final arrangement of all living species in such an order as that each shall be always inferior to all that precede it, and superior to all that follow it.[58]

Let us notice that these principles, and especially the first, would also be used in sociology to account for the progress of humanity.[59]

Such a conception of the animal hierarchy led Comte to reject one of the most influential models at the time: Cuvier's "embranchement" system. Although he agreed with Cuvier, against Lamarck, on the issue of the fixity of species,[60] Comte nonetheless fully endorsed the serial conception of the animal hierarchy advocated by Lamarck and, later, Blainville. The influence of fixism over that conception bore only on "the continuity or discontinuity of the organic progression."[61] The kind of discontinuity characteristic of that scale was specific: although it did not display a straightforward continuity, for the species did not transform into one another, the progressiveness of the scale was still preserved, in accordance with the continuism characteristic of Comtean philosophy. As emphasized by Comte, the "progressive course of the animal organism," which fixism prevented from being a "real natural law," nonetheless remained "a convenient abstraction."[62]

This last remark leads us to think about how the subjective perspective of the *Système* caused Comte to reconsider the question of the animal series. Contrary to the realist position he had adopted in the *Cours*, in the *Système* Comte finally came to emphasize its "instrumental" character: "Discussions suggested by it [i.e., the biological series] can have no issue so long as the series is regarded as an absolute expression of some external reality, instead of what in truth it is, a subjective logical instrument intended to assist us in investigating the more difficult problems."[63] This analysis, which was indeed quite close to Lamarck's positions on the artificial character of the series,[64] provided the proposal already suggested by Comte in the fortieth lesson with a more consistent framework: the introduction of fictitious species in order to facilitate transitions.[65]

The high esteem in which Comte held the comparative method showed that he belonged to that age of biology characterized by the domination of comparative anatomy. His endorsement of the idea of a hierarchy of beings, the scope he ascribed to it (from plants to man and mankind) and the place he gave man as a superior type and summary of the "series" brought him closer to a certain kind of Romantic biology.[66] For Comte, this method indisputably was one of the greatest achievements of the human mind: "No philosopher can contemplate without admiration the eminent art by which the human mind has been aided to convert into a potent means what appeared at first to be a formidable difficulty. I know no stronger evidence of the force of human reason than such a transformation affords."[67] Accordingly, one should not be surprised by the fact that the proper meth-

Vegetal and Animal Physiology

od of sociology, the historical method, which was supposed to rule over all the other methods, would also be the ultimate mode of the comparative method.

The organization of Comtean physiology was laid out in the forty-third lesson:

> Though all vital phenomena are truly interconnected, we must, as usual, decompose them for purposes of speculative study, into those of greater and those of lesser generality. This distinction answers to Bichat's division into the organic or vegetative life, which is the common basis of existence of all living bodies; and animal life, proper to animals, but the chief characters of which are clearly marked only in the higher part of the zoological scale. But, since Gall's time, it has become necessary to add a third division—the positive study of the intellectual and moral phenomena which are distinguished from the preceding by a yet more marked specialty, as the organisms which rank nearest to Man are the only ones which admit of their direct exploration.[68]

Although this outline followed a principle of decreasing generality, it was not altogether consistent. The division between organic and animal life rested on the distinction between two kinds of life or functions that were fundamentally different, whereas, properly speaking, cerebral phenomena belonged to animal life to the extent that they were relative to sensibility. That is the reason why Comte sometimes talked about a provisional separation, motivated merely by the degree of complexity.[69] Yet Comte finally stuck to this tripartition, and that would be the one that also appeared in the *Système*. Because of its importance, I will study cerebral physiology in an independent section and will focus here on vegetal and animal physiology.

As the previous quotation indicates, both the outline and the claims upheld in the lessons on physiology were dominated by the distinction between organic and animal life as introduced by Bichat in his *Recherches Physiologiques sur la vie et la mort* (1800). Bichat described that distinction as follows:

> It may be said that the vegetable is the rough sketch, the *canvas* of the animal; and that to form this last, nothing more is necessary than to display upon this canvas the external organs proper to establish its different

relations.... The functions of the animal form two very distinct classes. The one is composed of a continual succession of assimilation and excretion; by these it is incessantly converting to its own substance the particles of surrounding bodies, and again ejecting these particles when they have become heterogeneous. It lives within itself only, by this class of functions; by the other, it exists, as it were, out of itself: it is the inhabitant of the world, and not, like the vegetable, of the spot which gave it birth. It feels and perceives what surrounds it, reflects its sensations, moves voluntarily according to their influence, and most generally has the power of communicating by voice its desires and its fears, its pleasures or its pains.[70]

Comte did not stray from that distinction and would keep on celebrating, as late as the *Système*, "Bichat's incomparable genius."[71] The only important modification Comte brought to Bichat's conceptions was to get rid of organic sensibility and contractibility. Comte had two reasons for making such a change: on the one hand, as shown previously, the aim was to rectify the correspondence between the ideas of organ and of function; on the other hand, Comte wanted the physiological principle of the division between the two kinds of life to be strictly upheld, for the ideas of organic sensibility and irritability clearly infringed on the division to the extent that irritability and sensibility defined animal life.

Finally, from a methodological point of view, the lessons on physiology avoided the pitfall of vitalism, building on "the true philosophical character of physiology," which "consists in establishing an exact and constant harmony between the static and the dynamic points of view,—between the ideas of organization and of life."[72] However, this principle, which governed the relations between structure and function, was applied quite differently depending on whether one dealt with organic or animal life.

Vegetal Physiology

The structure of the forty-third lesson strictly followed the dual anatomical and physiological hierarchy described above. The chart below summarizes the outline of the lesson.[73]

Two somewhat different sets of results originated from vegetative life: those relative to composition and decomposition and those relative to reproductive phenomena. The defining feature of universal life (i.e., the life common to all being)—"the double interior motion, general and continuous, of composition and decomposition"[74]—attracted a fair share of

Table 4.3

Properties (cellular tissue)	Functions	Results
Hygrometricity	Absorption	Composition /decomposition
Capillarity	Exhalation	Vital heat
Retractility		Electrization
		Reproduction
		Organic development / decline

Comte's attention. This dual motion was dependent on the functions of absorption and exhalation. Both the structural basis—the cellular tissue—and the physiological properties it was associated with—hygrometricity, capillarity, and retractility—also contributed to the production of that main result. One could therefore say that organized structures were responsible for the most important organic phenomenon. However, Comte kept emphasizing the physicochemical nature of these functions: "the essential acts vegetative life is made out of are, by their nature, mere physicochemical phenomena."[75] This analysis raised the problem of the autonomy of biology that Comte, because of the "antimaterialistic" stance he had adopted, wanted to defend at all costs, even though it revealed a weakness in his conception of organic physiology, which was supposed to be balanced by his presentation of animal physiology.

What might surprise the reader in the considerations of the *Cours* dedicated to the phenomena of reproduction and development is the fact that Comte did not seem willing to defend any specific claim relative to those. As he himself pointed out, "Still, the preliminary requisite for the formation of doctrine—a fundamental analysis—remains unfulfilled; and the ascertainment of the laws of production and development is not, therefore, to be attempted at present."[76] This reluctance by an author who attempted, in sociology, to study the development of mankind, which he considered as some sort of organism, might seem puzzling. Such a stance is also problematic for those who claim, drawing on serious textual evidence, that Comte had a preformationist conception of history.[77] Here again, the absence of explicit references to this embryological claim remains quite puzzling with respect to an author like Comte, who was quite wary of the use and precision of the models he resorted to.

Animal Physiology

As Comte pointed out, "nowadays, all the general phenomena of animal life are . . . associated with irritability and sensibility, which are both considered to be the characteristic features of a strictly defined tissue."[78] Accordingly, the phenomena of irritability and sensibility were the objects of animal physiology. At the outset of the forty-fourth lesson, Comte emphasized the absolutely original and unexplainable character of these phenomena: just like gravity and heat, sensibility and irritability could not be reduced to any other category. One might argue that the previous claim was a second line of defense against any reductionist or materialistic argument. The first line of defense was premised on the idea that living beings, apart from "all the phenomena, whether mechanical or chemical, which occur in inorganic bodies," exhibited "a wholly special order of phenomena, the vital phenomena, properly so-called, which belongs to organization."[79] Hence the idea of a specificity of the organized bodies. However, the emphasis on the very physicochemical character of vegetative processes had weakened that first line of defense. The assertion of the radical originality of irritability and sensibility was thus conceived as a reinforcement for this first line of argument and made more explicit the fact that the specificity of biology was primarily due to animal physiology.

In order to study these phenomena, Comte followed the same anatomophysiological progression he had adopted for vegetative life, except that he inserted, between functions and results, considerations relative to the "mode of action" of animal functions.[80] The following charts try to render the conceptual structure of the forty-fourth lesson.

Let us focus on two points: the distribution of the functions of sensibility and the general theory of modes. As for the former, the striking feature of Comte's analysis of sensibility was the fact that he took up the categories put forward by Pierre Jean Georges Cabanis (1757–1808). The threefold distinction between external sensations, internal sensations, and internal sensibility indeed mirrored the fundamental division introduced in the *Rapports du physique et du moral de l'homme*, which contrasted external impressions, internal impressions, and the impressions directly produced by the action of the cerebral organ itself.[81] That distinction had been devised to complement the sensualist paradigm that, since Étienne Bonnot de Condillac (1715–1780), merely restricted itself to stating that our ideas

Table 4.4

Tissues	Properties	Functions
Muscular	Irritability	Voluntary movement
		Involuntary movement
Nervous	Sensibility	Sensation
		Internal sensation
		Internal sensibility (affective and intellectual functions)

Table 4.5

	Modes of action (common to the phenomena of irritability and sensibility)		Results
Relative to a function separately conceived	Mode of the animal phenomenon: intermittence [Bichat]		Sleep [Bichat]
			Habit [Bichat]
	Degree of the animal phenomenon	inferior degree	Boredom [Leroy]
		intermediate degree	Health, welfare, happiness
		superior degree	Pleasure and pain
Relative to the association of the animal functions	Normal state: *synergies* [Barthez]		General sense of the Self.
	Pathological state: *sympathies* [Bichat]		

originate from external sensations. Cabanis claimed that instincts derived from internal impressions (those produced by the organs of vegetative life) and that spontaneous cerebral phenomena were the third source of impressions.

Comte took up these distinctions, but he developed them in a distinctly antisensualist context that was under the spell of the phrenological claim of the innateness of cerebral functions. Accordingly, in these pages on sensibility, there was no trace of any account of the genesis of knowledge, nor even the lineaments of an associationist theory: this was a serious departure from the way French philosophy, through Condillac, Antoine Destutt de Tracy (1754–1836), or even Cabanis, had conceived the theory of knowledge. It was for sociology to take charge of the genesis of our beliefs, whereas cerebral physiology was to provide the theory of our intellectual faculties.

Finally, let us remark that, even though, in the forty-fourth lesson, internal sensations had just been mentioned, they would play a crucial role later, for they would be connected with the selfish instincts, as opposed to altruism. This was why Comte considered ethics, which the second volume of the *Système* turned into the ultimate science of the positive encyclopedia, to be aimed at limiting the influence of those internal sensations and at regulating the relations between the moral and the physical in man. By doing so, he was aware that he achieved "the doctrine that Cabanis had so nobly sketched."[82]

The theory of the modes of action is the second noticeable feature of the forty-fourth lesson. Its principle is quite straightforward. If the equivalent of such a theory was absent from the forty-third lesson, it was due to the fact that vegetal economy was conceived as the continuous fulfillment of a single function, whereas animal functions were intermittent and diverse. Hence the necessity of a study that focused on intermittent phenomena and on the issue of the association of functions. Such a theory originated from the description Bichat gave of animal life in the *Recherches*, in which he also emphasized the importance of irregularity and habit. However, Comte also modified that theory to a certain extent, most notably by establishing a connection between intermittency and habit that did not exist in Bichat and by adding to it some theses borrowed from Charles Georges Leroy (1723–1789) and Paul-Joseph Barthez (1734–1806) (see table 4.5). Finally, to be complete, it must be pointed out that Blainville's lectures constituted the secondary source of that study of modes.[83]

That theory gave pride of place to a series of concepts that would prove to be crucial for Comtean anthropology: the phenomena of irregularity and habit would provide the idea of progress with its biological foundation;[84] boredom would be understood as an accelerator of social evolution; and the feeling of the self would be the ultimate result of the associations of the animal functions. The latter point enables us to return to the issue of the unity of the living and to qualify what has been said previously. From the anatomical point of view, the unity of organized bodies was to be considered a premise: it indeed accounted for the irrelevance of experimentation. From the physiological point of view, the fact that living bodies functioned as wholes was merely a precarious outcome, achieved through the balancing of all animal functions: "a necessary result of a harmony between its various chief functions."[85] "There is, in this direction, . . . no other real subject of positive investigation than the study of equilibrium of the various animal functions,—both of irritability and of sensibility,— which marks the normal state, in which each of them, duly moderated, is regularly and permanently associated with the whole of the others, according to the laws of sympathy, and yet more of synergy. The very abstract and indirect notion of the I proceeds from the continuous sense of such a harmony; that is, from the universal accordance of the entire organism."[86]

Such an analysis, which turned the self on which metaphysics rested into "a purely fictitious state,"[87] showed that the living and society faced the same problem: that of unity. All the higher organized bodies, be they mammals (Comte took the example of the cat), individual man, or humanity, were taking part in a process of unification. In physiology, unification was carried out by the nervous system. As Comte stated, "The physiologists of our time seem to be all agreed as to the nervous system being the necessary agent of all sympathy."[88] In other words, *in fine*, unification was dependent on the brain, which was considered to be the central apparatus of the nervous system: cerebral unification would then perfectly mirror the unification of the living being.

Biology in the *Système*

The first volume of the *Système* contains the second general presentation of positive biology.[89] Most of the theses elaborated in the *Cours* are taken up in the *Système* and, as seen previously, the outline of the biological curriculum suggested by Comte follows the structure set in the lessons of the *Cours*, except for the openly finalistic perspective. However, the presentation of chapter 3 of the *Système* was quite original, for the perspective and arrangement it adopted were very different from those of the *Cours*.

The Subjective Point of View

The perspective adopted by the *Système* was subjective: the very aim of the "introductory principles, scientific and logical"[90] was to reconsider the whole encyclopedia, sociology excluded, from the subjective point of view. The idea on which the subjective point of view was premised was that unity of knowledge was impossible to achieve from an objective point of view. The sciences that formed the encyclopedia were too heterogeneous. The only way to satisfy the deep need for unity felt by the human mind was to consider knowledge as a whole from the point of view of humanity: "One must henceforth conceive . . . but one single science, human science, or more exactly social science, whose principle and goal is our existence."[91] Once the creation of sociology had completed the encyclopedia, an achievement carried out in the *Cours*, it became possible to consider all the sciences from that point of view and to go through the encyclopedia the other way around: from man to world (subjective order), and not from world to man (objective order).

Within this framework, biology had a particular role, to the extent that it was the science that, in the objective order, immediately preceded sociology. This explained why, from the subjective point of view, it had been the first science to be reconsidered. The chapter on biology of the *Système* showed that, contrary to the other sciences, biology depended on sociology not only for its definitive logical and scientific constitution but also for its completion.

The "abstract theory of life"[92] was the main originality of the biological outline of the *Système*. The subjective renewal of the encyclopedia revealed, as a consequence of the "radical contrast between Life and Death," a sharp division between the cosmological sciences and biology or the theory of life. For it was indeed a theory of life, in the broadest sense of the term, that the *Système* introduced. It consisted of three kinds of vitality, which all rested on three fundamental laws. The following chart summarizes the gist of the *Système*'s presentation.[93]

The transition from vegetative to animal life was made possible by the modification of the mode of nutrition: contrary to plants, animals, which could not construct their tissues directly from the inert environment, were either herbivorous or carnivorous. The transition from animal life to humanity was made possible by the development of the internal functions of the brain, which was restricted to only one species.

The Completion of Biology

The inclusion of the study of humanity in the abstract theory of life meant that biology as such could not cover the full scope of the object it considered. It was for sociology to study the last mode of vitality, as was illustrated by the fact that the three laws that Comte laid out were the three funda-

Table 4.6

Vegetality	Animality	Humanity
Law of the constant renewal of structure	Law of the alternation of action and rest	Law of the three stages
Law of death	Law of habit	Law of the classification of abstract conceptions
Law of reproduction	Law of improvement by exercise	Law of the evolution of activity

mental laws of positive sociology. Only the subjective point of view, which emanated from sociology, could remedy the shortcomings of biology.

It is possible to give two examples of these shortcomings and of the ways Comte dealt with them. The first example concerned vegetal life. Comte emphasized the fact that the three laws of elementary life were independent and did not imply one another. From an objective point of view, it was therefore impossible to claim the unity of that first kind of life. It was only possible subjectively: "the only unity really possible is the subjective; this alone lies within the compass of both our capacities and our wants."[94] The second example bore on the study of cerebral functions, which was a part of biology. In the *Cours*, it had been presented as the most complex part of physiology so-called. Now Comte claimed that the knowledge of the cerebral faculties, the development of which allowed for the third mode of vitality, belonged to biology. The direct consequence of such a statement was that biology would be definitively incomplete without sociology and the subjective perspective: "We see more clearly than ever the impossibility of regarding the science of Life as an isolated study, since its highest phenomena form the subject of a distinct science. Nor would it be feasible to limit the field of pure Biology to the two lower phases of vitality, abstracting the highest phase.... Animal life as a whole would be unintelligible without the higher attributes which Sociology alone can estimate."[95] This last example clearly showed that the study of cerebral functions was a crucial element in the constitution of biology. Comte's final position consisted in claiming that this study was dependent on sociology. We will address below the problems that such a stance created for the theory of human nature.

The Biocratic League

The subjective point of view, when applied to biology, not only resulted in epistemological and encyclopedial effects but also led to a somewhat original and ecological thought: the theory of "biocracy," which was the outcome of a struggle in which humanity, the only species that was to form the Great Being, prevailed: "Each race of animals has in fact been struggling for exclusive dominion over the earth, as each people has been struggling to subdue all others. But both these contests necessarily cease at the same time."[96] When the "true Great Being" succeeded in unifying itself its preponderance over the Earth would then be definitive, so that the advent of "sociocracy" would go hand in hand with that of biocracy: "In the conception of this association between Humanity and the races capable of domestication we have the scientific basis for the widest and most permanent aspect of Positive Polity; the combination of organic nature against inorganic nature for the purpose of developing all the resources of our planet."[97] We are not far from the modern idea according to which man has become responsible for the living world. However, such a modern construal of Comte's thought needs to be qualified in light of the fact that the biocratic association concerned only sociable races—Comte had no objection to the eradication of the races that were not sociable. Yet his position remained one of the strongest expressions of a vitalism that could be traced back to Bichat.

Physiological Phrenology and the Theory of Human Nature

Physiological phrenology, or, as indicated by the title of the forty-fifth lesson, "the positive study of the intellectual and moral, or cerebral functions,"[98] certainly was one of the most striking manifestations of Comtean naturalism. The claim according to which intellectual and moral faculties were cerebral faculties appeared as early as 1824 in Comte's writings.[99] Cabanis stated the principles, but Gall was to be praised for achieving what remained mere "advice" in Cabanis's works. It cannot be denied that Comte had been an extremely devoted, though lucid, advocate of Franz-Joseph Gall (1758–1828). As late as the *Système* he described himself, even though he had built his own cerebral theory, as the successor of both Gall and Condorcet.[100] Nor can it be denied that Comte, throughout his intellectual evolution, never stopped thinking about and working on phrenological physiology.[101] Together with the specifically biological problems it raised, that theory indeed dealt with most of the biological factors that allegedly determined sociological evolution. As its encyclopedic situation suggested, phrenological physiology, at the juncture of biology and sociology, held a strategic position.

In the reminder of this section, after an outline of the main principles, which would remain the same in the various versions of the theory, on which rested physiological phrenology, I will study its evolution, which will introduce us to the "theory of human nature."

The Fundamental Principles of Physiological Phrenology

Cerebral physiology rested on the following philosophical principles: "the innateness of the fundamental dispositions, affective and intellectual, and the plurality of the distinct and independent faculties, though real acts usually require their more or less complex concurrence."[102] Both principles, which were physiological in the proper sense of the term, found their anatomical counterpart in the idea that the brain was divided "into a certain number of partial organs, symmetrical like those of the animal life, and, though more contiguous and mutually resembling than in any other system, and therefore more adapted both for sympathy and synergy, still distinct and mutually independent."[103]

Comte fully endorsed these principles, which were, together with their anatomical interpretation, the basis of Gall's doctrine.[104] Innateness fit perfectly with both Comte's antisensualism and, to a lesser extent, his moderate Lamarckism: the environment, the circumstances, were totally deprived of the shaping power they had been granted by the materialistic philosophy of the eighteenth century.

However, both philosophically and biologically, it was the assertion of the plurality of cerebral organs that was crucial. Philosophically, it destroyed the metaphysical and theological claims relative to the unity of the self. Moreover, it prevented the characterization of man by a single attribute, such as intelligence: in man as in animals there existed intellectual but also affective and instinctual faculties. Furthermore, unlike the classic definition of man as a "rational animal," phrenology showed that affective faculties were the most numerous and active. Finally, this principle enabled one to conceive, within the fixist framework characteristic of Comtean philosophy, the true evolution of cerebral functions. For what had evolved during the course of human history was not, strictly speaking, the mental endowment—there had been no creation of new faculties—but the synergies between the faculties and the degree of energy associated with those faculties, whose anatomical translation was the increase or decrease of the size of the organs.

To adapt the cerebral plasticity presupposed by Gall's doctrine to his historical evolutionism, Comte introduced a "scientific principle" that was not to be found in Gall:

> The scientific principle involved in the phrenological view is that the functions, affective and intellectual, are more elevated, more human, if you will, and at the same time less energetic, in proportion to the exclusiveness with which they belong to the higher part of the zoological series, their positions being in portions of the brain more and more restricted in extent, and further removed from its immediate origin,— according to the anatomical decision that the skull is simply a prolongation of the vertebral column, which is the primitive center of the nervous system. Thus, the least developed and anterior part of the brain is appropriated to the characteristic faculties of humanity; and the most voluminous and hindmost part to those which constitute the basis of the whole of the animal kingdom.[105]

What this principle claimed was crucial: a model of how a normal brain would operate, with some possibility of marginal variation, in accordance with the moderate Lamarckism advocated by Comte.

In Gall's work, the coefficients of energy and volume were randomly allocated among the various individuals of a given species.[106] Such a stance corresponded to the fact that Gall's psychology was primarily one of individual differences—what was also called, in the nineteenth century, a "characterology."[107] This differential perspective, which was the distinctive feature of the reception of phrenology at the time, appeared only marginally in Comte.

Comte used cerebral physiology in quite a different manner, as a means to construct a general type whose variations were primarily historical, and only secondarily, and to a lesser extent, individual. It was in this latter respect that the scientific principle played a crucial part: through the determination of the faculties' coefficients of energy, it balanced the differential effect of the plurality of faculties; it also allowed for the elaboration of a model of how the brain operated that was both rigid enough to define a normal state for the whole species and flexible enough to accommodate a series of different configurations that would not correspond to an idiosyncratic state but to a stage of civilization. For Comte, phrenology was a tool that enabled him to think about the identity of human nature and to argue for the uniform character of the cerebral modifications that resulted from social development. Hence a fundamental difference between Comte and Gall as to the very spirit of cerebral physiology: the philosopher took up the categories Gall had elaborated in a differential perspective but intended to use them within a universalist and historical perspective.

The Physiological Models of Social Development

With regard to the classification of phrenological faculties as such, it is worth noticing that Comte did not take as his starting point Gall's list of faculties but the classification drawn by Spurzheim (1766–1832).[108] Such a choice might well have been motivated by the importance Comte ascribed to classification as a method, siding with Spurzheim against Gall.[109] But Comte also borrowed Spurzheim's model for the early outlines of the cerebral chart, the properly Comtean categories being progressively elaborated afterward. From the outset, as early as the *Cours*, the distinction between penchants and feelings was reinterpreted sociologically: the former were relative to the individual, the latter to the species. It was of course from that distinction that the opposition developed in the *Système* between selfish instincts and altruistic feelings derived. The other Comtean innovation, of much less importance, amounted to introducing practical faculties, which started to appear very gradually in lesson 56[110] and which would be addressed as such only in the *Système*. The final positive classification would consist in a threefold division: affective, intellectual, and practical faculties; and a crucial subdivision: the division of affective faculties into selfish and altruistic ones.

On the basis of that classification, and with the help of the scientific principle previously mentioned, Comte elaborated successively two ideal types of synergic cerebral functioning. The first model was introduced in the lessons of the *Cours* dedicated to social statics and dynamics. The purpose of social development "is to subordinate the satisfaction of the personal instincts to the habitual exercise of the social faculties, subjecting, at the same time, all our passions to rules imposed by an ever-strengthening intelligence, with the view of identifying the individual more and more with the species."[111] In a sense, this model put intellectual faculties and social feelings on a par. They all had to exert a unifying action in order to promote their own development and to moderate the "personal instincts" that drew their strength from their relation to the needs characteristic of vegetative life.

The second model, primarily elaborated in the *Système*, is very close to the first, except that it ranks altruistic feelings above intellectual faculties, the only legitimate role of the latter being that of serving the former. In this model, intelligence is merely a tool in the service of altruism, whose value lies in the fact that it achieves the unification of both society and the individual.

It is certain that most of the importance Comte ascribed to phrenological physiology derived from the possibility it offered of modeling social development on a biological basis. As suggested above, with the consideration of phrenological physiology, the difficult question of the articulation of biology and sociology had to be raised.

The Theory of Human Nature

There are several ways to tackle the issue of the theory of human nature. For instance, one might question its actual role within sociology;[112] or one might also consider the biological status of that theory. Although the two issues are linked, I will focus on the second.

The phrase "positive theory of human nature" was introduced by Comte in the forty-eighth lesson to designate the "preparatory conceptions of the biological theory": "Thus every law of social succession disclosed by the historical method must be unquestionably connected, directly or indirectly, with the positive theory of human nature; and all inductions which cannot stand this test will prove to be illusory, through some sort of insufficiency in the observations on which they are grounded."[113] The theory of human nature thus referred to a set of biological conceptions relative to man, constructed independently from sociology, and whose function was to control the results delivered by sociology with the help of the historical method. It was allegedly intended as a description of "human nature." It consisted chiefly of the theory of intellectual and moral faculties, because, as demonstrated in the *Système*, the rise of humanity was primarily due to the development of those faculties.

More precisely, the relationship between this biological corpus and sociology could take two forms. With respect to "the most elementary terms of the social series,"[114] the theory of human nature could govern sociological analysis. However, when the progress of civilization was advanced enough to allow for the application of the historical method and prevented a direct deduction from biological data, the theory of human nature merely served as control, for only the historical method was able to account for the "chief phenomenon in sociology,—the phenomenon which marks its scientific originality,—that is, the gradual and continuous influence of generations upon each other."[115]

The idea of a biological control ultimately rested on a fundamental assumption: the idea that human nature was fixed and that social development could not run counter to that nature. That is also one of the possible interpretations of the fundamental, but nonetheless ambiguous, maxim according to which "progress is the development of order." Finally, that idea underlay the preformationist interpretation of Comtean sociology: the course of history would merely be the development of the "seeds" present within human nature.[116] But the evolution of Comte's views on the epistemological status of phrenological physiology would deprive this biological control of its primary function and would eventually render it unworkable.

For how was one to establish such a "theory of human nature"? Comte suggested at least three different scenarios. The first is given in the forty-eighth lesson and carried out in the forty-ninth and fiftieth. Drawing on the results of physiological studies, Comte assumed that it was possible to elaborate an autonomous theory of human nature. The sections of social statics dedicated to the individual and the family were instances of the way biological assumptions shaped sociological theories.

However, the possibility of building a theory of cerebral functions independently from sociology was questioned as early as the end of the *Cours*. That criticism, clearly formulated in the forty-eighth lesson of the *Cours*, rested on the exclusive primacy of the collective over the individual advocated by Comte: "Man is a mere abstraction, and there is nothing real but Humanity, regarded intellectually or, yet more, morally."[117] The prominence of the collective irrevocably ruined the legitimacy that still remained associated with the claims of biology regarding the study of mental phenomena: for Comte, the science of the living could not outgrow the individual point of view. At the same time, such an assertion confirmed the role of sociology as a hegemonic science. That scenario, as previously seen, found its culmination in the *Système*: it was for sociology to take over the study of intellectual and affective phenomena.

The third scenario, which is merely outlined in the fourth volume of the *Système*, is in fact a variation of the first. It consists, on the one hand, in dividing the study of the cerebral functions between three sciences.

Biology clears the ground for the study, sociology deals with the intellectual faculties, and ethics (the seventh and final science of the encyclopedia that had been introduced in the second volume of the *Système*) accounts for feelings. On the other hand, it consists in entrusting ethics with the institution of the theory of human nature.

Without delving into the complex meaning of positive ethics, it seems to us that this third scenario clearly indicates the direction taken by Comte's views on the issue of human nature. From an epistemological problem relative to the biological control of the results achieved through the sociological method, it became a moral problem, that of enforcing upon the individual a certain kind of unification compatible with social life.

To conclude, let us return to the first question raised by the theory of human nature. The window of opportunity available for those wishing to claim or deny the biologization of Comtean sociology is in fact quite narrow: it corresponds to what may be labeled the "classical" formulation of the problem, the one that appears in the forty-eighth and fifty-first lessons of the *Cours* and which was the basis of the controversy with Mill.[118] Afterward, it is difficult to address that very issue, for it seems that we are, rather, dealing with a massive movement of subjectivization of biology and of an integral sociologization or moralization of the theory of human nature.

★ ★ ★

As is the case for all the important topics covered by Comtean philosophy, the results of that overview are mixed. The trajectory of the theory of human nature mirrors quite well that of biology as a whole. On the one hand, there was an attempt at building an autonomous science, with its own laws and its own methods, whose relations with the preceding sciences were mediated by the notion of *milieu*, which borrowed nothing from sociology (which followed it in the encyclopedia) but provided this new social science with a basis for its deductions and means of controlling them. On the other hand, there was a subordinate and incomplete biology, subjected to the social and moral agenda of positive politics, which, through the biopolitical notion of "biocracy," revived the antagonism between life and death that Comte had previously denounced as metaphysical. The fame and recognition enjoyed by Comte's biological philosophy was primarily due to the rigorous and knowledgeable exposition to be found in the *Cours*. Yet as seen later in the *Système*, the theoretical inventiveness of a thought that flouted both objective regulations and current scientific developments might have confused and surprised many, if only because of the contrast with the architectonics that Comte had previously developed. It nonetheless remains true that, each in their own proper style, the lessons or chapters on biology constitute some of the most striking pages of Comte's works.

5

Comte and Social Science

Vincent Guillin

There was plenty for a social philosopher in the first half of the nineteenth century to think about: the French Revolution, which both spelled the fall of the Ancien Régime and sparked the hectic advent of a new social and political order in France and well beyond its borders; the spread of an ever-growing scale of the effects of the Industrial Revolution in the West and the conflicting development of the social forces (most notably, the bourgeoisie and the proletariat) on which it fed; and the tumultuous rise of nationalism in Europe and the universal hopes of freedom, independence, and happiness it originally carried with it.

Among the few thinkers who felt the urgency of elaborating a theoretical framework that could capture that jumble of apparently uncoordinated events, Auguste Comte undoubtedly stood apart. Convinced that mankind was merely going through a critical phase of its evolution, Comte declared that the time was ripe for the articulation of a systematic knowledge of the goals of human development, of the various conditions on which it depended, and of its historical path. This system of knowledge would provide an intellectual and moral compass for the harmonious growth of modern societies.

This imposing project, the *philosophie positive*, culminated in Comte's *physique sociale*, for which he later coined the infelicitous neologism *sociologie*, i.e., a genuine science of society. This science conceived of society as an organic whole and maintained that its historical development could be explained and predicted through laws that would be as objective as those of the natural sciences and that would nonetheless reveal the very specificity of this development, namely that the main driving force behind it was the progressive unfolding and diffusion of the "positive spirit."

In this chapter, I offer a comprehensive view of positive sociology as elaborated in the *Opuscules de jeunesse* of the 1820s and then developed in the *Cours de philosophie positive* (1830–1842), of the historical and political importance Comte ascribed to it, of its main concepts, of the methods upon which it relied, and of the conclusions it was supposed to deliver. I will pay particular attention to Comte's complex analysis of the "encyclopedic" position of sociology, which testified both to its dependence on the other sciences (most notably, biology) and to its epistemological autonomy. I will also question some of the peculiarities of Comte's understanding of the social (his proscription of psychology and political economy from the encyclopedic scale of the sciences, his rejection of any attempt at mathematizing social phenomena, and his reliance on the pseudoscience of phrenology).[1]

Of course, from a methodological point of view, one might question the very usefulness of such an overview, since, except for the somewhat ritual celebration of Comte as a "forerunner" or "founding father" of sociology, there are but very few vestiges of the Comtean legacy that survive in the actual practice of social science. For instance, most contemporary "positivist" sociologists seem to have lost any taste for the kind of grand historical schemes favored by Comte, just as they are wary of avoiding the conflation of descriptive statements with normative claims typical of Comte's sociological analyses, not to mention their rejection of the politically conservative and authoritarian implications of positive politics. Yet a reassessment of Comte's views on social science might nonetheless prove beneficial from a different perspective.

First of all, it might serve as a reminder that, historically, sociology as an academic endeavor has emerged in a certain political, moral, religious, and cultural context in which the resilience of the very fabric of society was at stake, and the contribution of the sociologist could have resulted not only in its understanding but also in its strengthening. For the establishment of sociology as an objective study of social phenomena was indeed conceived, by Comte and some of his contemporaries, as the only appropriate intellectual response to the growing threat of the dissolution of social relations and the correlative dangers associated with the rise of atomistic individualism.[2] In that respect, although the problems we are facing today are somewhat different from those diagnosed by Comte, one might argue that a sociological understanding of their nature is still widely considered a necessary, if not sufficient, condition for their resolution. In other words, sociology as an objective science remains, as was the case in Comte's time, an offshoot of the value-laden reaction to the various ills of modernity.

Secondly, a look back at Comte's sociology might also be interesting to the extent that it provides a vivid illustration of the ways in which the creation of a scientific discipline always presupposes a critical relation to its past. For Comte's sociology is historical through and through, not only in the sense that it aims at discovering the historical laws of social evolution, but also in its attempt to locate its own emergence in the broader course of the intellectual development of mankind. Such a reflexive approach, in which the writing of the history of a science is part and parcel of its theoretical elaboration, might still be worthy of interest, and, although one is perfectly entitled to doubt the scientific value of Comte's contribution to sociology, it would seem rash to conclude that his conception of science, of its historical evolution, and of the kind of narrative most suited to its depiction have also lost all their significance. On the contrary, Comte's "sociology of sociology," i.e., his account of the development of the various conceptual elements constitutive of "positive" sociology, is a perfect instance of a distinctive philosophical understanding of scientific rationality that acknowledges both its epistemic primacy and its historical nature.

Finally, it is worth insisting on the distinctiveness of Comte's sociological approach, for, as Robert Scharff has forcefully argued, his idea of a "historicocritical defense of science" clearly demarcates him from other versions of positivism, such as those developed by Mill or the Vienna Circle.[3] Because he was acutely aware of the various obstacles that had impeded the growth and diffusion of the positive spirit, and despite his progressive endorsement of an increasingly "dogmatic and ahistorical scientism,"[4] Comte remained convinced that if positivism were to prevail as a systematic worldview, it had to rely on a "historically minded reflectiveness"[5] as to whence it came, where it was leading, and how its development could be furthered or hastened. In other words, since, as Comte put it, "no idea can be properly understood apart from its history,"[6] those who believe that science still has a crucial role to play in the advancement of mankind would certainly benefit from a closer reappraisal of the ways in which Comte himself, for better or for worse, theorized the historical contribution of philosophy to intellectual, ethical, religious, aesthetic, and social progress.

The Political Destination of Positive Philosophy

Although latter-day positivism has been sanitized as a pure philosophy of science, whose endorsement of the fact/value distinction was considered a protection against ideological ravings, Comte's "positive philosophy" was through and through a political endeavor. As his early writings of the mid-1820s show, Comte first conceived it as the only proper intellectual answer to the practical issues faced by postrevolutionary societies. And as the later publication of the *Système de politique positive* (1851–1854), demonstrates, Comte's mature concerns mostly focused on the political impact of "positive philosophy." In between, the writing of the *Cours de philosophie positive* provided Comte with the theoretical means for achieving the political and social aims he was pursuing.

The outlines of Comte's distinctive approach were first elaborated in what he later came to regard as his "fundamental opuscule," his 1824 *Plan de travaux scientifiques nécessaires pour réorganiser la société*. The urgent need for such reorganization, Comte claimed in his pamphlet, was prompted by the state of moral and political anarchy in which advanced Western societies found themselves and which was primarily due to the coexistence of two opposite long-term social trends: on the one hand, the inescapable collapse of the feudal and "theological" system, the demise of which had been hastened by the repeated assaults of the "critical" or "metaphysical" tendency characteristic of modernity as exemplified by the Enlightenment and the French Revolution, which had challenged all the convictions buttressing the old world; on the other hand, the inevitable advent of a new system of beliefs that was supposed to reorganize society on new grounds, but that had so far not prevailed. Comte ascribed the latter's inability to establish itself both to the deleterious influence of the "critical" spirit, whose constant questioning hampered the emergence of an "organic" set of social, moral, and political principles, and to the unending and unproductive confrontation between the feudal and "theological" system and revolutionary ideas, which merely perpetuated the crisis without providing any hint at its resolution.

According to Comte, the symptoms of that social crisis, which also testified to the accurateness of his diagnostic, consisted both in the fact that the Revolution, after having destroyed the Ancien Régime, had only been able to produce successive constitutional arrangements that never lasted (which suggested that the purpose of "critical" ideas was purely negative and transitional) and in the fact that the various attempts at reinstating the old order of things, either by way of monarchical restoration or imperial fiat, always resulted in new revolutionary upheavals. This endless reshuffling of political regimes demonstrated that traditional principles were not, to use a typically Comtean phrase, "in tune with their times"[7] and thereby failed to meet the current demands of mankind, and that the "critical" spirit was unable to provide the new system with proper foundations, for "weapons of war cannot, by a strange metamorphosis, suddenly become building instruments."[8] But how was one to escape the hopeless alternation of barren destructions and short-lived revivals? By elaborating, Comte argued, a doctrine that would facilitate "the formation of the new social system, the final object of the crisis, and the goal of which everything that has occurred hitherto is just the preparation."[9]

Drawing on a contrast between the "kings," who contented themselves with searching in the past for solutions to present evils without acknowledging the reality of historical development, and the "peoples," who mistakenly focused all their efforts on mere political and institutional reforms, Comte argued that those truly willing to contribute to the reorganization of society had first to identify the "new principle according to which social relations must be coordinated."[10] Failing this "theoretical" or "spiritual operation," there would be no hope of ever succeeding in the "practical" or "temporal" task of "determin[ing] the mode of distribution of power and the system of administrative institutions which are in close conformity with the spirit of the system as settled by the theoretical operations."[11]

Given this logical dependence of the "practical" on the "theoretical," there was no difficulty in singling out those who were to carry out the speculative work required: "whenever, in any particular direction, society needs theoretical works, it is acknowledged that it must appeal to the corresponding class of scientists."[12] This clearly implied that, since what was needed was a social theory, a new class of scientists specializing in the study of social phenomena had to be formed, which would consist of "men who, without devoting their lives to the special cultivation of any science of observation, possess an aptitude for science, and have made a sufficiently close study of the general shape of positive knowledge to be penetrated by its spirit, and to have become familiar with the principal laws of natural phenomena."[13]

But scientific expertise would not be the only asset of that new breed of scientists, for they would also enjoy a moral authority that would enable them to secure approval for the new social order: "Independently of the fact that they are alone competent to form the new organic doctrine, they are exclusively invested with the necessary moral force to secure its acceptance. . . . This is the habit which society has gradually acquired, since the foundation of the positive sciences, of submitting itself to the decisions of scientists for all particular theoretical ideas, a habit which the scientists will easily extend to general theoretical ideas, when they are entrusted with their co-ordination."[14]

Accordingly, the very next step to be made by that new class of scientists would be to apply the objective outlook already adopted in the study of natural phenomena to that of social phenomena, so as to "treat politics in a scientific manner."[15] That transformation called for a threefold evolution: the submission of imagination to observation, with politics giving up "the absolute search for the best possible government"[16] and focusing on actual social facts; the endorsement of the idea that "social organization [is] intimately tied to the state of civilization and determined by it;"[17] and the acknowledgment that "the progressive development of civilization is subject to a natural and irrevocable course, derived from the laws of human organization, which in turn becomes the supreme law of all political phenomena."[18]

Comte quite surprisingly argued that the adoption of the last two principles—the former stating that social organization depended on civilization, that is, on "the development of the human mind, on the one hand, and, on the other hand, on the development of man's action on nature which is its consequences," respectively giving rise to "the sciences, the arts, and industry,"[19] the latter claiming that "the progress of civilization develops according to a necessary law"[20]—was the precondition of the scientificization of politics, whereas one would have rather expected that those claims would have been, in a straightforward inductivist manner, the results of observation.

Such a puzzling reversal nonetheless points toward two very distinctive features of Comte's sociology: firstly, its "intellectualism," that is, the idea that the development of the mind was the primary explanatory factor of human progress;[21] secondly, its "historicism," that is, the belief that this intellectual development consisted of a long-term historical trend that could be captured, together with its social consequences, by laws similar to those governing natural phenomena. This in turn explained the intricate relationships existing between Comte's theory of knowledge and his sociology, historically, conceptually, and methodologically.

Sociology in History: The Maturation of the Social, the "Law of the Three States," and the "Encyclopedic Scale"

Comte's contextualization of the emergence of "positive politics" provides a prime example of the "historically minded reflectiveness"[22] characteristic of Comtean positivism. For Comte not only diagnosed the underlying causes of the ongoing social and political crisis and exposed the equal futility of theological-monarchical and metaphysical-revolutionary remedies but also explained both why any attempt at solving the crisis had so far been premature and what conditions had to be fulfilled for doing so. Such a rationale, for which the *Plan des travaux* provided a blueprint that was later developed in the *Cours*,[23] focused both on the historical timeliness of the sociological enterprise and the epistemological maturity it required.

As to the first aspect of Comte's explanation, it rested on the idea that a "positive" social theory could emerge only after "the definitive social state of the human race, the one which best suits its nature, that in which all its means of prosperity are to receive their fullest development and their most direct application," had reached a stage of development sufficiently advanced to allow for the identification of its main characteristics.[24] As long as this social maturation had not taken place, "positive politics" lacked an "adequate experimental basis": "A system of social order had to be established, accepted by a numerous population and composed of several nations; and this system had to have lasted for an extended period, so that a theory could be founded on this vast experience."[25] Comte, for all his dramatization of the critical state in which modern Western societies found themselves, nonetheless judged that this empirical condition had been met, most notably because the widespread acceptance of "positive" or scientific ideas and the ever-growing spread of industrial activity clearly indicated the very nature of the last stage of human development. But social maturation was not on its own a sufficient condition for the constitution of "positive politics"; it also demanded that the human mind itself had reached a stage of epistemological development that would enable it to grasp scientifically or "positively" the social reality it was supposed to explain. It was at this critical juncture of the argument that Comte's "law of the three states" entered center stage.[26]

As is well known, the canonical version of the law outlined in the first lesson of *Cours de philosophie positive* claims that "each of our principal conceptions, each branch of our knowledge, passes in succession through three different theoretical states"—the theological, the metaphysical, and the positive—giving rise to "three methods of philosophizing, whose characters are essentially different and even radically opposed to each other."[27] In the *Plan des travaux*,[28] the law of the three states, which Comte held to be vindicated by the actual historical course of the four fundamental sciences—astronomy, physics, chemistry, and physiology[29]—is applied to politics and is intended as a guarantee for its imminent "positivization." For, according to Comte, it was clear that politics had already passed through the theological (as illustrated by the "doctrine of kings" with its ideas of "divine right" and of "an immediate supernatural direction" of human events) and the metaphysical states (as expressed by the "doctrine of the peoples," with its "abstract and metaphysical assumption of an original social contract" and its invocation of "rights, viewed as natural and common to all men to the same degree, which it guarantees by this contract"),[30] and that it was about to enter the last and final positive state. And if the past was any indication of the future, Comte confidently concluded that "there has therefore never been a moral revolution at once more inevitable, more mature, and more urgent than that which is now to elevate politics to the rank of the sciences of observation."[31]

Yet Comte also seemed to realize that his very insistence on the unavoidable coming of age of "positive politics" raised a new question: compared to the other sciences, why did it take so long for the positive spirit to expand its reach to the study of social phenomena? To explain this backwardness, Comte invoked another element of the positivist philosophical paraphernalia, the "encyclopedic scale" of the sciences, which, combined with the law of the three states, accounted for the order in which "the different human theories have successfully attained, first, the theological; then the metaphysical; and, finally, the positive state."[32] According to this classificatory principle, which is introduced in lesson 2 of the *Cours de philosophie positive*[33] and which structures the reminder of the book, the development of the various sciences had been determined by the degree of simplicity or generality of the phenomena considered, those studying simpler or more general phenomena evolving more rapidly toward positivity than those dealing with more complex or specific phenomena. This resulted in what Comte called an "encyclopedic," or hierarchical, scale, which ranked the positive sciences in order of the decreasing generality and increasing complexity of the phenomena studied, with mathematics first, astronomy second, physics third, chemistry fourth, and physiology fifth. For Comte, this scientific series not only captured the objective dependence existing between the various kinds of phenomena (for instance, biological phenomena would be impossible in the absence of certain astronomical, physical, and chemical conditions) but also clarified both the logical connections between the various sciences and their historical succession: since social phenomena were "the most special, complicated, and concrete phenomena," which made them "depend more or less upon all the preceding phenomena without, however, exercising any influence upon them,"[34] it was normal that their study had "necessarily progressed more slowly than preceding orders"[35] of phenomena. But their complexity, albeit responsible for the backward state of "positive politics," also marked out their originality and called for the establishment of a "distinct class"[36] of inquiry. The time was ripe, concluded Comte, for the positive conception of social phenomena: "Now that the human mind has founded celestial physics, terrestrial physics (mechanical and chemical), and organic physics (vegetal and animal), it only remains to complete the system of observational sciences by the foundation of social physics."[37]

What is interesting in Comte's historicist justification of the timeliness of the sociological enterprise and his epistemological contextualization of its emergence is that it sheds light on two key assumptions of the positivist conception of sociology: on the one hand, in accordance with the idea of an encyclopedic classification, the claim that sociology was dependent, both scientifically and methodologically, on the other sciences that had preceded it and that this twofold dependence had to be clearly acknowledged and theorized; on the other hand, in line with Comte's scientific pluralism, which held that each fundamental science dealt with a specific kind of phenomena and developed its own specific methods and concepts, the claim that sociology was an autonomous science because the phenomena it considered were truly original and irreducible to those studied by the other sciences, and therefore it needed to elaborate methods of inquiry and theoretical constructs of its own. As we will see below, encyclopedic dependence and epistemological autonomy were the two complementary principles that structured Comte's philosophical and methodological appraisal of positive sociology.

What is also worth emphasizing in Comte's plea for the establishment of a positive science of social phenomena is its somewhat problematic reliance on the law of the three states. For what Comte regarded as the "fundamental law of mental development"[38] was indeed, despite the misleading psychological ring of the phrase, a sociological law, and undoubtedly the most important one, since it provided the key to the understanding of human evolution.[39] But if so, would not Comte be guilty of an illegitimate attempt at bootstrapping his demonstration of the necessary advent of sociology to an argument that itself rested on the sociological law of the three states, thereby assuming what was to be proved, namely, the possibility of a positive sociology? To that objection, one might reply that Comte viewed the impending emergence of sociology as a very reasonable bet to make given the advance of the positive spirit in the natural sciences, the progress of which is documented in the first three volumes of the *Cours*, which deals with the slow but steady positivization of mathematics, astronomy, physics, and physiology, and which certainly paved the way for further sociological developments.

Yet such a deflationist interpretation would miss the historicist dimension of Comte's argument, which right from the start assumed, as the title of lesson 46 tellingly puts it, "the Necessity and Timeliness of *Social Physics*."[40] Comte's approach to social phenomena, which conceived both institutions and beliefs to be dependent on the state of intellectual development reached by the human mind, indeed qualified his positivism as a philosophy of history, that is, as an attempt at a comprehensive teleological interpretation of the course of human evolution intended to provide the series of historical events with both a direction and a goal. However, contrary to theological or idealistic versions—which viewed history either as governed by a divine plan or as the earthly realization of an absolute spirit—Comtean historicism considered historical development as the natural and spontaneous result of dispositions or potentialities inherent in human nature, which required some historical maturation to manifest themselves, and the knowledge of which therefore presupposed that a certain stretch of history had already elapsed. In that regard, positive sociology was indisputably deterministic;[41] but whereas Marxian or materialistic historicism argued for the determination of the ideological superstructure by the productive infrastructure, Comte's intellectualism went for a dependence of the "temporal" on the "spiritual" that ascribed ideas, beliefs, and mores a causal role in the development of mankind, thereby turning material, social, and political organizations into consequences of the progress of the human mind. But if it was indeed the case that the course of history was governed by laws that were themselves premised on a law of intellectual development, then it followed that the study of history itself had undergone the very same intellectual transformation that the law of the three states described in general terms for the human mind at large. Accordingly, if "no science could be thoroughly understood until its true essential history has been considered,"[42] this was all the more true of sociology and explained why Comte felt it necessary to provide a historical account of its progressive development.

The History of Sociology: "From Aristotle to Me"

A distinctive feature of Comte's attempt at historicizing sociology was that it not only aimed at discovering the laws of the historical development of social phenomena but also tried to define the historical conditions of possibility of the knowledge of these laws. The second aspect of Comte's project, what might be called his "sociology of sociology," insisted both, as argued previously, on the historical maturation required for an adequate grasp of human evolution—as Comte himself put it, "only the present generation . . . which has experienced the full impact of the revolutionary crisis, can at last find, for the first time, in the social past as a whole, an adequate basis for a rational exploration"[43]—and on the methodological assumptions called for by the establishment of a positive social science—only minds already acquainted with "the diffusion of the positive spirit in all the other fundamental branches of natural philosophy" could "apply the positive method directly to the general study of social phenomena."[44] Such a reflexive approach, which took as its epistemological standard the main characteristics of the positive spirit—the subordination of imagination to observation, the subjection of all phenomena to natural laws, the proscription of the search for primary or final causes[45]—provided Comte with a benchmark against which to assess the respective claims to scientific soundness of the "main philosophical attempts carried out so far in order to constitute social science."[46] It allowed both for a teleological appraisal of the historical development of sociological knowledge (to the extent that the law of the three stages enabled one to judge the various attempts at establishing social science as "normal," "premature," or "retrograde") and for practical guidance, because the law of the three stages and the encyclopedic hierarchy of the sciences specified the epistemological steps still to be taken and the methodological requirements that needed to be satisfied before the foundation of sociology could be properly proclaimed.

This manner of writing the history of science structured Comte's various accounts of the development of sociology, the most paradigmatic of which undoubtedly being the one offered in the forty-seventh lesson of the *Cours*.[47] There, Comte started by tracing back the origin of positive social science to the *Politics* of Aristotle, this "immortal father of philosophy," whose minute observations on the constitution of the Greek *polis* strikingly contrasted with the "dangerous fancies of Plato and his imitators about joint ownership."[48] However, and despite his willingness to subordinate imagination to observation, Aristotle's focus on the "principle and form of government," which derived from his inability to grasp the "progressive tendencies of mankind" and the "natural law of civilization,"[49] testified to the hold of the metaphysical state over his sociological conceptions. Accordingly, as Comte put it, a social theory could be considered positive only when "the final preponderance of the positive spirit in the study of the less complex phenomena allowed for the real understanding of natural laws and . . . the true notion of human progress, either partial or total, has eventually assumed some real consistency."[50]

In the light of these criteria, Comte singles out two main contributors to the positivization of social science. The former was Montesquieu, whose *Esprit des lois*—drawing on the progress of the positive spirit in the inorganic sciences—clearly conceived "political phenomena as being necessarily subjected to invariable natural laws, like all the other phenomena."[51] But this intuition never developed into a systematic body of social laws, with Montesquieu instead losing himself in a "barren accumulation . . . of facts, indifferently borrowed, often without any philosophical criticism, from the most opposed states of civilization."[52] In other words, although Montesquieu rightly acknowledged the nomological nature of sociological knowledge, he failed to identify "the great general fact which dominates and is the true regulator of all political phenomena, the natural development of civilization,"[53] the law of which social science should aim at discovering. Because he lacked this "indispensable fundamental notion of the general progress of mankind,"[54] Montesquieu erred, in his otherwise positive "climate theory," in ascribing environmental causes a decisive role in social development, whereas "the influence of climate on political phenomena has only a modifying influence on the natural course of civilization, which retains its character as a supreme law."[55] Accordingly, "Montesquieu felt the necessity of treating politics in the manner of the sciences of observation; but he did not conceive the general work which should stamp it with this character."[56]

"Since Montesquieu," Comte concluded, "the only important step taken by the fundamental conception of *sociology* is due to the famous and unfortunate Condorcet."[57] As Pierre Macherey rightly underlines, it was no coincidence that Comte chose to coin his most famous neologism at this point of his historical narrative: with Condorcet, social science had indeed reached a new epistemological threshold, since, "for the first time, the truly paramount scientific notion of the social progress of mankind has been clearly and directly introduced, with the full universal preponderance it must exert over such a science as a whole."[58] Condorcet was the first, Comte argued, "to see clearly that civilization is subjected to a progressive course all of whose steps are rigorously connected to each other according to natural laws which can be unveiled by philosophical observation of the past."[59]

Such a perspective is mirrored in the very structure of Condorcet's greatest positive achievement, the *Esquisse d'un tableau historique des progrès de l'esprit humain* (1794), which is divided into ten "Epochs" summarizing the various stages of improvement that led to the current state of mankind. The last chapter is dedicated to a "sketch of the future," which, far from being utopian, testifies to the scientific orientation of Condorcet's approach, since "the determination of the future by philosophical observation of the past would seem . . . a very natural idea with which all men have become familiar for other classes of phenomena."[60] Prediction was indeed one of the distinctive features of the positive spirit, for it made possible the verification of theories and hypotheses, and Condorcet perfectly understood that positive sociology had to submit to the same methodological standard that governed scientific practice in the natural sciences.[61]

However, although Condorcet benefited from an even greater diffusion of the positive spirit in the natural sciences than did Montesquieu—most notably with respect to advances made in biology, especially in anatomy and taxonomy[62]—and had been a firsthand witness of the fall of the theological regime and of its revolutionary aftermath,[63] which allowed him to be "the first to conceive the true nature of the general task, which is to raise politics to the rank of the sciences of observation," "he [nonetheless] executed it in an absolutely pernicious spirit in the most essential respects."[64] On the one hand, because, like Montesquieu in his climate theory, he could not draw on the "fundamental laws of human nature," Condorcet stuck to an outdated 'sensationalist' theory that rested on an erroneous empiricist understanding of the workings of the mind and led him to endorse "vague and irrational conceptions about indefinite perfectibility."[65] On the other hand, Condorcet's epic narrative of how knowledge prevailed over superstition and finally resulted in an age of Enlightenment that put an end to the dark times of the Middle Ages also reveals his "metaphysical" or "critical" bent. Like Goethe's "spirit that always denies," Condorcet "allowed himself to be blindly dominated by . . . prejudices; he condemned the past instead of observing it; and as a consequence his work was no more than a long and tiring declamation, from which no positive instruction really resulted."[66]

Lacking historical judgment, Comte ruled, Condorcet could not adopt the objective outlook appropriate for a fair appraisal of the past and therefore failed to understand that "any social force that has been active for long has necessarily contributed to the general production of human development."[67] Neither admiration nor condemnation, any more than fatalism or optimism, should have their say in sociology. What mattered was the impartial assessment of the respective contributions of the different ages of mankind to the advent of the positive era. And, Comte argued, the time was ripe to pass that judgment because, on the one hand, the positive spirit had prevailed in all the other domains of knowledge, and, on the other, because the social situation itself called for the establishment of a sociological theory.

Sociological Dead Ends: Psychology and Political Economy

However, to many, Comte's encyclopedic hierarchy of the sciences seemed problematic or incomplete, for it appeared to create a gap between biology and sociology. What happened to human individual phenomena? Were they just absorbed by biology? And what about the abstract laws of thought followed by human individuals? Were there not anthropological invariants immune to historical evolution? Other disciplines were already trying to answer these questions while Comte was laying the foundations of sociology: whereas logic and psychology, as distinctive philosophical subfields, aspired to state the laws of thinking and the norms of reasoning, political economy—the science of human actions prompted by the desire to possess wealth—intended to show that social interactions were also governed by regularities. Strikingly enough, Comtean sociology built itself in explicit opposition to both psychology (and logic, which Comte lumped together) and political economy, which might have otherwise seemed to be straightforwardly positive.

Various motivations led Comte to reject the traditional architectonics of the "moral sciences." Firstly, he proscribed psychology and political economy from the encyclopedic hierarchy because he considered them to be flawed attempts at explaining human phenomena. Secondly, Comte's holistic approach to the social prompted him to discard the individualistic perspective characteristic of psychology and political economy for ethical and social reasons. Finally, that dual proscription served a highly polemical function, since, within the French context, psychology and political economy indeed were, in the first third of the nineteenth century, the moral sciences that strove, each in its own way, to rule over the study of human knowledge, the former in the philosophical and moral domain—mainly through Cousinian eclecticism, with the Sorbonne as its stronghold[68]—the latter in the social and political domain. (French liberal economists diffused their ideas from the Académie des Sciences Morales et Politiques and the *Journal des economists*.[69]) The eviction of psychology and political economy was therefore a way for Comte to mark out the specificity of sociology as a distinctive intellectual endeavor, although it could also result in its institutional marginalization.[70]

Comte's opposition to psychology can be traced back to his early writings and was further developed in his more mature works.[71] In its most elaborate form, it consisted of an argumentative scheme that articulated a threefold criticism: methodological (the methods resorted to were contradictory or barren), architectonical (psychology infringed the encyclopedic scale of the sciences), and scientific (the results and doctrines at which psychologists arrived were either false or illusory). Although his targets varied with time, one might argue that Comte took issue with psychology not only as a specific discipline but as a more general approach that considered the human mind in itself, independently of the body and of the world to which it belonged, and that intended to deliver conclusions as to its properties and nature. This approach in turn resulted in logic, understood as the inquiry into the norms of reasoning, metaphysics as the study of the powers and essence of the mind, and ideology as the analysis of its perceptual and reflexive capacities.[72] Despite their venerable philosophical pedigree, Comte had no qualms discarding those attempts, for, as he told his friend Pierre Valat, "these alleged observations made on the human mind considered in itself and a priori are mere illusions; and . . . all that is called logic, metaphysics and ideology is a chimera and a fantasy, when it is not nonsense."[73]

Besides being methodologically flawed,[74] psychology, according to Comte, also infringed the encyclopedic hierarchy since that "pseudo-science independent of physiology, superior to it, and to which alone would belong the study of the phenomena specially called *moral*" refused to view "the study of intellectual and affective functions as inseparably linked to that of all other physiological phenomena, and necessarily pursued by the same methods and in the same spirit."[75] Instead of clinging to an outdated Cartesianism,[76] one would be better off adopting the same outlook that resulted in the positivization of the study of mental phenomena, as illustrated by the achievements of Cabanis (who emphasized the visceral determination of thought), Broussais (who elaborated a physiological theory of psychopathology), and Gall and Spurzheim (whose phrenology opened up the way for an analytical science of the brain ascribing specific cerebral localizations to the various intellectual and affective functions).[77] However, one should also be wary of reducing moral science to physiology. For, as Comte clearly put it, although the study of the human species was undoubtedly founded on the latter, "it nonetheless constitutes a separate science, having its own special methods and getting its own observations from the history of human society."[78] For positive sociology, epistemological autonomy was no less important than encyclopedic dependence.[79]

Finally, Comte argued that all psychological schools erred in their depiction of human nature. First, they all concurred in asserting the primacy of intelligence over affective functions, whereas "daily experience shows . . . that affections, inclinations, passions constitute the main springs of human life; and that, far from resulting from intelligence, their spontaneous and independent impulse is indispensable for the first awakening and continuous development of the various intellectual faculties."[80]

So, even though intelligence would finally dominate the human mind in the positive state, affective tendencies provided the impetus for its development. Furthermore, when proper attention was paid to affective functions, they were generally reduced to one single principle, like sympathy according to the Scots or egoism in Helvétius's protoutilitarianism.[81] This myth of a domineering intelligence ruling over a simple mind Comte regarded as an unfortunate offshoot of both the belief in a radical distinction between men and animals and the metaphysical idea of the unity of the self. On the contrary, a positive approach to mental phenomena should adopt a naturalistic perspective and consider the self as a somewhat precarious outcome of the organic consensus.

Eventually, Comte insisted on the deleterious social consequences that were drawn from that mangled and individualistic view of human nature. On the one hand, the French Idéologues, prompted by their extreme empiricism to exaggerate the power of education, ended up representing man as "essentially a ratiocinative being, continually carrying on, unbeknownst to him, a multitude of imperceptible calculations, without almost any spontaneity of action" and reducing "social relations to vile coalitions of private interests."[82] On the other hand, "German psychology"—i.e., French eclectic spiritualism—conceived of the self as "essentially ungovernable, due to the erratic freedom that truly characterizes it" and was forced to give in to "some sort of universal mystification, in which the alleged permanent disposition to govern one's conduct exclusively in accordance with the abstract idea of duty would finally result in the exploitation of the species by a handful of clever charlatans."[83] The Scottish school of Hume, Smith, and Ferguson, "which acknowledged both sympathy and egoism, certainly was closer to reality,"[84] but failed to develop a systematic view that could really translate into an influential conception of human nature. Accordingly, because it failed methodologically, encyclopedically, and scientifically, Comte held that psychology could be legitimately banned from the hierarchical scale of the sciences and that sociology could be directly grafted onto biology.

To a great extent, Comte's criticism of the "alleged science of economics" and of the "necessary futility of the scientific claims" of its proponents mirrored his attack on psychology. Already present in his early writings,[85] it was repeated in the *Cours* and reappeared sporadically in later works, including the *Système*.[86] Drawing on a similar argumentative scheme, it blended a negative appraisal of political economy's epistemological credentials with a stern denunciation of its deleterious practical implications. But Comte's critical engagement with political economy also had more idiosyncratic and contextual motivations: on the one hand, it publicly enacted his definitive break with his former mentor Saint-Simon,[87] whose *industrialism* echoed some of the ideas developed by classical political economists (most notably Jean-Baptiste Say); on the other hand, it intended to displace an institutionally well-established theoretical rival with clear and explicit political and social ambitions, most notably that of imposing "laissez-faire" liberalism.[88]

Although Comte readily acknowledged the positive contribution of political economists, and primarily that of Adam Smith, both theoretically (particularly their insistence on seeing the division of labor as a distinctive feature of the modern social order) and practically (especially their involvement in the dissolution of the feudal system and the transition from a military to an industrial society that would favor not birth but talent), he nonetheless criticized the narrowness of their views, which prevented them from realizing that specialization was not a phenomenon restricted to the productive sphere but applied to society as a whole and from grasping its various drawbacks—the disappearance of synthetic views, the weakening of solidarity, and the dissolution of the general interest. Furthermore, economists also fell for the misleading picture of the *Homo œconomicus* as an "essentially calculating being, moved by the sole motive of private interest."[89] Hence the erroneous belief in the spontaneous convergence of interests, the mischievous call for the limitation of state intervention, and the sheer indifference to the moral needs of populations. On the contrary, Comte thought that material relations had to be regulated, that a less repressive temporal government had to be supported by the instauration of a spiritual authority that would reassert the demands of the common good[90] and inculcate upon individuals the private virtues necessary for the reaping of public benefits.[91]

But Comte also took issue with political economy for its methodological shortcomings. While emphasizing the lack of scientific training of the jurists and publicists parading as social scientists, Comte also drew attention to the metaphysical state of the alleged science of economics, what Thomas Kuhn would have called its preparadigmatic stage, within which the very meaning of its most basic concepts such as *value*, *utility*, or *production* still gave rise to bitter controversies that closely resembled "the strangest debates of medieval scholastics regarding the fundamental attributions of their purely metaphysical entities."[92] More fundamentally, Comte objected to the analytical method favored by political economists, which focused exclusively on the productive dimension of human activity, thereby infringing the encyclopedic order of the scale of the sciences and ignoring the holistic nature of social phenomena. As Comte put it,

> In social studies, like in all studies relative to living bodies, the various general aspects are necessarily linked to one another and rationally inseparable.... When one leaves the world of entities to approach real speculations, it becomes certain that the economic or industrial analysis of society could not be positively carried out in abstraction of its intellectual, moral, and political analysis, either for its past or even its present: so much so that, conversely, that irrational separation provides the indisputable symptom of the truly metaphysical nature of the doctrines that take it as their foundation.[93]

Just as the psychologists mistakenly conceived of man as a disincarnated solipsistic subject, the egoistic ratiocinative agent of the political economists seriously misrepresented human nature. Accordingly, what Comte really discarded with his dual proscription of psychology and political economy from the encyclopedic hierarchy of the sciences was the very idea of a social science that could start from the consideration of individuals and explain, through the study of their interactions, the emergence of collective phenomena and their historical development. That would not be the way sociology would proceed.

The Spirit, Methods, and Encyclopedic Position of Sociology

Right from the start, Comte's attempt at establishing a genuinely positive science of society faced an "extraordinary fundamental difficulty."[94] For, in accordance with the principle he had first stated in a 1819 letter to his friend Valat and then endorsed throughout the *Cours*, if "methods" could not be grasped independently from "doctrine"—i.e., if the study of the processes of discovery and justification of scientific theories and hypotheses was inseparable from their actual historical development, then the founding of sociology would prove almost impossible, since the laws from which sociological methods were supposed to be drawn were still to be discovered.[95] But positive philosophy, Comte argued, had the resources to break from this "vicious circle":[96] on the one hand, the encyclopedic scale perfectly warranted the borrowing of methods already in use in the natural sciences, granted their application to that new field of inquiry respected the specificity of social phenomena; on the other hand, the law of the three states, which was indeed the first sociological law, provided a blueprint on which to rely for the elaboration of the method of positive sociology. In other words, it was the same kind of methodological bootstrapping we previously encountered that enabled Comte to deal with sociology while it was establishing itself.

In the *Cours*, Comte dedicated four lessons to the basics of positive sociology (lessons 48–51), before turning to a grand narrative of the evolution of mankind from the theological (lessons 52–54), through the metaphysical (lesson 55), to the positive state (lessons 56–57), and finally concluding on the various prospects of "positive philosophy" (lessons 58–60).[97] In line with the general scheme of development he had previously identified in the history of the other sciences, Comte argued that sociology would definitively become positive when observation takes over imagination and social phenomena are reduced to invariable laws, thereby allowing for "rational prevision," "the most unquestionable criterion of scientific positivity."[98] But these "essential conditions" did not specify "the precise subject and the proper character of these laws."[99] What was the general spirit of positive sociology?[100]

Drawing on a distinction borrowed from his mentor the physiologist Henri Ducrotay de Blainville,[101] Comte conceived of sociology as a dual attempt to approach social phenomena *statically* and *dynamically*. Explicitly referring to biology, which divided itself into the anatomical study of the structure of living beings and the physiological appraisal of their functioning, Comte argued that sociology could focus either "on the fundamental study of the conditions of existence of society" or on "the laws of its continuous motion."[102] Just as anatomy dealt with "organization" and physiology with "life," "social statics" investigated "order" while "social dynamics" explored "progress," these two departments of sociological inquiry being complementary to the extent that order is the basis of progress and progress is nothing but the development of order.[103]

Pushing further the biological analogy, Comte argued that "social life necessarily manifests to the highest degree . . . that inevitable universal consensus that characterizes all the phenomena of living beings," ascribing to social statics "the positive study, both experimental and rational, of the mutual actions and reactions the various parts of the social system exert on one another."[104] Through his extension of the age-old medical notion of *consensus* to the social sphere, Comte decidedly emphasized the holistic nature of the sociological perspective, going so far as to advocate some sort of Cuvierian "social anatomy" in which the functional dependence existing within the "social body" would rationally warrant the deduction of "organs" of society that had not yet been actually observed. According to

Comte, what was blatantly obvious in the intellectual sphere (the encyclopedic dependence existing between the different sciences and the various interactions between the sciences and the arts), in addition to a somewhat casual observation of human history, gave good grounds for believing in the pervasiveness of "social consensus," the "key idea of social statics":[105] "not only must political institutions and social mores on the one hand, and mores and ideas on the other, be always mutually connected; but . . . this whole system, by its very nature, must also always be connected to the corresponding state of the integral development of mankind, considered in all its various modes of activity, be they intellectual, moral, or physical."[106]

This synthetic character of social phenomena in turn marked a methodological shift, in fact already initiated by biology, that favored synthesis over analysis in the explanation of complex phenomena, where consideration of the whole should take precedence over the examination of the parts: "since social phenomena are indeed thoroughly connected, their real study should never be rationally separated, hence the permanent obligation . . . to always consider simultaneously the various social aspects, either in social statics or . . . in social dynamics."[107] Although Comte pointed out that the inorganic sciences also took into account connectedness (as illustrated by the use of the concept of "system" in astronomy, for example), he nonetheless suggested that biology and, a fortiori, sociology, induced a transformation of our understanding of phenomena. It was as if, above a certain level of complexity, natural and social phenomena gave rise to a new kind of order in which the whole was not merely the sum of its parts but a system in which the parts primarily existed in their relation to the whole. This explains why Comte, in the fiftieth lesson on social statics, claimed that, since "any system [has to be] necessarily made out of elements that are essentially homogeneous to it, the scientific spirit [could not] allow one to consider human society as being genuinely made out of individuals" but found in the family "the true social unit," which "spontaneously exhibits the true seed of the various essential dispositions that characterizes the social organism,"[108] i.e., a unity of purpose, a hierarchical organization, and a rudimentary form of specialization. At the global level of society, whose advancement depended both on the "separation of labors" and the "cooperation of efforts,"[109] human association extended well beyond the limits of family and gave social consensus its full development.

That progressive emergence of a synthetic order also resulted, Comte argued, in a new way of articulating the division of scientific labor, with the study of complex phenomena calling for the development of a synoptic perspective. Sociology should therefore be driven by an *esprit d'ensemble*, not an *esprit de détail*, and "those who try today to dismember further the system of social sciences, through a blind imitation of the methodic fragmentation characteristic of the inorganic sciences, involuntarily fall prey to the fundamental aberration of considering as an essential means of philosophical improvement an intellectual disposition that is radically incompatible with the fundamental conditions of such a subject."[110] In other words, the unwarranted adoption of the analytical method misled psychologists and political economists into developing an individualistic science of the social, whereas a positive appraisal of the prevalence of "social consensus" unmistakably suggested the primacy of a synthetic outlook.

With "social statics" properly defined, Comte then turned to "social dynamics," which he considered the truly original contribution of positive sociology to the understanding of social phenomena, most notably because "it gives direct precedence to the notion that distinguishes most clearly sociology properly so-called from mere biology, that is the key-idea of continuous progress or, rather, of the gradual development of mankind."[111] Comte indeed regarded historical evolution, understood as "the action of each generation on the next,"[112] as the distinctive feature of social phenomena considered dynamically. And there again the synthetic perspective prevailed, but, unlike social statics, which emphasized the synchronic coexistence of the various aspects of social life—or "solidarity," as the *Système de politique positive* would later put it—"social dynamics" focused on the diachronic succession—or "continuity"—of the various stages of humanity, hypothetically conceived as a single subject that had undergone "the consecutive social modifications actually observed among distinct populations."[113] Social dynamics as a component of positive sociology aimed at discovering the laws of succession of these social states, each of them being conceived as "the necessary result of the previous one and the indispensable driving force of the next, according to the illuminating axiom of the great Leibniz: *The present is pregnant with the future*."[114]

As with social statics, Comte argued that intellectual progress provided primary evidence that the historical evolution of mankind was law-governed, since "the continuous development of the scientific spirit, from the early works of Thales and Pythagoras to those of Lagrange and Bichat" exhibited a definite pattern captured by the "law of the three stages" and the encyclopedic scale of the sciences, so much so that "no enlightened man could doubt today that, in that long succession of efforts and discoveries, human genius had always followed an exactly determined course."[115] Although he acknowledged that, so far, artistic and political phenomena seemed to have escaped such a strict determination, Comte nonetheless claimed that the synoptic outlook characteristic of sociology would allow for a better grasp of the general course taken by the evolution of mankind, while also emphasizing, in line with his idea that the advent of sociology was linked to some definite historical circumstances, that "fundamental laws also become, of necessity, all the more irresistible, and therefore more easily observable, when they apply to a more advanced civilization, for social change, at first vague and uncertain, must naturally accentuate and consolidate itself further as it goes on, overcoming, with increasing energy, all adventitious influences."[116] That was exactly what Comte intended to prove in the sociological lessons that would follow in the *Cours*.

What is puzzling about Comte's conception of social dynamics is that it both implicitly relied on a biologically inspired scheme that would seem to belie his endorsement of a nonreductionist epistemological stance and had an unmistakeably "ideological" ring to it that did not chime well with the proscription of the individualistic approach of psychologists and political economists. Indeed, after having discarded the notion of progress because of its association with the metaphysical idea of unlimited perfectibility and its absolutist overtones, Comte opted for the use of the concept of "development"[117] that was inspired by comparative anatomy and the outdated embryological theory of preformation—as the reference to Leibniz's "axiom" suggested—and which he thought "designate[d], without any moral appreciation, an unquestionable general fact."[118] What this dynamic perspective concretely amounted to was the claim that "humanity constantly develops itself throughout the gradual course of its civilization, particularly in the most eminent faculties of our nature, be they physical, moral, intellectual, or political; i.e., these faculties, at first numb, reach, through an ever more extended and regular use, an ever fuller development, within the general limits set by the fundamental organism of man."[119]

Given the premium Comte put on intellectual evolution, it surely is difficult not to read his definition of development as "the simple spontaneous development, gradually aided by an appropriate cultivation, of the preexisting fundamental faculties that constitute our nature"[120] as a straightforward psychological claim. However, Comte would certainly have vigorously objected to the subjective reduction of the "states" described by the law of the same name to private and subjective mental representations. As to the threat of "physiologizing" sociology through the use of the notion of development, it did not seem to worry Comte, perhaps because, as we will see shortly, he thought he had made clear the encyclopedic distinction between sociology and biology.

Comte's emphasis on the connectedness and developmental nature of social phenomena thoroughly influenced his discussion of sociological methods. In line with his idea of encyclopedic dependence, Comte specified in what ways the procedures used in the natural sciences could be taken up in sociology. As for observation, while dismissing "pyrrhonistic" attacks on the objectivity of historical data, he contrasted the narrow approach of historians, who specialized in a given figure, country, or period, and the analytical method of political economists with the synthetic perspective of sociologists, who could therefore benefit, because of "the universal connectedness of the various social aspects," not only from "the immediate inspection or direct description of all sorts of events, but also [from] the consideration of what might appear the most insignificant customs, the appreciation of all sorts of monuments, the analysis and comparison of languages, etc."[121] Comte also discarded any attempt at deriving static or dynamic inductions from a mere collection of facts and reasserted the need for observation to be guided by a definite theoretical framework, considering the law of the three states and its various sociological corollaries as the best candidates for such a role.[122]

Although deliberate experimentation could not be directly used for social phenomena (because of both their intimate connectedness and their magnitude), Comte nonetheless argued that one could still rely on "natural experiments" for a better understanding of social laws. Drawing on what he coined "Broussais's principle,"[123] which stated that pathological manifestations were mere quantitative modifications of the normal state of the organism, Comte thought that certain extreme situations could help determine the exact limits of variation that the social order could endure.[124] Even though he acknowledged that "it has not yet been possible to actually apply that procedure to any inquiry in political philosophy,"[125] Comte himself did not hesitate to consider the social situation prevalent in postrevolutionary Western European countries as pathological,[126] thereby further reflecting the hold of organicist ideas on his thought and the teleological dimension of his sociological conceptions.[127]

Comparison, the proper method of biology, could also be extended to the study of society in at least two respects: on the one hand, interspecifically, with the comparison of mankind with other animal species, both with regard to mental dispositions and social organization; and on the other hand, intraspecifically, with the comparison of the respective degree of development reached by the various human populations. Since Comte held that human groups, because of race, climate, or political institutions, instantiated at a different pace the potentialities of human nature, he argued that "from the unfortunate inhabitants of Tierra del Fuego to the most advanced peoples of Western Europe, one could not imagine a social nuance that would not be actually realized in certain parts of the globe, and even almost always in various localizations clearly separated from one another."[128] However, this synchronic or geographic comparison, which sometimes seemed to verge on racialist differentialism, was rooted in Comte's belief in "the necessary and constant identity of the fundamental development of humanity, because of the irresistible preponderance of the common type of human nature, in the midst of all the differences in climate or even race, the real differences only being able to affect the actual speed of each social evolution."[129]

But the comparative method still lacked one key sociological element: the sense of history. It could lead us astray because it failed to grasp what was most essential in the evolution of mankind, namely, the direction of its development. That shortcoming explained, Comte argued, why "the necessary influence of the various human generations on those that follow them . . . soon becomes the primary consideration in the direct study of social development,"[130] and resulted in the promotion of what Comte called the "historical method" in sociology, that is, the attempt "to form the whole of human events into coordinated series that clearly show their gradual connection."[131] Eager to sharply distinguish social dynamics from current historical practice, Comte emphasized the "synthetic" and "social" dimension of the "historical method" and refused to reduce history to a primarily political narrative, focusing on treatises, battles, individual figures, and unanticipated turns, which would merely amount to "a vain accumulation of incoherent monographs, in which the very idea of the true filiation . . . of the various human events would necessarily be lost in the midst of this barren jumble of confused descriptions."[132]

Once again drawing on the parallel with scientific progress, in which the history of one science can only be understood in relation to "human progress as a whole,"[133] Comte advocated a bird's-eye view of human evolution, in which the development of the human mind was the prime mover of the whole process and the key to the understanding of its unfolding. Accordingly, Comte conceived of sociological analysis as a "successive appraisal of the various states of humanity that shows, based on the whole set of historical facts, the continuous increase of whatever physical, intellectual, moral, or political disposition, combined with the indefinite decrease of the opposite disposition, from which would result the scientific prevision of the final domination of the former and the definitive demise of the latter."[134]

In the *Cours*, such a method would both allow for postdictions (i.e., the interpretation of the historical record in the light of the law of the three stages, as for instance in the explanation of the shifting balance between temporal and spiritual powers throughout history) and predictions (the imminent triumph of the industrial spirit over military proclivities in modern societies), whereas the *Système* would rely on it to prove the gradual prevalence of altruism over egoism in the course of human history. However, the historical method was not without dangers. Because it focused on long-term transformations, it was too prone to "mistake a continuous decrease for a tendency towards total extinction or, conversely, . . . to conflate continuous variations, be they positive or negative, with unlimited ones,"[135] just as if one would argue from the fact that the daily intake of food had steadily decreased during the course of history that men would give up eating at one point. But how was one to avoid such pitfalls?

Comte's answer lay in the specific conception he developed of the architectonic relations between sociology and the other sciences.[136] Since social phenomena were influenced, to a greater or lesser extent, by astronomical, physical, chemical, and biological conditions, sociological explanations were logically dependent on the laws of the "inorganic sciences," which accounted for the "external conditions of the human existence," and on those of biology, which "alone introduces to the real laws of human nature."[137] Comte particularly emphasized the latter connection, with regard to both statics and dynamics.[138] As to the first, Comte argued that it fell on biology to provide sociology with an inventory of the inborn capacities responsible for "human sociability and the various organic conditions that determine its specific character."[139] As the forty-fifth lesson of the *Cours* had made clear, Comte thought that a positive version of Gall and Spurzheim's phrenology, i.e., the naturalistic attempt to identify man's innate intellectual and affective endowment by way of an empirical study of cerebral conformation, could be used to ascertain the spontaneous basis from which social relations originated and developed. For instance, in the fiftieth lesson, which introduced the basics of social statics, Comte's description of the fabric of society rooted human sociability in an "instinctive drive toward communal life,"[140] family relations in the inborn sympathetic tendencies of mankind, and social cooperation in the intellectual functions.

On the other hand, Comte considered phrenology a means of verifying the dynamic inferences drawn by sociologists, since he required that "no law of social succession" be accepted if it had not been "rationally connected . . . to the positive theory of human nature"—that is, if it postulated the existence of an intellectual capacity, affective tendency, or moral trait not acknowledged by the "science of human nature" or if it referred to the disappearance of a well-attested mental faculty. According to Comte, neither the "primitive" nor the "continuous" modes of dependence, as he called them, endangered the autonomy of sociology, since its encyclopedic position testified to its being "a perfectly distinct science, directly established on bases of its own, but thoroughly connected, either in its point of departure or in its continuous development, to the entire system of biological philosophy."[141] In other words, biology merely identified the human faculties that made social existence possible, but it was for sociology alone to determine the historical laws of their development.[142]

In this last regard, Comte made it clear, at the outset of his sociological lessons, that the distinctive contribution of social dynamics was that it proved "intellectual evolution" to be "the principle that necessarily governed human evolution as a whole."[143] Even though he both acknowledged that intelligence "necessarily needed the primitive awakening and the continuous stimulation impressed on it by the appetites, the passions and the feelings," and underlined the mitigating effects of various circumstances, either "primary"—like race, climate, or political institutions—or "secondary"—boredom, life expectancy, and demographic growth[144]—on the speed of its development, he nonetheless reaffirmed that "it was . . . under the necessary guidance [of intelligence] that the whole of human progression has always been achieved."[145] Even more important, social dynamics unambiguously specified the nature of the changes that mankind was undergoing, by showing that "the general result of our fundamental evolution" primarily consisted in developing "our most eminent faculties, either by constantly diminishing the empire of physical appetites and by stimulating further the various social instincts, or by continually exciting the development of the intellectual functions, even the highest, and by spontaneously increasing the daily influence of reason over the conduct of man."[146]

It is worth insisting on the extent to which the social evolution depicted by Comtean sociology was infused by the biologically loaded concept of development and the idea that historical change was a process in which, so to speak, nothing was lost and nothing was created but everything was transformed. For, in the positive understanding of history, there was no place left for the emergence of radical novelty; as Comte put it while asserting that the "seeds" of theological and positive philosophy had forever coexisted in the human mind, "human life [could] never exhibit any true creation whatsoever, but always a mere gradual evolution."[147]

Even more strikingly, but unsurprisingly given the role he ascribed to phrenology as a "theory of human nature," Comte claimed that historical evolution was exactly mirrored in the increasing anatomical prevalence of the "various organs of the cerebral apparatus, as they recede from the vertebral region and get closer to the frontal one," that is, the exact localization of the intellectual faculties.[148] For all its epistemological autonomy, sociology indeed drew heavily on biology, both conceptually and empirically. However, with the direction and goal of human evolution clearly set, that is, with a genuine science of social phenomena definitively established, the reorganization of society Comte had called for since the *Plan des travaux* could really begin: positive politics could at last rely on a positive sociology.

The "Intellectual Sovereignty" of Sociology: The Final Encyclopedic Primacy of the "Social Point of View"[149]

Comte was perfectly aware that the hold of theological and metaphysical beliefs would continue to impede the positive reorganization of society for a long time. Nevertheless, he also had the hope that sociology could already exert a positive influence over science itself and further hasten its advancement. Encyclopedic dependence was indeed a process that worked both ways: just as the positive spirit developed itself through the successive sciences of the encyclopedic hierarchy and thereby provided sociology with its logical and scientific foundations, sociology, because it proved that social phenomena did fall under the purview of scientific inquiry, gave the positive spirit its full scope and could then react positively on the other sciences. Just as the political destination of positive philosophy was to offer a shared system of beliefs that would help avoid the anarchy of conflicting opinions, sociology, Comte argued, could similarly contribute to a "total reorganization of the speculative system" that would regulate the intellectual division of labor between the sciences and mitigate the deleterious effects of the "irrational dispersion of scientific endeavors"[150] resulting from excessive disciplinary specialization. For he was thoroughly convinced that society and science were both exposed to the same threat: division. And since the positive spirit gave birth to a systematic conception of the social, it lay with sociology to systematize human existence, theoretically and practically, spiritually and temporally.

Originally voiced toward the end of the first lesson of the *Cours*, and then fully developed in the forty-ninth, fifty-eighth, and fifty-ninth lessons, the search for a "philosophical unity," conceived as the "primary fundamental condition of the intellectual and moral reorganization of the most advanced populations,"[151] finally took the form of a contest. According to Comte, only mathematics and sociology could legitimately claim "rational supremacy" over the other sciences, the former on account of its generality (since all phenomena, inorganic and organic, natural and social, were subject to the laws of geometry and mechanics), the latter because sociological laws equally governed the development of all the sciences. Prima facie, Comte's alternative might seem a bit strange, but it made perfect sense, with respect to both his own conception of the encyclopedic scale and the current organization of the scientific field. For it was indeed the case that the positive spirit had first appeared and thrived in the context of mathematical thinking, that the "geometers" (and most notable among them Descartes, with his *mathesis universalis*) were the first to attempt a unification of the natural sciences through the use of mathematical means, that they had achieved domination over the most prestigious scientific institutions (such as the French Académie des Sciences), and that they have even tried to develop—under the various guises of "moral arithmetic" or "social mathematics"—a positive science of social phenomena. However, for Comte, "mathematical conceptions [were], by nature, essentially unable to direct the formation of a real and complete philosophy."[152]

Comte's radical demonstration of the "philosophical incapacity"[153] of mathematics to unify the encyclopedic scale focused on both its "logical" and "scientific" shortcomings. On the one hand, although he acknowledged that "universal positivity" should be studied "at its genuine primitive source," Comte nonetheless claimed that mathematicians merely had the "most imperfect notion" of the positive method since they "only conceived it in its most rudimentary state."[154] Lost in their ideal abstractions, the analysts could not benefit from the methodological improvements achieved by the natural sciences, whereas sociologists, in contrast, could sharpen their understanding of the positive method by drawing on the various procedures elaborated throughout the history of the natural sciences.

On the other hand, Comte made it very clear that the analytical methods characteristic of mathematics could not be applied to complex phenomena, on account of their variability and connectedness: "Just as any idea of actual number or of mathematical law is already forbidden in biology . . . it must also be, and even more so, radically excluded from the even more complex speculations of sociology."[155] In that last regard, Comte was particularly adamant in denouncing "the vain conceit of a large number of geometers to render social studies positive by way of a chimerical subordination to the illusory theory of chances."[156] This viciously individualistic attempt at providing a probabilistic appraisal of human actions,[157] which could still be ascribed, by Jacob Bernoulli[158] or even Condorcet,[159] to a misguided manifestation of the "premature instinct of the necessary renovation of social studies,"[160] had become an unquestionable sign of the faulty positive education of the contemporary geometers, like Laplace,[161] who advocated such a "philosophical monstrosity."[162] For by doing so, they both infringed the encyclopedic order, reducing the "superior" to the "inferior,"[163] and belied the very idea at the heart of the sociological enterprise, namely that social phenomena were governed by deterministic laws: "Would it be possible to imagine a more radically irrational conception than the one consisting of giving to social science as a whole, as its basis or primary means of final elaboration, an alleged mathematical theory in which, while usually mistaking signs for ideas, according to the common character of purely metaphysical speculations, one tries to subject the necessarily sophistic notion of numerical probability to calculus, which directly amounts to giving our real personal ignorance as the natural measure of our various opinions' degree of likelihood[?]"[164]

Quite surprisingly, what we now call 'sociological positivism,' that is, the application of quantitative methods to the explanation of social phenomena, is the exact antithesis of Comtean sociologism. Given these various "logical" and "scientific" shortcomings and his conviction that "a cradle could not be a throne,"[165] as he pompously phrased it, Comte concluded that mathematics was not qualified to govern the encyclopedic hierarchy: like all the other sciences, it had to submit to the rule of sociology.

According to Comte, the extent and diversity of the "necessary philosophical reaction of social physics on all the preceding sciences, regarding either doctrine or method,"[166] testified to "the inevitable rational supremacy of the sociological spirit over any other mode, or rather degree, of the true scientific spirit."[167] With respect to "doctrine," Comte claimed that "the human, or rather social, point of view" characteristic of sociology would favor the development of a genuine synthetic perspective in all the natural sciences, which would in turn better respond to the needs and problems of mankind.[168] Furthermore, since sociology, through the law of the three stages and the encyclopedic scale, had clearly identified the epistemological capacities of the human mind and the respective domain of each science, it could easily "regulate the natural progress of the various sciences,"[169] by banning theological or metaphysical investigations, suggesting proper topics and methods, or adjudicating the claims of the various sciences over a given kind of phenomena.

In fact, Comte boasted that the *Cours de philosophie positive* had already provided examples of the "importance and fecundity of the many general reactions [of sociology]"[170] on the other disciplines, most notably the reduction of astronomical research to the solar system and the formulation of a positive theory of hypotheses.[171] Concretely, the advent of a positive sociology could lead—although Comte knew it would be difficult to challenge the current "speculative domination of geometers"[172]—to an institutional reform of existing scientific bodies: for instance, the Parisian Académie des Sciences would see the "introduction of a new and prevalent section, specially dedicated to social physics and positive philosophy" that would foster the diffusion of the synthetic spirit. Comte added that the "just rational supremacy of that additional section [would be] regularly acknowledged by the exclusive prerogative of always providing the annual President and the perpetual Secretary of the Academy and by its resolute participation in the partial deliberations of the other sections."[173]

As for "logic," Comte claimed that, since each discovery was itself a social phenomenon, a sociological understanding of the development of the human mind as provided by social dynamics could allow for a positive anticipation of scientific progress. On the one hand, "the various scientific discoveries become, to a certain degree, amenable to a true rational prevision, by way of an exact appraisal of the previous movement of science, duly interpreted in accordance with the fundamental laws of the real course of the human mind."[174] On the other hand, "the art of discoveries" could be subjected to "some sort of rational theory, which might usefully direct the instinctive efforts of individual genius."[175] So, although it was hard, as Mill rightly emphasized,[176] to find in Comte a proper "*organon* of proof"—what we would now call a "logic of justification"—there was no question that sociology provided Comte with a powerful "logic of discovery." And this last "reaction" of sociology over the encyclopedic scale really marked out the crucial epistemological role he ascribed to social science: whereas more recent versions of positivism have advocated the "naturalization of epistemology," Comte, on the contrary, because he considered sociology to be the true epitome of the positive, i.e., scientific, spirit, called for the "sociologization of science." It is not sure that one would find many natural scientists still sharing that belief nowadays. Nor is it very likely that that many sociologists of science do conceive of their scientific contribution as Comte did.

★ ★ ★

Comte was convinced that the *Cours de philosophie positive* had definitively established sociology as a positive science. Of course, the very evolution of his own ideas and the various historical and personal events he was confronted with throughout the remainder of his life would prompt Comte to insist or, on the contrary, to downplay certain aspects of his sociological thought in his later works. For instance, the *Système*, which bore the imprint of his romantic encounter with Clotilde de Vaux and was mostly written in the aftermath of the revolutionary upheavals of the late 1840s, both emphasized the social role of affective tendencies (while relativizing the importance of intelligence) and gave greater significance to statics—that is, the science of social order—than had been the case in the *Cours*. But, as could be expected from somebody as obsessed with cohesion as Comte was, these elaborations were mere developments—the apt word to use here—of "seeds" that were already present in him before he chose to dedicate most of his efforts to concrete social transformation. To be sure, Comte had realized that sociology could not provide direct guidance for practical interventions, especially when dealing with individuals (hence the introduction of a seventh science in the encyclopedic scale, morals). But he remained convinced that a social understanding of the structure of human communities and historical evolution of mankind was the necessary key to securing humanity's happiness. Most importantly, one might well argue that Comte's methodological proscription of the individualistic attempts at explaining human phenomena associated with psychology and political economy and his insistence on the connectedness characteristic of social existence to be found in the *Cours* were simply the theoretical counterparts of the Religion of Humanity and the *Système*'s infatuation with continuity and solidarity. In a way, positive politics was merely the continuation of sociology by other means.

Part II

Comte's Social and Political Thought

6
Comte's Political Philosophy

Michel Bourdeau

> You ask me what institutions I want to see established? That is not the point at all, for I do not speak of establishing any. For me, what matters is the spiritual reorganization of society, and in no way its temporal reorganization; that is, I am looking forward to the establishment of doctrines, not of institutions.
>
> **To Pierre Valat, December 25, 1824**

According to the *Discours sur l'ensemble du positivisme* (1848), positivism essentially comprises two parts: on the one hand, a philosophy, i.e., a philosophy of science, to be found in the *Cours de philosophie positive*; on the other hand, a politics, to be found in the *Système de politique positive*. Their respective success reveals a sharp contrast. Whereas philosophers of science have always acknowledged Comte as one of the founders of their discipline, it is hard to find within contemporary political philosophy the slightest reference to positive politics, just as if it had never existed. However, as will be shown in the concluding chapter, the situation was quite different at the end of the nineteenth century: at that time, Comte's political ideas were taken seriously, and his disciples were active almost all over the world.

Moreover, whatever the judgment passed on this politics, it is impossible to ignore it when studying Comte, for it inaugurated his intellectual career as well as closed it. In 1854, wishing to underline the unity of his thought, he reprinted some of his early works in an appendix to the *Sys-*

tème, and the most important among them, which he always referred to as his "fundamental opuscule," bears a title that leaves no doubt as to its political aim: *Plan des travaux scientifiques nécessaires pour réorganiser la société*. For Henri de Saint-Simon's young secretary, the question was: how is society to be reorganized? And science was merely a means to achieve that goal.

The shortest answer Comte ever gave to that question can be found on the Brazilian flag: *Ordem e progresso*. Usually, politicians ask us to choose between order and progress without realizing that by doing so they are heading to failure, and that "as long as Progress tends toward anarchy, so long will Order continue to be retrograde."[1] Assuming that progress is nothing more than the development of order, Comte displayed an enduring ambition to reconcile both. One might think that he never achieved that goal and that he had rather changed sides. Initially, positive politics relied almost exclusively on the advocates of progress and, as late as 1848, Comte still pinned his hopes on the working class. Yet in 1855 he launched an *Appel aux conservateurs* (Appeal to conservatives) and, closer to his death, he went so far as to declare that progress "must be systematically considered to be characteristic of early development, and even incompatible with the final state, the economy of which it would disturb."[2]

However, if one is willing to understand the almost complete lack of interest in Comte's political philosophy, another answer to the question raised above needs to be considered. To reorganize society, it is necessary to establish a new spiritual power. This idea, which stems from Saint-Simon, was upheld by Comte when he was young, as early as 1826, in the *Considérations sur le pouvoir spirituel*, and it remained the focal point from which positive politics later developed. That is certainly the main reason why positive politics seems as strange today as it did two hundred years ago. Baruch Spinoza and Thomas Hobbes founded modern political philosophy by freeing politics from the domination of theology, which quickly resulted in rendering the distinction between temporal power and spiritual power obsolete. To be willing to reinstate such a distinction would amount to questioning what seems, to many of us, some of the most valuable achievements of modern times. "Metaphysical politics," to use the term by which Comte referred to seventeenth- and eighteenth-century political philosophy, was charged with being a threat to order and an obstacle to further progress. It was conceived as a way of overthrowing the Ancien Régime—but this goal had been reached and the mission had changed: the time was ripe for the reorganization of society. The advocates of metaphysical politics mistook critical principles for organic principles and did not realize that "weapons of war cannot, by a strange metamorphosis, suddenly become building instruments."[3]

From a historical perspective, this latter claim makes sense. It cannot be denied that the French Revolution put an end to the Golden Age of political philosophy. In the nineteenth century, with the exception of John Stuart Mill, the most important contributions to political philosophy came from sociologists: Karl Marx, Alexis de Tocqueville, and Comte. But the significance the latter ascribed to the spiritual power is so contrary to our firmest beliefs that, unlike his two contemporaries, it is as if there existed no common ground on which to engage with him. Accordingly, positive politics discredits itself and appears to be some sort of nostalgia for a past that it is neither desirable nor possible to revive. As Thomas Huxley put it, positive politics is "Catholicism minus Christianity." Mill's final judgment was even harsher: It is "the completest system of spiritual and temporal despotism which ever yet emanated from a human brain, unless possibly that of Ignatius Loyola."[4]

This overview will be divided in three sections. The first two introduce the two axioms of positive politics: there is no society without a government, and no society can last in the absence of some sort of priesthood.[5] The last section will then detail how these principles have been put into practice, or at least how they were supposed to be applied.

Even if it took some time for this to become explicit, spiritual power was right from the start conceived in relation to religion, as illustrated by the numerous references to Joseph de Maistre's *Du Pape* in Comte's early writings. How is one to justify so seemingly retrograde a move? What meaning can we ascribe to a claim such as: "Man becomes more and more religious"?[6] Part of the problem had already been emphasized by Mill, when he noted that "[to] have no religion, though scandalous enough, is an idea [our contemporaries] are partly used to: but to have no God, and to talk of religion, is to their feelings at once an absurdity and an impiety."[7] The collapse of the theologicopolitical, which Comte did not deny (he wanted to reorganize "without God or King"), does not mean that we are done with religion. A certain form of spiritual power is indeed needed if we want to account for the formation of an enlightened public opinion. To make sense of the idea of a spiritual power, we can also consider the functions that it is supposed to fulfill: education, diplomacy, the resolution of social conflicts, and social classification. From that point of view, Comte's proposal appears to be, at least in part, an unusual way of describing familiar phenomena.

Despite its centrality, it should be always remembered that spiritual power always follows temporal power, as indicated by the terminological change introduced in the *Système*, where "moderating power" is given as a more appropriate phrase.[8] That is why an overview of positive politics must begin with the general theory of government. To the extent that this theory relies on the study of social forces, it belongs to social statics, which explains how men are "naturally disposed to submit to government."[9] Division of labor, understood more broadly than political economists usually define it, is one of the characteristic features of society. Between the various members of society, there exists a spontaneous, albeit fragile, consensus that government is intended to strengthen. Social forces can be material, intellectual, or moral, but these three forms are not on a par, and ordinary language hits it right when it uses "force" *simpliciter* to refer to material force. Positive politics acknowledges this primacy and maintains that, ultimately, social order depends on force, "the indispensable foundation of any social organization."[10] This claim requires qualification in at least two respects. First of all, as a general rule, sheer force is not enough, and that is why we call on the spiritual power to assist. Secondly, the very possibility of a social science presupposes that, like any other natural phenomena, social phenomena are governed by underlying regularities from which we cannot escape. In line with the principles of positivism, positive politics therefore contains an analysis of the limits set to the power of government, not by man, as in the liberal tradition, but by the very nature of social phenomena themselves.

The last section will address more specific topics, related to what could be called the cosmic aspect of positive politics. Actually, despite its numerous antimodern features, positive politics might, quite surprisingly, meet some of the expectations of contemporary readers. For example, politics must take into account the intimate relations humanity sustains with Earth. Both powers entertain a specific relation with space, and the very nature of spiritual power implies that its domain extends to the whole surface of the globe.[11] As for temporal power, it is common to oppose central to local powers, a French Jacobin tradition to a British Liberal tradition, Comte being usually associated with the former. However, things are not that straightforward. Why did an advocate of centralization propose that

France be divided in seventeen intendancies? Positive politics reminds us that "local" is not so much opposed to "central" as to "global." Regarding colonization, Comte's position is more consonant with ours than Tocqueville's, for instance. Comte's conception of world history, like that of most of the authors of his century, was indeed Eurocentric: Europe embodied the elite of humanity and was in charge of spreading civilization all over the planet. But Comte was one of the few who saw that military colonization is incompatible with such a mission: he stated publicly that he wished that Algeria be given back to the Algerians, whereas British positivists defended similar positions with respect to Gibraltar or India. Positivism also acknowledged the need for an environmental policy. (On the notion of *milieu*, see chapter 4.) As a former Saint-Simonian, Comte was an industrialist, but he was also aware of the fact that in the long run, the consequences of industrialization could turn out to be devastating. As early as 1841 he called for the creation of "a department of foreign relations."[12] Later, in the *Système*, he paid particular attention to our relations with animals and suggested enlarging the sociocracy (the political system he was willing to establish) into a wider biocracy.

The Positive Theory of Government

The subtitle of the *Système de politique positive* introduces the four-volume opus as a *Traité de sociologie, instituant la religion de l'Humanité*.[13] What might first appear as a category mistake was in fact the expression of the very intimate relationship between society and government, that is, between sociology and politics, established by the first axiom mentioned previously. Government and society are correlative notions: government is by definition, one might be tempted to say, the government of a society. Conversely, there is no society without government. Putting government at the heart of political philosophy implies two things. First of all, it amounts to refusing to ascribe this central position to the concept of the state, suggesting that the latter is quite overrated. Moreover, if, as Tocqueville claims, "in the estimation of the democracy, a government is not a benefit, but a necessary evil," then democracy errs.[14] If no society can do without a government, on what basis should we judge that it is an evil?

In order to introduce the positive theory of government, we will start from the idea of society and, more precisely, from its most characteristic feature, the division of labor, also called the "distribution of functions" by Comte.[15] The development of the latter is the sine qua non condition of progress, but it also induces some harmful effects, which need to be corrected. Such is the function of government. To grasp what can be and what cannot be expected from it, one must remember that we cannot govern phenomena as we please and that we need to develop a theory of the limits of political action.

Social Statics, or, the Theory of Order: Aristotle's Principle, the Division of Labor, and Dispersive Specialization

The positive classification of the sciences starts with a distinction between the sciences of unorganized bodies (astronomy, physics, and chemistry) and the sciences of the organized bodies (biology and sociology). Among the latter group, biology studies individual organisms, and sociology focuses on collective organisms. Unorganized bodies have parts, whereas organized bodies have members. Collective organisms are characterized by the independence of their members: when their members are severed from the whole to which they belong, they are able to keep on living. Comte's antiindividualism is well known, and claims such as "man as such does not exist, only humanity has a real existence," and his hostility toward human rights did not contribute to his fame.[16] Positivism, however, fully acknowledges the necessity of respecting individuality. The first objection that the 1848 *Discours sur l'ensemble* addresses to communism is precisely its "dangerous tendency to suppress individuality. Not only do [communists] ignore the inherent preponderance in our nature of the personal instincts; but they forget that, in the collective Organism, the separation of functions is a feature no less essential than the cooperation of functions. . . . in social life, individuality cannot be dispensed with. It is necessary in order to admit of that variety of simultaneous efforts that constitutes the immense superiority of the Social Organism over every individual life."[17]

From that perspective, cooperation (Comte more readily speaks of *concours* or *consensus*) between individual agents becomes the fundamental social phenomenon. The two main overviews of the positive theory of government, both of which follow a progression that is already present in the *Considérations sur le pouvoir spirituel*, begin with a discussion of the division of labor, understood broadly and not exclusively from an economic point of view, which the *Système* traces back to Aristotle.[18] If society exists, it is not only because individuals are independent but also because they are interdependent. This interdependence results from a certain order that is both social (it is related to social phenomena) and natural (like the vital consensus within the organism, it appears spontaneously and nonintentionally).[19] For the same reason, progress, to the extent that it is the development of order, is essentially the development of the division of labor.

If division of labor is indeed, according to positivism, the sine qua non condition or, so to speak, the engine of progress, all its effects are far from being beneficial. The analysis of the various drawbacks it creates is a recurring topic in Comte, and one might even think that all his works constitute a resolute attempt at reducing as much as possible the cost of what he called "dispersive specialization." Its deleterious effects indeed have an impact on social life as a whole. No matter whether you spend your life making pinheads or classifying insects, tightening bolts or solving equations, the results are the same: the hypertrophy of an organ, the atrophy of all the others, and an indifference toward the general course of the world as long as there are pins to make or insects to classify. Intellectually speaking, the mind thins as it sharpens: "Similarly, from the moral point of view, as everybody is closely subjected to the mass, it is natural that everyone be diverted by the very development of one's special activity, which constantly reminds one of its private interest, whose true relation with the public interest is only vaguely discerned. . . . The very same principle has allowed for the development and extension of society as a whole, but it also threatens, in another respect, to dismember it into a multitude of incoherent corporations, which almost seem not to belong to the same species."[20]

It is difficult to overestimate the importance of such considerations, which appear as early as the first lesson of the *Cours*, in the development of Comte's thought. If the *Cours* is a course on positive philosophy, and not on the positive sciences, as Comte explains, it is because it aims at reconciling positivity and generality. Now, it is true that the division of labor is and will remain the necessary condition of scientific progress. Specialization was needed in order to constitute the various positive sciences and nowhere is it suggested that it should be abolished, but this stage may be considered as over, the time being ripe for claiming again the rights of the comprehensive mind. To the extent that social phenomena include all other phenomena, they are more general, and the sociologist, as a specialist, becomes by the same token the specialist of generalities. The general conclusions of the *Cours* will return to that question, when they address the issue of which science is to head the encyclopedic scale.

On the political level, the situation is similar: "The social destination of government seems to me to consist primarily in containing and preventing, as much as possible, the deadly tendency of the fundamental dispersion of ideas, feelings, and interests, which is the necessary outcome of the very principle of human development and which would unavoidably lead, if left to its natural course, to the stopping of social progression in all its important aspects. This conception constitutes, to my mind, the first positive and rational basis of the elementary and abstract theory of government as such."[21]

The Art of Governing: Social Science and Political Art, Natural Order, and Artificial Order

By giving government the function of refocusing minds on the general interest, from which they are prone to diverge, positive politics recaptures a commonplace of political philosophy: government speaks for the general interest. In accordance with the etymology of the term, it must set the course; it must direct and repress or, to use a phrase from the *Catéchisme positiviste*, it must "repress divergences and foster convergences."[22] As in biology, function creates the organ, and the *Système* takes pains to describe how, early on, the principle of cooperation spontaneously gives rise to a differentiation such that, although men are naturally governable, the domination of certain families, which "have been disciplined, educated, and so placed as to have the means and the responsibility of exercising power rightly, . . . forms in the heart of every society a government of a kind." In this context, Comte remarks that "men who ignore the existence of unselfish instincts naturally attribute obedience to fear, which is not true even of the animals; and [that] they have gone so far as to make a virtue of revolt," whereas, by contrast, "metaphysical ideas about the vices of submission have given a kind of sanction to the thirst for power, which the ambitious represent to be devotion to the public good." Yet Comte adds that, given the weakness of altruistic tendencies, if "men's egoism did not create a more active incentive to exertion, societies would often find it difficult to obtain a government, or to replace it when lost."[23]

Social science not only provides us with a theory of government but also helps those in power with their actions. Governing is an art, and the political art is to sociology what medicine is to biology. Political philosophy leads us back again to philosophy of science, namely, to the principle "knowledge, hence foresight; foresight, hence action." Science enables us to know the laws that govern nature; its applications allow us to replace this natural order by an artificial order more consonant with our needs. The same principle applies in politics. If there is a social science, that is because social phenomena follow an ordered course that science aims at describing. If politicians want to alter that course, they first have to know it. In order to grasp Comte's position more precisely, let us contrast it with the idea of a natural spontaneous order that Friedrich Hayek takes to be the foundation of social science and liberalism. As discussed previously, division of labor is at the heart of Comte's thought: it is common to insist on his criticisms of economics while omitting that he praised it for having been the first to acknowledge the existence of natural laws governing social phenomena. This should not be so much of a surprise if one considers that Comte started his career reading Adam Smith and Jean-Baptiste Say. He always held in high regard the Scottish Enlightenment and, as Mary Pickering remarks, what we generally ascribe to the influence of de Maistre—for instance, the criticism of social contract theories—might stem from Adam Ferguson, whom he readily quoted.[24]

The disagreement with the economists and the liberals exclusively bears on the attitude to adopt when confronted with this spontaneous order. Acknowledging its existence, advocates of "laissez-faire" conclude that it is impossible, or at least useless, to try to modify it. Comte ascribes their position to a deficient scientific education. If they had been more familiar with scientific thinking, economists would have seen that social phenomena's spontaneous order is just a special instance of what is observed in all natural phenomena; they might also have learned that even in the cases in which this order is incomparably more stringent, it can be modified, either by natural causes or through human intervention. The more complex phenomena are, the more modifiable is the order they exhibit. Accordingly, there is no reason to forbid a regulatory action of man upon social phenomena, even though the limits of such an action have to be acknowledged.[25]

The Theory of the Limits of Political Action[26]

The idea that a theory of government must include a theory of the limits of governmental action is not very original, but the way Comte addresses the issue differs greatly from what we are used to and illustrates once again the intimate relationship between political philosophy and philosophy of science in positivist thought. It is indeed quite common to consider, as Tocqueville does in the quotation given previously, that government represents a constant threat to individuals: "All force is liable to abuse."[27] Accordingly, it is necessary to set limits to the power of government. That is exactly the rationale of the system of "checks and balances" proposed by Montesquieu: "Power should be a check to power."[28] Although the creation of a spiritual power is an answer to that very concern, Comte rejects the means proposed by classical political philosophy to achieve such an end, for he thinks it is still caught in a theologicometaphysical mode of thought: in the absence of limits we set, the sovereign would be, like God, an omnipotent being that could impose its most arbitrary wishes on us. Such measures have been conceived to fight against absolute power, but there is no more an absolute in politics than anywhere else. Actually, even independently of our actions, politicians face limits that are due to the nature of things, and room for maneuver is severely restricted.

These limits are what the Comtean theory intends to describe, for the limits of political action are just a specific case of the limits imposed on the actions of man on nature and, in that respect, they are to be dealt with by the general theory of the modifiable order, the principle of which is the following: we do not govern phenomena, we can only modify them.[29]

Accordingly, the limits of political action are the limits of modification of the social order, which in turn depend on the theory of modifiable order, the basis for the Comtean theory of the applied sciences or industry and an offshoot of the classification of the sciences. The encyclopedic scale starts with the simple and ascends toward the complex. To arrive at the theory of the modifiable order, one just needs to add: the more complex the phenomena, the more modifiable they get. Since social phenomena are the most complex phenomena, they are also the most modifiable—so modifiable, in fact, that for a long time people thought they were not subject to any law. For that reason, the limits imposed on political action are less stringent, even though they nonetheless remain quite narrow. As Comte remarks, all politicians have had the experience of the stubborn resistance of facts. So, what he says about the universal order also applies to political life: "We may regret that the order of things is not more within man's power to alter. But true wisdom forbids our wishing it to be in any part open to indefinite modification. As we advance, so far from shrinking from this inevitable yoke, we extend its range by paying to human institutions the obedience we cannot refuse to the laws of nature."[30]

The contrast between the two conceptions of the limits of governmental action is striking. On the one hand, the limits are legal conventions, the study of which belongs to political philosophy; on the other hand, the limits are facts the existence of which does not depend on us and whose study draws together the art of governing and social science, political philosophy, and the philosophy of science. However, in practice, the distance between positive politics and liberalism of laissez-faire is less than it appears. Comte is no constructivist in the Hayekian sense of the term, and he often ends up endorsing conservative positions. "Political intervention can have no efficacy unless it rests on corresponding tendencies of the political organism or life, so as to aid its spontaneous development.... It would be exaggerating the scope of such an art to suppose it capable of obviating, in all cases, the violent disturbances which are occasioned by impediments to the natural evolution."[31] This principle applies all across the board, for instance within social classification.

Spiritual Power

Despite the importance of the theory of government, the idea of spiritual power remains the core of positive politics. From his early writings to his later works, the lasting goal of Comte was the instauration of a new spiritual power. Accordingly, this notion provides us with the best way of approaching what Comte aimed at and a comprehensive view of positive politics. If Comte deals more extensively with spiritual power than with temporal power, it is just because the establishment of the former is incomparably more urgent.[32] Of course, temporal power also needs to be reformed; but these reforms can wait, and, in the meantime, we can rely on temporary measures. One of the principles of positive politics indeed is that reforms must follow one another according to a strict order. Given that all our evils result primarily from the reigning intellectual anarchy, reorganization has to start there. It will then be possible to alter mores and, finally, institutions. This emphasis on spiritual power is undoubtedly the main reason for the lack of interest in positive politics. As Comte himself remarked, political modernity started with the refusal of any spiritual power.[33] The relations between politics and religion established during the Middle Ages have been so radically destroyed that Comte's proposal sounds like an unacceptable step backward and would provide enough of a reason for dismissing positive politics as a whole.

The objection is somewhat embarrassing, and it is possible that there is no satisfying answer to it. However, if one is willing to make sense of such an idea, maybe it is worth using a different term and speaking instead, as Comte started doing in 1848, of a "moderating influence."[34] This change of terminology indeed allows for the clarification of some misinterpretations. First of all, with the disappearance of the tacit reference to the Middle Ages, it is easier to grasp that the argument rests on a conceptual analysis of the workings of society. Moreover, *spiritual power* "tends to suggest a false idea of a power even more moral than intellectual."[35] It also emphasizes the fact that the relation between temporal and spiritual powers is asymmetrical. The former could exist on its own, but the latter, which always presupposes a preexisting power, could not. Even if the primacy of the temporal power might be challenged, it does not call into question the fact that political power as such belongs to force. This terminological change also explains what might otherwise appear as a slight incoherence, namely, the recurring criticisms addressed to the doctrine of the philosopher-king, which is easily confused with the idea of a spiritual power. Actually, there exists a wide gap between Plato and Comte on that issue. If the "alleged absolute domination of the mind" must be rejected, it is not only because it contradicts the separation of powers but because it also tends to entrust the mind with a function it is not able to fulfill; that is why the Platonic doctrine is deemed not only a "dangerous utopia," but also a "radical absurdity."[36]

Two Features of the Moderating Power: Popular and Public

Such an emphasis on the moderating nature of spiritual power provides a key to understanding not only its relation to temporal power but also two of its most characteristic features: its popular nature and its role in the formation and expression of public opinion.

It is generally assumed that the goal ascribed to spiritual power is to "govern opinions and mores."[37] But this description is unfortunate to the extent that it suggests that the spiritual power would intend to get some sort of grip on minds and impose on them preconceived ideas. It is more enlightening to characterize spiritual power in contrast with temporal power: the latter commands actions; the former modifies wills. Accordingly, the separation between them rests on the distinction between two types of action: "One is material, the other moral, and they are entirely heterogeneous, either with regard to their sources or their modes, although they always coexist. The former bears directly on actions, so as to determine some of them and prevent others; it ultimately rests on force or, in other words, on wealth. . . . The latter consists in the regulation of opinions, feelings, and wills, that is, of tendencies; it rests on moral authority, which ultimately results from the superiority of intelligence and enlightenment."[38]

In other terms, the most characteristic feature of spiritual power is negative: it eschews the use of force and chooses instead to address the mind. In that respect, it is important to note that the theory of spiritual power has its foundation in social epistemology, whose principles were very clearly identified by positivism. So, it is a mistake to restrict, as is usually done, the concept of faith to the analysis of theological beliefs: there also exists a positive faith, defined as a disposition to "take on trust the principles established in sciences by competent men."[39] Similarly, it is also necessary to make room for the concept of authority within intellectual life; if this authority, which "is not liable to be, by its nature, an object of decree or prohibition," is characterized as *spiritual*, it is because it also has a moral dimension, without which trust is impossible.[40]

> Whereas temporal power ultimately depends on a certain material preponderance, either based on force or wealth, whose unavoidable empire is often felt as a nuisance, spiritual authority, which is both softer and more intimate, always rests on some trust spontaneously granted to intellectual and moral superiority; it presupposes a free and continuous assent, either resulting from conviction or persuasion, to a common elementary doctrine, which simultaneously regulates the exercise and the conditions of such an influence that is destroyed as soon as this faith disappears. But the philosophical nature of such a doctrine must deeply influence its elementary features.... The theological faith, which is always linked to some form of revelation in which the believer cannot participate, is assuredly very different from the positive faith, which can be studied by everybody in certain conditions, although both equally result from this universal disposition to trust.[41]

Secondly, "any spiritual power must be, by its very nature, essentially popular."[42] One might be tempted to think that sheer opportunism prompted Comte to endorse that idea. He indeed quickly realized that, contrary to what he first believed, scientists were not really interested in forming a new spiritual power, and he then turned to the popular classes in the hope of finding political support for his endeavor.[43] However, the main reason why Comte believed that spiritual power is popular has to do with its moderating nature. Force manifests itself in two forms: concentrated or diffuse, as wealth or number. Spiritual power influences both, but not in the same way. It indeed turns out that power is not equally divided between the two types of force. Temporal power belongs to the rich, and it is primarily their manner of behaving toward the greatest number that the spiritual power will have to moderate. Furthermore, "the characteristic mission [of the spiritual power] being primarily that of ensuring, as much as possible, the domination of universal morals on the social movement as a whole, its most extended duty is toward the constant protection of the most numerous classes, which are generally more exposed to oppression, and with which, through common education, it has more daily contacts."[44]

The theory of public opinion shows that the relation between spiritual power and the popular classes is even closer: spiritual power is the voice of the popular masses, the support of which it can count on in return.[45] Positive politics aims at establishing "the empire of public opinion." Actually, public opinion pursues the same goal as spiritual power—the regulation of social life—and, like it, eschews force. To fulfill its mission, it first needs "the establishment of fixed principles of social action; secondly, their adoption by the public, and its consent to their application in special cases; and lastly, a recognized organ to lay down the principles, and to apply them to the conduct of daily life."[46] In that respect, the existence of a spiritual power results from the necessity of having, between the doctrine and the public, "a philosophical organ, without which their relation would otherwise abort on almost every occasion." Otherwise, it "would be difficult to conceive that a system of moral and political principles should be possessed of great social influence, and yet at the same time that the men who originate or inculcate the system should exercise no spiritual authority."[47]

Spiritual Power's Main Functions

Three words suffice to describe the main prerogatives of spiritual power: it must advise, sanction, and direct. But one might perhaps get a better idea of its main function by focusing on education and social classification.[48] Education has always been considered the primary task of the spiritual power and, as early as 1822, Comte had planned a *Traité de l'éducation universelle*, which still appeared on his "to-do list" when he passed away.[49] Together with labor, education is one of the two great social needs, and the main point of contention between positivists and communists bears precisely on their relative importance: whereas communists think that people primarily need work, positivists give education priority.⁵⁰ The latter is first and foremost a family matter, but it also concerns society, and that is why it is included in the governmental sphere of competence.

By entrusting the spiritual power with education, positivism acknowledges its intrinsically political character as well as ensures its independence from the state, which might take the opportunity of using it to subject individuals to its own will. Education must be universal, i.e., it must be aimed at everybody. In the Middle Ages, the church taught catechism to all, but instruction remained the privilege of but a few. In the positive state, the possibility of attending school must be open to all, poor and rich, men and women. Comte spent a good part of his spare time delivering popular lectures on astronomy, and his disciples contributed extensively to the creation of a secular, free, and compulsory system of education. Given the openness of the educational system, the spiritual power is in touch with all the classes of the population, and the way it contributes to the formation of the individual is the main source of its influence outside of school.

The empire of opinion mentioned above only makes sense for an enlightened opinion. The primary goal of education is not to train qualified workers but rather to enable everyone to fulfill properly his duties as citizen. The best way of achieving such a result is to familiarize oneself with the sciences. The last volume of the *Système* describes in detail the syllabus of positivist education: through the study of the seven fundamental sciences (ethics having been added to sociology), pupils will discover the order that governs both the external and the human world, and they will learn to submit to the strength of demonstrations, thereby gaining enduring beliefs that will protect them from skepticism.⁵¹ In accordance with the law of the human order, which states that "the living are always and increasingly dominated by the dead," education also aims at developing the feeling of continuity, which provides a new way of distinguishing the spiritual and the temporal powers: the former ensures continuity, whereas the latter takes care of solidarity.⁵²

Social classification is also one of the main functions of the spiritual power. Tocqueville taught us how to appreciate the way democracy tends toward the equality of conditions; positive politics, on the other hand, reminds us that the inequality of conditions constitutes the basis of any social structure.⁵³ An adequate political philosophy must not only include a theory of order but also provide a theory of subordination. This decidedly antimodern stance is to be traced back not so much to the influence of the retrograde school, but rather to that of Ferguson. When he condemns the "antisocial notion of Equality,"[54] Comte refers not to the Catholic hierarchy de Maistre had praised but to the division of labor dear to economists,[55] and if one is still willing to link Comte's ideas to counterrevolutionary thought, it would be better to turn to military life, which he considered to have been superior to industrial life for a long stretch of history.[56] The contrast between military and industrial societies was so important to him that he returned to it in all of his main political works.[57]

Classification is a task the scientist is familiar with, and the sociologist, when he classifies, has to apply the methodological principles used in the sciences. Of course, classifying and ordering are two distinct activities, but they are close to one another; most of the time, classification brings about order. That is indeed the case here, for social classification is mostly a description of the hierarchical structure of society. It consists mainly in distinguishing two kinds of classification: a "concrete" classification, which considers social position, and an "abstract" classification, which bears on individual merit. The former is objective and describes social structure, with its various relations of dependence resulting primarily from the division of labor, without paying attention to who is in charge of such-and-such function. Once this functional analysis is completed, one still needs to guarantee "the harmony between the function and its organ."[58] Hence one comes to the second classification, which is more difficult to carry out because it also considers persons, a consideration that forces it to divide itself in two, depending on whether it judges the ability of somebody to fulfill certain functions or her intrinsic merit, independently of her social function.

Let us focus on two points. First of all, although it is true that the sociologist aims at describing social structure, it is difficult to see how he could avoid such a work of classification, however unpopular it is. Of course, these structures are imperfect and liable to change, but it is absurd to attempt to abolish, as a matter of principle, all the inequalities. Between the various social positions, there will always be relations of subordination. To pretend, as the levelers do, that it could be otherwise just amounts to deluding ourselves. Accordingly, Comte is quite explicit in his wish to "ennoble obedience and strengthen command."[59] Secondly, it is also true that social positions are often not occupied by the most apt person; we are then entitled to act so that the position will be entrusted to a more qualified individual and, more generally, so that everybody occupies the function that best fits her talents. Part of such a requirement rests on the educational system, one of its goals being precisely the providing of the means to "put everybody in the position most compatible with their abilities, independently of the rank they have been given by birth"; this system thereby ensures social mobility and the renewal of elites. However, our possibilities in that domain should not be overestimated.[60] For it is indeed the case that "most men do not have definite vocations and . . . most social functions do not call for it."[61] Moreover, in most cases, it is even more advisable to accept these imperfections, the very attempt at correcting them leading in general to making the situation worse: "Well-marked personal superiority is not very common; and society would be wasting its power in useless and interminable controversy if it undertook to give each function to its best organ, thus dispossessing the former functionary without taking into account the conditions of practical experience."[62]

Abstract classification not only aims at judging the abilities required for the fulfillment of certain functions; it also meets a deeper need. Most of the time, only the powerful enjoy social consideration, men being judged according to the positions they occupy. That is why we need another kind of classification, which is more difficult to establish and which is ordered not by reference to power but by reference to dignity.[63] Saint Bernard of Clairvaux submitted to the authority of the pope, just as Saint Paul submitted to Saint Peter, although they were both superior as individuals.[64] Now, it is true that moral dignity is not an easy thing to appreciate. Here again, Comte warns us against an improper use of that second kind of classification and suggests that time be the judge: the sacrament of incorporation will be posthumous, just like the cult of great men.

Spiritual power provides us with a good illustration of the problems raised by the idea of subordination. As a moderating power, it presupposes temporal power, which then has pride of place. But their relation can be understood in the opposite direction, so that the moderating character belongs to temporal power. One just needs to consider its specific task (ordaining actions) as an indirect way of modifying wills. From this point of view, government becomes a mere auxiliary of the spiritual power.[65] Any classification falls prey to this sort of predicament, which can be dealt with by distinguishing between objective and subjective generality. As complexity increases, subjective generality does also, whereas objective generality decreases. For instance, social phenomena are, when considered objectively, the less general (they only constitute a subset of biological phenomena, which in turn . . .), but they are subjectively the most general (they relate to beings that are also governed by biological laws, chemical laws, etc.). Classification can be read from top to bottom or from bottom to top. Accordingly, levels that are objectively superior are subjectively subordinated to levels that are objectively inferior; as a result, one may characterize "the higher functionaries as the involuntary servants of their voluntary subordinates."[66] Similarly, the degrees of independence and dignity vary in opposite directions: "The Spiritual power surpasses all others in subjective generality and in social dignity; on the other hand, material power is the more general objectively and the most independent. Material force therefore forms the essential base of the political construction, spiritual power being the apex."[67]

In the preceding section, I have tried to make sense of the positive theory of spiritual power. For that reason, I have set aside the eminently religious character it took after 1848. Even so, numerous problems remain. For example, one might accept most of what has been previously said but still reject the very idea of a spiritual power. It is not the necessity of having an educational system or of relying on an enlightened public opinion that is called into question, but do we really need, in order to achieve such results, a single authority fulfilling these various functions, one distinct from the temporal power as such? If so, what form must it take within society? Does it require, as Comte has it, the formation of a new class, of a clergy? If spiritual power must give rise to the constitution of a church, is it not the case that Benjamin Constant's fears are justified?[68] Leaving all these questions open, let us analyze instead a few examples of the manner in which Comte developed his program.

Further Insights: The Cosmic Character of Positive Politics

In order to provide a more precise view of the way positivism was planning to put its ideas in practice, let us examine briefly its positions on the issue of the relation between central and local powers, on colonialism, and, finally, on what might be called, following Foucault, biopolitics. Comte's take on these problems illustrates another distinctive feature of positive politics, namely, its cosmic dimension. Comte was indeed acutely aware of the manner in which humanity depends on astronomical conditions for its existence: just imagine minor alterations in the elliptical orbit of the Earth or in the inclination of the ecliptic and life—at least life as we know it—

would have been impossible. In accordance with the principle of the subordination of the organism to the milieu, humanity, the proper subject of sociology, is closely related to Earth through its "twofold fluid envelope."[69] Despite the Copernican revolution, the human planet remains for all of us the firm, unshakeable ground on which everything else rests. It is in this feeling that one can find the source of the attachment to the homeland, even for nomads, for whom "the Tent, the Car, or the Ship, [is] . . . a sort of moveable Country."[70]

A striking feature of the concrete application of positive politics is the way it juxtaposes two kinds of remarks that appear to be, at first sight, incompatible: on the one hand, Comte emphasizes the eminently provisional nature of political measures; on the other hand, many of his proposals seem totally utopian (actually, although the *Cours* readily criticizes the utopian movement that was thriving at the time, the *Système* ends up making significant room for positive utopias). This apparent anomaly is due to the specific relationship that positive politics, and more generally Comte's thought, has to the future. As early as 1822, the *Plan* claimed that the chronological order is not the logical order according to which epochs should be considered: one should not say "past, present, future," but "past, future, present." "It is in effect only when we have, through the past, conceived the future that we can usefully return to the present—which is only a point—in such a way as to grasp its true character."[71] Accordingly, the "synthetic presentation of the future of man," included in the last volume of the *Système*, starts by dedicating three long chapters to a description of a so-called normal state toward which humanity is supposed to move, before turning, in a final chapter, to an analysis of the stages that still separate us from it and that should be covered within a few generations. Alongside these proposals, which may sound vain and which Mill ascribes to the bent for regulation typical of the French (the number of future intendancies into which France would be divided, the number of positive schools in the Occidental Republic, the number of teachers and the level of their salary, etc.), the presentation of the normal state also includes a description of ideal limits, which are explicitly introduced as utopian but are nonetheless positive.[72] As long as the normal state has not been reached, the appropriate steps to take can be only transitional. This might explain the admiration of positivists for Danton and the members of the *Convention*, who clearly acknowledged the provisional nature of their actions.[73]

Central, Local, and Global: The Withering Away of the State and European Integration

After 1851, Comte proposed to dismember France into seventeen "intendancies."[74] According to the received view, which holds Comte to be an advocate of centralization, this fact is difficult to understand. It is true that when, in the historical lessons of the *Cours*, he referred to the classical opposition between two great political traditions (on the one hand, a French tradition, Jacobin and centralist; on the other hand, a British tradition, liberal and in favor of local powers), Comte decidedly sided with the first: the situation of France is the normal one, England being only an exception.[75] In 1848, the Société Positiviste was set up following the model of the Club des Jacobins, and the report on government drafted on its behalf shortly afterward granted the Parisian people a power totally out of proportion with its representativeness. Only a superficial reading of his works could conclude that Comte was an advocate of centralization, since it amounts to ignoring what he had always considered paramount, namely, the distinction between the temporal and the spiritual powers. He indeed complained that he had been misunderstood on that issue, and the case at hand is a good example of the effects of such a misunderstanding. According to the power one considers, the situation changes dramatically, for material and spiritual forces do not have the same relation at all with space. Centralization applies only to spiritual power (Comte clearly had in mind the papacy here). The mind ignores borders: accordingly, spiritual power can be only catholic and universal, and the whole planet is its natural domain. On the contrary, temporal power is by its nature local, for it is important that it be close to the public it serves.[76] Therefore, the normal size of a state should not exceed that of Belgium or Sardinia.[77] Moreover, one should consider that the states as we know them are historical products that did not exist before the sixteenth century and that there is thus no reason to grant them an unlimited duration. One should also remember that positivists believed that, along with the advances in civilization, spiritual power was poised increasingly to replace the temporal power. Given these conditions, it makes sense to talk of a withering away of the state in positive politics.[78]

Comte's approach points to the notable interest, uncommon at the time, for international relations and most notably for European integration. Although this issue, which was on the agenda in 1815 just after the fall of Napoleon, had been quickly set aside, the long discussion of the new politics that concludes the lessons on sociology closes with a call to create an occidental positive committee, which would include members from the great European nations (France, Great Britain, Germany, Italy, and Spain) and which would be responsible for the establishment of the Republic.[79] Toward the end of his life, Comte also paid close attention to the Crimean War and to the conflict between Russia and Turkey; he even went so far as to write letters to the czar and the sultan (or his vizier). Since neither of these countries belongs to the Occidental Republic, the question of the exact extension of Europe arises. In this struggle for hegemony in Eastern Europe, religious questions were important. The sultan also acted as caliph; Russia, as an Orthodox country, presented itself as the heir of the Eastern Roman Empire and claimed that it would protect Slavic peoples, who were also Orthodox, against Ottoman oppression. This gave Comte the opportunity of discussing at length the respective merits of Christianity and Islam. First he sided with Russia, but then backtracked and adopted a position somewhat more favorable to Turkey. He even went so far as to suggest that the European capital be moved from Paris to Constantinople so as to bring closer together the East and the West. Through Ahmed Rıza, the Young Turks were influenced by positivism, the separation of church and state providing them with a solution to the numerous problems that plagued the Ottoman Empire.

Colonization and Eurocentrism

Positivists also paid attention to another aspect of foreign affairs, colonization. Comte was resolutely opposed to colonial empires, the development of which he witnessed during his lifetime. He prided himself for having wished in his youth for the victory of the Spanish people over the Napoleonic armies, he publicly called for the restitution of Algeria to the Algerians and of Gibraltar to Spain, and his English disciples were among the first to denounce the imperial policies of their country. At the time, such an attitude was exceptional: Tocqueville supported the conquest of Algeria, and the Saint-Simonians actively contributed to the colonization of Northern Africa. Yet to criticize colonialism does not amount to forsaking a Eurocentric vision of history. The positivists were so confident about the superiority of European civilization that they believed it had been entrusted with a civilizing mission. And that is indeed why they considered colonization unacceptable. It resorts to methods (war and constraint) that are incompatible with its intended purpose. The mission of the West is indeed spiritual: it consists in spreading positive thought (that is, after 1848, the Religion of Humanity), which requires a free espousal from the mind and the heart.

What is at stake in these discussions, which take place within a framework provided by the theory of progress or social dynamics, is nothing less than the issue of the relationship Europe has with humanity considered as a whole. The positive theory of humanity underwent a notable evolution. In the version of positivism given in the *Système*, even though it remains the proper subject of sociology, the Great Being has now become an object of worship. As a result, that which relates to Europe and the West is approached in religious terms: religion is universal and that is the reason why the whole planet is summoned to embrace it. In fact, this religious dimension merely builds on a doctrine elaborated in the *Cours*, where the notion of humanity already serves to define the subject of sociology. The core idea, borrowed from Condorcet, is that of a single people.[80] Now, Comte did not ignore the diversity of races or cultures, but he claimed that the emphasis put on such a diversity obscures the fact that the inhabitants of the Earth have a lot in common, and that what they share increases with progress. Accordingly, within the positivist calendar, Comte took pains not to restrict himself to Europe and introduced figures from all civilizations (not only Buddha, Muhammad, and Confucius but also Manco Capac and Toussaint Louverture, to quote just a few names).

With this unitary background in place, the situation of Europe appears even more peculiar. It is indeed the case that "the past is divisible in two great periods: the one, common in its essential features to all nations, includes Fetishism and Theocracy; the other, peculiar to the Western nations, effects as a spontaneous process the transition from Theocracy to Sociocracy."[81] It is for this reason that Europe is to lead the way for the other peoples. The last pages of the *Système* describe the order in which they will adopt the Religion of Humanity. Once the West has been regenerated, it will be the turn of the other great monotheisms: first the Turks, or western Muslims, then the Russians and the other Orthodox peoples, and finally the Persians, the eastern Muslims. They will be followed by the polytheists (India, China, and Japan) and, eventually, the fetishists of Africa, Oceania, and the Americas, who have been spared by theologism and who could directly adopt positivism. (In 1848, the *Discours sur l'ensemble du positivisme* gives a slightly different order.)

Sociocracy and Biocracy

The cosmic dimension of positive politics also derives from its desire to grasp political relations in their most general forms. It requires that one consider an "eminent type of institutions, whose importance has not been adequately felt so far, and which is intended to regulate properly the relations . . . of humanity with the world, and most notably with the other animals." In that respect, Comte praised fetishism for having initiated such a perspective and went so far as to suggest the creation of "a department of foreign relations"![82] In this context, Comte was led to develop views on the relations between man and animal that Lévi-Strauss later acknowledged as "prophetic."[83]

The discussion was sparked by the consideration of the transition from sociocracy (a neologism coined by Comte to refer to the political way of life that is consonant with the teachings of sociology) to biocracy, conceived as the horizon of positive politics. To understand this final extension, it is useful to start from a comparison of humanity and animality, since it provides "the true explanation of human greatness. Varying the candid expression of a hero who knew what ambition was [Julius Caesar], we may say that it is better to be the first animal than the lowest of angels."[84] If one regards man as poised to develop more and more "his natural ascendancy over the animal kingdom as a whole," that domination needs to be regulated, as fetishism already anticipated.[85] To do so, Comte suggests that animals be divided into three main groups: pests, which Comte, insensitive to biodiversity, thinks are to be destroyed; the "laboratories of our food"; and the "affective auxiliaries." For the former of these last two cases, Comte calls for the improvement of their living conditions and, even more importantly, of the way they are killed; he also considers transforming herbivores into flesh eaters, in the hope that they would become "more active, more intelligent, nay even more devoted" and therefore ascend in the animal hierarchy.[86] But biocracy concerns primarily our "affective auxiliaries," for they are the animals with which we have the closest association. Just as we progressively replace man by the animals as sources of mere physical power, we must aim at freeing animals from the most mechanical tasks in order to entrust them with others that are more consonant with their dignity.[87] We should turn to biology to get "the point of view [of] the policy of the human race, nay of the whole animal kingdom."[88] If sociocracy leads to biocracy, it is because "the conception of this association between Humanity and the races capable of domestication provides the scientific basis for the widest and most permanent aspect of positive politics: the combination of organic nature against inorganic nature for the purpose of exploiting all the resources of our planet."[89]

★ ★ ★

Of course, more could be said, for instance about the close relations established by positivism between politics and religion on the one hand and politics and morals on the other. In the former case, the omnipresence in the *Système* of a religious vocabulary that was absent in previous works is certainly one of the main reasons for the limited success enjoyed by the book. In the latter case, the topic appeared right from the start. Comte often uses the term "moral power" to refer to the spiritual power, as opposed to political power as such—that is, temporal power—and he always considered the subordination of politics to morals one of the main advances achieved by the Middle Ages.[90] However, in the foregoing, we have tried to focus on the essentials and attempted to provide an overview of positive politics true to Comte's project—an overview that would allow for the identification of its strengths and weaknesses.

It should be pointed out that positive politics derives entirely from two axioms: no society can exist without government, which amounts to acknowledging that the two notions are correlative; and a society cannot last without some sort of spiritual power. The first axiom locates government at the center of political philosophy. The distinctive feature of an organism indeed lies in the separation of functions and the combination of efforts or, to put it concisely, in the independence of its members and their cooperation. It is true that, in certain respects, these two components counterbalance one another. Therefore, ways should be found to prevent them from destroying one another; such a situation explains why a society must be governed. The division of labor is not only what distinguishes the political order, which is based on cooperation, from the domestic order, which is based on sympathy; it is also the main engine of progress to the extent that it enables society to expand. Yet as the division of labor develops, it causes an increasing gap between private interests and the common good, which multiplies the opportunities for conflict. Government is here to guarantee social cohesion through the alleviation of the adverse effects generated by the division of labor and through the development of cooperation. Once the function of government has been defined, the nature of its action is still to be specified. Conceived in this second sense, the theory of government states that politics is not a science, but an art, which depends on sociology the way medicine depends on biology. The knowledge of the laws of social phenomena, to the extent that it allows for the prediction of the future course of events, can indeed help rulers to take the right decisions. This illustrates the manner in which positivism links political philosophy to philosophy of science, and it is not improper to liken this perspective to an engineering approach, which suggests that governmental action be included in a more general theory of the applied sciences.

The second axiom is much more problematic. On that point, Comte failed to convince. Although the spiritual power described in the early writings was not religious, the increasing importance of this dimension and the somewhat extravagant form it often took provided the enemies of positive politics with easy targets. Many of those who, like Mill and Émile Littré, had placed great hopes in the author of the *Cours*, refused to follow him in that direction; but even in its initial, secular formulation, according to which scientists were to be entrusted with the mission of replacing the clergy, the idea was not more successful. However, even though it is hard to accept the solutions suggested by Comte, it is no less difficult to set aside the problem he was trying to solve. As Tocqueville had noted, "the question is, not to know whether any intellectual authority exists in the ages of democracy, but simply where it resides and by what standard it is to be measured."[91] Comte's position originates in a belief he expressed in the opening pages of the *Cours*, one that separates him radically from Marx's historical materialism: "The world is governed and overturned by ideas, or, in other words, . . . the whole social mechanism rests finally on opinions."[92] It follows that the evils that plague society are caused by intellectual anarchy and that politicians err when they focus their action on the institutional level. To achieve lasting results, reforms must proceed in an unalterable order: one must first transform minds, then mores, and only after that institutions. A society cannot survive without a common doctrine, a minimal consensus on the rules of community life. In the positive era, this doctrine has to be positive. Classical political philosophy, which triumphed with the French Revolution, does not meet this requirement, and the works of Comte as a whole constitute one single effort to provide a doctrine more suited to our societies.

What also needs to be emphasized is the difficulty of passing a fair judgment on positive politics. It challenges most of our intellectual schemes, almost to the point of becoming subversive. Michel Houellebecq provides a good description of the situation when he observes that "everything, in Comte's political and moral thought, seems intended to infuriate the contemporary reader."[93] It would not make sense to try to salvage positive politics as it is presented in the *Système*, and nobody is even considering doing so. Yet the silence that surrounds positive politics and the few clichés that save us from a closer analysis do not make more sense. The source of our difficulties is quite easy to identify. It lies in Comte's irreducible hostility toward the metaphysical politics we are still massively endorsing. When he attacks the parliamentary regime and human rights, Comte rejects entirely what we take to be the most fundamental values. By refusing to accept this very basis of discussion, Comte excludes himself from the democratic debate. The denunciation of the "occidental disease" goes along with an almost complete indifference to law, this great regulator of social life that is considered to be the political equivalent of the philosophical concept of cause.[94] Despite his admiration for Rome, Comte forgets that military polytheism also bequeathed to us Roman law. The same omission also undoubtedly explains some of Comte's misappreciations, such as when he attributes to Aristotle the "the famous aphorism on the necessary subordination of laws to mores," whereas Aristotle in fact held that mores depend on education and, through it, on laws.[95]

Listing the weaknesses of positive politics would be tedious and would not solve a question that still needs to be addressed: how is one to explain that minds as eminent as Mill and Littré could, at least for a time, recognize themselves in it? Comte prided himself for having remained a republican all along but, contrary to Littré, he conceived the Republic as dictatorial, not parliamentary. On this account, one might be tempted to suspect him of totalitarian inclinations. Nothing could be more unfair. Because he believed that "no combination of men can be durable, if it is not really voluntary," Comte decided to get rid of "all artificial and violent bonds of union, and retain only those which are spontaneous and free,"[96] and he demanded that "the question of moral and intellectual reorganization shall be left to the unrestricted efforts of thinkers of whatever school."[97] As Jean-Claude Casanova remarks, "his posterity never contributed to any of the political crimes of the twentieth century"; moreover, one cannot dispense with the analysis of his philosophy to the extent that "it challenges liberalism by identifying its limits and by trying to think order and progress, outside of the parliamentary régime and the capitalist market."[98] Hayek said of Comte: "So far as [he] is concerned, it is true that I strongly disagree with most of his views. But this disagreement is still of a kind which leaves room for profitable discussion because there exists at least some common basis."[99] It seems that this discussion had not taken place, but everything suggests that it would be no less profitable today.

7 Art, Affective Life, and the Role of Gender in Auguste Comte's Philosophy and Politics

Jean Elisabeth Pedersen

Auguste Comte made his first public statements about the connections between art, affective life, and the organization and reorganization of society in the first piece that he published under his own name, the work that he would later characterize as his "fundamental opuscule," the *Plan des travaux scientifiques nécessaires pour réorganiser la société*, which he circulated in 1824 as the first volume of a projected *Système de politique positive*.[1] "For a new social system to be established," he asserted, "it is not enough for it to have been conceived properly; it is also necessary for the mass of society to become impassioned about constituting it."[2] Comte was especially interested in establishing an enduring new social system because he had experienced the social upheaval of the French Revolution and observed the difficulties of creating or restoring any sort of social order in its aftermath. Contrasting the emotional power of art with the intellectual power of science, he argued that the necessary passion for social change in the service of social stability could come only from the work of pioneering artists who used their creative imagination to produce inspiring visions of the future society that his new science of "positive politics" would predict.

Comte incorporated significant discussions of gender into his discussions of art, science, and social change when he presented his new positive philosophy and its associated political and religious recommendations at greater length in the more substantial publications that scholars have identified as his most important works: the *Cours de philosophie positive*, published from 1830 to 1842; the *Discours sur l'ensemble du positivisme*, published in 1848; and the *Système de politique positive*, published from 1851 to 1854 with a new version of the *Discours* at the beginning of the first volume and a new version of the *Plan* in the appendix to the last. As he outlined his new "social science," "positive philosophy," and "positive politics" in the *Cours*, for example, he praised men for their superiority in reason, identified women with a "secondary superiority" in emotion, and completed his recommendations for the public organization of politics and intellectual life with recommendations for the private organization of marriage and family life.[3] When he reviewed his intellectual trajectory in the *Système*, he claimed that the "necessary influence" of the *Plan* and his other "early sociological works," while "especially concerned with the political order properly speaking," had already "extended implicitly as far as the domestic order"; described the *Cours* as "the great philosophical elaboration" that had already offered "some decisive explanations on the subject"; and promised that his complete *Système* would finally provide "the true theory of the family, founded on an exact knowledge of human nature."[4]

As Comte moved to refine his theory of the family, introduce the study of morality, and codify the Religion of Humanity in the *Discours sur l'ensemble du positivisme* and the *Système de politique positive*, he elaborated his ideas on the interconnections between art, affective life, and gender in an especially visible way, for he shifted the focus of his analysis from the nature of science to the nature of art, from the importance of reason to the importance of emotion, and from the intellectual abilities of men to the emotional influence of women. The common preamble that he wrote for the *Discours* and reprinted in the *Système* asserted that "only the feminine point of view would allow positive philosophy to embrace the true ensemble of human existence," described how "the aesthetic appreciation of positivism" would follow "as the natural result of the explanation[s] related to women," and proclaimed that his new work on the complete male and female "ensemble of human relationships" was what had finally enabled him to develop for the first time "a true general theory of the fine arts."[5] The "final invocation" that he inserted just after the "total conclusion" of the *Système* itself indicated the ultimate significance of the arts, the emotions, and emotional women in his new religious and philosophical system by stressing that "the final expansion of a religion [that would be] more aesthetic than theoretical" was possible only with "the solemn sanction and intervention of that [female] sex whose sympathy best disposes it to the synthetic state."[6]

Social theorist Mike Gane has characterized the work of Auguste Comte as a philosophy that was born in heresy, and Comte's changing ideas about how to understand the interconnected political and social significance of art, affective life, sex, gender, and family life would become some of the most controversial aspects of his philosophical system.[7] Comte had drafted the *Plan* while he was working as the secretary for the social theorist Henri de Saint-Simon, and he initially expected it to appear in Saint-Simon's series *Du système industriel* in 1822.[8] When Saint-Simon actually published the *Plan* as an installment of his *Catéchisme des industriels* in 1824, however, he added a preface that criticized both Comte's decision to prioritize scientists over industrialists and his decision to focus on science itself without according enough importance either to feeling or to religion.[9] Comte's anger at what he saw as Saint-Simon's refusal to recognize his intellectual independence led both to his resignation as Saint-Simon's secretary and to his decision to develop and popularize his new social theories on his own.[10] Six months after Comte started his public lectures on positive philosophy in January 1829, Olinde Rodrigues read him out of the Saint-Simonian intellectual tradition on the grounds that Comte had privileged reason, "cold calculation," and the constraints of science over imagination, generous "devotion," and the power of poetry.[11]

As Comte added a gender dimension to his continuing explorations of the proper relationship between reason and emotion in the *Cours*, the *Discours*, and the *Système*, his evolving ideas about men's and women's respective roles in the ideal positive polity became just as controversial as his early ideas about the respective roles of science and art in the service of social change. The conflict between Comte's initial insistence on women's intellectual inferiority to men in the *Cours* in 1838 and John Stuart Mill's enduring commitment to women's rational equality with men, eventually elaborated most influentially in Mill's *Subjection of Women* in 1869, was at the heart of the conflict that destroyed the budding collaboration between the two social theorists after an exchange of almost one hundred letters in the period from 1841 to 1847.[12] Comte's later valorization of women's emotional superiority to men in the *Discours* and his elevation of women to the status of priestesses in the Religion of Humanity in the *Système* were two of the most visible aspects of the shift in the focus of his philosophical work that eventually led republican intellectuals such as sociologist Émile Durkheim and philosopher Émile Littré to reserve their admiration for Comte's earlier positive philosophy alone while inspiring antirepublican figures such as conservative novelist Maurice Barrès and monarchist journalist Charles Maurras to adopt the ideals of Comte's later positive polity instead.[13]

While historians, philosophers, social theorists, and literary critics have returned time and time again to the question of how to understand and evaluate Comte's changing views on the significance of women and their place in the family and society, they have paid much less attention to the closely connected question of how to assess Comte's changing views on the significance of art, its relationship to science, and its role in the new Religion of Humanity.[14] For Comte himself, however, these two sets of topics were all intricately interconnected. When he asserted women's "fundamental inferiority" to men in the *Cours*, for example, he cited "the irrecusable organic subalternity of the feminine genius . . . even in the fine arts."[15] When he introduced "the Worship of Humanity" in the *Discours*, he praised the anonymous "misunderstood angel" that he would later identify in the *Système* as his beloved Clotilde de Vaux not only for inspiring "the incomparable private affection that had reanimated his public life" but also for enabling him to complete his earlier work on reason with his new work on "sentiment, and even imagination."[16] When he published the *Système* itself, he summed up his new ideas on the complex relationships between art, affective life, and gender when he dedicated the first volume to de Vaux's memory, prefaced his philosophy by reprinting her fiction, and stressed the complementarity between the "two solidary but independent functions" that the two of them had developed together: "one tending to establish active masculine convictions through the scientific route; the other, to develop profound feminine sentiments through the aesthetic route."[17]

In order to analyze the larger intellectual and political implications of Comte's changing views on the ideal relationships between art and science, emotion and reason, and women and men, I have divided the body of my essay into four sections. The first, focusing on Comte's ideas about art and imagination, explores the ways in which his invocations of the fine arts and aesthetics related to his larger philosophical, scientific, and political projects. The second, focusing on his ideas about emotion and the affective life, explores the ways in which he ultimately sought to change not only the practices of philosophy, science, and social science but also the practices of religion, morality, and ethics. The third, focusing on his ideas about the proper relationship between rational men and emotional women, explores the ways in which he connected his abstract philosophical ideals to a series of practical political recommendations, especially the ways in which he intended his new social science to counter the competing claims of the utopian socialists who were active at the same time, particularly those in the Saint-Simonian school. The fourth and last section, finally, returns to the vexed question of how to balance assessments of continuity and change between Comte's two most important multivolume publications, the *Cours de philosophie positive* and the *Système de politique positive*, by highlighting the parallel and simultaneous development of Comte's interconnected ideas about the power of art, the significance of affective life, and the social, scientific, artistic, political, and religious roles of women and men.

Science and Art: Aesthetics in Comte's Philosophy and Politics

Comte used the term *art* in three different, though interconnected, ways over the course of his work. In the oldest of these meanings, which harkens back to the original Aristotelian distinction between doing and knowing, art appears as the practical extension of science.[18] So, for example, when Comte introduced his "general considerations on the hierarchy of the positive sciences" in the first volume of the *Cours de philosophie positive* in 1830, he connected the basic distinction between the work of human "speculation" and human "action" first to the distinction between "theoretical" knowledge and "practical" knowledge, then to the distinction between "science" and "art." While he briefly mentioned the "fine arts" in this section of the *Cours*, he spent more time talking about "the true theory of agriculture," which he took as "the most essential case," the best example of a complex art that depended on several "corresponding sciences," including physiology, chemistry, physics, "and even" astronomy and mathematics.[19]

In a second pair of definitions, "art" still appears in distinction to "science," but without such a clear sense of the subordination of the first term to the second. When Comte published the last of the six volumes of the *Cours* in 1842, for example, he promised that the positive ascendancy of "the sociological spirit" over "the mathematical spirit" would ultimately eliminate the "radical opposition" that "ancient philosophy" had established between "the active point of view and the speculative point of view."

As a sign of the benefits that would flow from "the most permanent intimate harmony" that his "new [positive] philosophy" would establish between action and speculation, Comte predicted an explosion of progress in "the two most difficult and most important arts, the medical art and the political art."[20] When he wrote the preamble to the *Discours sur l'ensemble du positivisme* that he published on its own in 1848 and converted into the first half of the first volume of his *Système de politique positive* in 1851, he proclaimed, similarly, that "the fundamental mission of positivism" would be "to generalize real science and to systematize social art," two projects that he characterized as "two inseparable faces of a same conception."[21] Presenting positivist philosophy and politics as the only solution to the persistent cycle of revolution and counterrevolution that had marked French history since 1789, he recommended a new focus on "the science and the art of sociability."[22]

While these binary comparisons, contrasts, and connections between "science" and "art" are intriguing, however, Comte's third and most common usage of the word *art* appears in the phrase "beaux arts" or "fine arts," the intermediate term in a series of shifting tripartite distinctions that became especially controversial as Comte increasingly explored the nature, significance, and interaction of the areas that he named over the course of his intellectual career as the "aesthetic life," the "affective life," and the "moral life." In his *Plan des travaux*, for example, he described the ideal relationship between "scientists [*savants*]," who determined a plan; "artists," who worked to encourage its adoption; and "industrialists," who put it into effect by designing the necessary "practical institutions."[23] In his *Cours de philosophie positive*, he presented the "aesthetic faculties" as an intermediary between the "affective life" and "the intellectual life," a link between the "purely moral faculties" and the "properly intellectual faculties," a force that was particularly important for his new educational projects and proposals because it had the power to work on both the "mind" and the "heart."[24] In his *Discours sur l'ensemble du positivisme* and his *Système de politique positive*, finally, he featured the "fine arts" in a third triple combination that identified "philosophy, politics, and poetry" as "the three great creations of humanity"; promised a "true general theory of fine arts whose principle consists in placing poetic idealization[s] in between philosophical conception[s] and political realization[s] in the positive coordination of the fundamental functions of humanity"; and clarified that poetry, as the single most important of the five branches of the fine arts, should have "a systematic position between philosophy and politics, emanating from one and preparing the other."[25]

Comte devoted increasing time and attention to the important social role of art as he moved from his *Plan*, through his *Cours*, to his *Discours* and his *Système*. Regardless of how much time and attention he devoted to the fine arts, however, and regardless of how much importance he accorded them in his history, philosophy, and politics, he consistently subordinated art and imagination to science and observation even when he claimed that he was elevating them to a new and more important position. Especially in his early works, but also even in his later ones, poetry might inspire action, but only if it respected philosophy first.

When Comte distinguished between "observation" and "imagination" in the part of the *Plan* where he discussed his proposed "political science," for example, he stressed the importance of the first of these two operations over the second when he specified not only that observation would guide the "formation" of the new social system, but also that imagination would provide only for its "propagation." Observation would discover the regular relationships between apparently isolated facts, while imagination would propose provisional connections only until observation could establish the real ones. Observation would determine "the general plan," he explained, and imagination would take on the "secondary function" of filling in the details.[26]

As Comte moved to the section of the *Plan* where he talked about how to put his "scientific politics" into practice, he designated the work of imagination as "equally indispensable" with that of observation on the grounds that only imagination could provide a view of the ideal positive society that would be appealing enough to inspire ordinary individuals to the cause of social reform by "driving back egotism," "pulling society out of apathy," and encouraging the internal "moral revolution" without which his new system could not succeed. Even as Comte promised that the imagination would "play a preponderant role" in the process of social reconstruction, however, he also specified that "it would exercise its action in the direction established by scientific research." Similarly, even as he proposed that the work of the imagination should be "left entirely to itself" in the cause of inspiring social action, he also specified that its only purpose was to support the goals of positive philosophy, "the adoption of [the system] . . . that will have been determined by positive politics."[27]

While Comte claimed that the role of imagination and the fine arts would be more substantial in his positive society than it had been in any of the earlier stages of civilization, then, the role that he envisioned for them in the *Plan* remained a decidedly subordinate one.

A similar dynamic of valorizing the fine arts while subordinating them to the influence of positive philosophy is also apparent in some of the most significant portions of the *Cours de philosophie positive*. As Comte outlined his method for the analysis of social dynamics that he published in volume 4 in 1839, for example, he argued that "the general history of *philosophy*" was "necessarily" more important than any other branch of the "intellectual history" of civilization, "even the history of the fine arts, including poetry."[28] As he introduced the history of the fine arts themselves into the more general history of civilization that he published in volume 5 in 1841, he began by going out of his way to reject the "irrational exaggeration" of certain unidentified contemporary thinkers who idealized the influence of the fine arts in classical antiquity both by stressing that "the great Homer . . . was certainly not a philosopher or a sage, still less a pontiff or a legislator" and by reminding his readers that Plato had excluded poets from his *Republic*.[29]

While Comte acknowledged that "poetic influences" had contributed to the extension and consolidation of civilization during the medieval, or "theological," period of history, similarly, he stressed that these influences alone would never have been able actually to "establish" it: "the faculties of expression have never been able directly to dominate the faculties of conception."[30] Finally, as he moved on to emphasize the ways in which the Enlightenment philosophers of the "metaphysical" period had destroyed the old systems of "theocratic" European civilization without successfully establishing new ones, his list of criticisms included the ways in which eighteenth-century intellectual leadership had passed from genuine "thinkers" to pretentious "literary figures," facile popular writers who had simply "set themselves up as philosophers" without undertaking the necessary "long and painful meditations" that valid philosophical claims would have required.[31] "Whatever the social importance of the fine arts . . . and although the future reserves an eminent mission for them, which I will indicate directly at the end of this volume," he concluded in his assessment of the laws of the "positive state" or "age of generalization" in volume 6 in 1842, "there is no doubt that the aesthetic point of view is less general and less abstract than the scientific or philosophical point of view."[32]

Unlike the conclusion to Comte's earlier *Plan*, with its purely instrumental view of the social role of art, Comte's promised conclusion to the *Cours* initially had more to say about how positive philosophy would serve poetry than about how poetry would serve positive philosophy. Seeking to refute those who had argued that "the positive spirit" was antiaesthetic, Comte claimed that once positive philosophy had moved from its "mathematical phase" to its "completely sociological systematization," it would be as good for art as it was for every other aspect of civilization, providing "the principal basis for an aesthetic organization no less indispensable than the mental and social renovation from which it is necessarily inseparable."[33] Comte predicted that positive philosophy would provide new subjects, new social functions, and new audiences for poets, who would be able to find a "fruitful source of new and powerful inspiration" in the sight of "the prodigies of man, his conquest of nature, and the marvels of his sociability," topics that would be particularly appealing to a modern public who had come to appreciate the whole series of purely human inventions that ran from "simple mechanical apparatuses" to "sublime political constructions." In addition to providing "modern art[ists]" with "an inexhaustible nourishment in the general spectacle of human marvels," positive philosophy and the society that it inspired would provide them with "an eminent social destination, [that of] making [people] better appreciate the final economy."[34]

While Comte's *Cours* gave more scope to the fine arts than his *Plan*, however, he still saw art ultimately only in terms of its social utility. In the same sentence in which he promised that modern art would be able to keep "the precious resources of its fictive beings," for example, he also reminded his readers that its "fundamental obligation" would nevertheless be to "subordinate all its conceptions to the ensemble of real laws."[35] Like the biologists who invented "imaginary organisms" to help them further their practical scientific experiments, in other words, artists could use the creative tools in their "aesthetic employ" to imagine modifications of the "human organism," but only if they stayed within the limits of what positive scientists had already determined to be "the fundamental laws of reality." "Aesthetic participation" might constitute "an important political office" in the preferred "positive future," but only if it followed the guidelines of "dogmatic philosophy" first.[36]

When Comte wrote the preface to his *Discours sur l'ensemble du positivisme* in 1848, he seemed to emphasize a new place for art in his philosophy when he identified "the decisive evolution" of the new *Discours* relative to the earlier *Cours* as "the happy ascendancy that positivism [now] accords directly to feeling, and even to the imagination."[37] Even though Comte devoted far more time and attention to the social role of art and the science of aesthetics in the *Discours* and the subsequent *Système de politique positive* than he had in the *Cours*, however, philosopher Annie Petit, who has produced one of the most recent critical editions of the *Discours*, has observed that the results are, at best, "ambivalent."[38] Despite their higher status in Comte's later works, poetry, art, and imagination actually remain just as subordinate to philosophy, science, and social science in Comte's second system as they did in his first.

Although Comte announced at the outset of his chapter on "the aesthetic aptitude of positivism" that "aesthetic functions have too much importance to be neglected in the final regime of humanity," for example, he started his discussion of the arts by repeating the same "immutable" law of social statics that he had already enunciated in the *Cours*: "Our faculties of representation and expression are necessarily subordinate to our faculties of conceptualization and combination."[39] Building on his earlier arguments about the limitations of art as a guide for politics, similarly, he criticized "the current delirium of poetic pride" that characterized contemporary literary figures and claimed that only his approach could both "explain" and "refute" the "anarchist utopias" of what he characterized in deeply derogatory terms as an "aesthetic pedantocracy."[40]

Although Comte now acknowledged that "the aesthetic genius" and "the scientific genius" were so closely connected that philosopher Denis Diderot could have been a poet and poet Johann Wolfgang von Goethe could have been a philosopher, he nevertheless continued to privilege the fields of philosophy and science in his assertion that "the true directors of the spiritual realm" in the new system of positive society and religion "would continue to be named philosophers more than poets because their ordinary attributes are more scientific than aesthetic." While he imagined an ideal future system in which the best philosophers might, if necessary, "pass from scientific activity to [a new] aesthetic activity" to become the best poets for the new society, he described no parallel reverse process through which the best poets might, if necessary, pass from aesthetic activity to scientific activity to become the best philosophers for it.[41] Indeed, as Petit has pointed out, while Comte raised the importance of the arts when he imagined a new educational system in which everyone would study the arts at a general level, he also diminished the importance of the arts when he hoped to create a new social system in which very few people would ever study them at a higher, more advanced, or more professional level.[42]

While Comte now promised that positivism would "call art to its best destination," give poetry "a dignity that had never been possible before," and privilege "aesthetic impressions" over "scientific impressions" in the elaborate series of public festivals and private rituals that he planned to feature in his new Religion of Humanity, finally, he closed his chapters on aesthetics in the *Discours* and the *Système* by explaining the social importance of art in the positivist project in the same instrumental ways that he had already indicated in the *Plan* and the *Cours*. "The principal function of art is always to construct the types with whose bases science furnishes it," he proclaimed. Even as he described the ways in which "positivist art," "the modern genre of poetic compositions," and "the new poetry" would inspire the peoples of Western Europe to create the positive new "normal state of humanity," he also stressed the ways in which these artistic initiatives would fulfill an "accessory office" to the work of science, take place "under sociological inspiration," and act "guided by philosophy."[43]

In his initial *Cours de philosophie positive*, Comte had repeatedly linked the triple series of the "moral," the "intellectual," and the "aesthetic" to the triple series of "the good, the true, and the beautiful," the same key triad that contemporary French figures such as philosopher Victor Cousin and poet Théophile Gauthier were using as the basis for their influential doctrine of art for art's sake.[44] Focusing on the importance of art for society's sake instead, Comte's first system subordinated the study of "the beautiful" to the needs of "the good" and, especially, "the true." While Comte's new emphasis on "the aesthetic aptitude of positivism" in the *Discours sur l'ensemble du positivisme* and the *Système de politique positive* seemed to reverse this pattern when it promised that the scientific study of "the true" would be limited to whatever the moral and aesthetic study of "the good" and "the beautiful" required,[45] the final pages of the conclusion that the *Discours* and the *Système* shared still announced that "the aesthetic cult" of the positivist future would be "worthily subordinated to the scientific cult."[46]

Reason and Emotion: The Affective Life in Comte's Philosophy and Politics

Comte made a close connection between his views on art and his views on emotions, especially when he explored the relative importance of the two conceptual categories that he labeled as "the intellectual life" and "the affective life" and the two attributes of the human person that he characterized as "the mind" and "the heart." In the *Plan des travaux*, for example, even as he stressed that reason and philosophy should guide imagination and art, he also acknowledged the unavoidable importance of emotion in setting up the ideal society when he stressed that "no one will ever impassion the mass of men for any system [just] by proving to them that it is the one whose establishment the progress of civilization has prepared from the beginning of time." While "a small number of minds" might accept "positive demonstrations" as convincing proof of the direction that had been laid out by "scientific politics" for them to follow, Comte argued that most people would require an emotional appeal instead, the sort of heightened attraction to an ideal that only the fine arts could inspire, the sort of "lively and seductive persuasion [that could only be] produced by ideas that move passions." "The only way to achieve the latter effect," he concluded, would be to use the emotional power of the imaginative arts "to show men a vivid picture of the improvements that the new system should bring to the human condition."[47]

Furthermore, just as Comte's discussions of art became progressively more extensive in the *Cours de philosophie positive*, the *Discours sur l'ensemble du positivism*, and the *Système de politique positive* than they had been in the *Plan*, so also did his discussions of emotion. When he initially published his first defenses of the emotional "feminine influence" and "aesthetic aptitude" of positivist philosophy in the *Discours* in 1848, for example, the title page of the volume still highlighted the political goals that had inspired his earlier philosophical work on "social physics" and the intellectual life in the *Cours* by featuring the original positivist slogan "Order and Progress."[48] When he reprinted the *Discours* at the beginning of the first of the four volumes of the *Système* in 1851, by contrast, he suggested his increasing interest in the importance of the affective life in a more visible way when he expanded the first epigraph on the new title page to read "Order and Progress; Live for Others."[49] He emphasized the importance of the emotions even more dramatically in a second epigraph that he added to the new title page of the *Système* at the same time, a phrase that he would identify in the second of the *Système*'s four volumes in 1852 as "the sacred formula of positivism"[50]: "Love as the principle, and Order as the base; Progress as the goal."[51]

While Comte's successive treatments of art consistently subordinated them to the needs of science, the discussion and development of his attitudes toward the emotions was more complex and less linearly straightforward. While the *Cours* largely privileged intellect over affect, for example, it also underscored the enduring importance of the emotions in key passages that suggested that "intellectual activity" and "the sympathetic instinct" would always share a "common social insufficiency" unless they worked together "mutually."[52] Even when the subsequent *Discours* and, especially, the final *Système* eventually claimed to privilege affect over intellect, similarly, they never failed to present the accomplishments of positive science and social science as a crucial prerequisite for the positive study of ethics and the Religion of Humanity's emphasis on emotions. While both the *Discours* and the *Système* went on to celebrate women for their especially close connections both to art and to emotion, they nevertheless still accorded most of the familial, political, and religious power to men in their varied capacities as husbands, fathers, philosophers, priests, and poets. In certain key ways, then, it seems that Comte persistently privileged thought over feeling in much the same way that he privileged science over art.

Whether Comte was talking about the individual, the family, the society, or the hierarchy of knowledge itself, for example, the *Cours de philosophie positive* almost always asserted the importance of reason over emotion. While he used and admired the work of physiologist Franz-Joseph Gall and phrenologist Johann Spurzheim, who had argued that "affective functions" took up most of the brain, for example, he also insisted that it was the "eminently reflective faculties, the most elevated of all" that were "the principal characteristic attribute of human nature."[53] When Comte subordinated women to men in his social statics, he explained that "positive biology" showed that emotional women existed "in a sort of continuous state of childhood" while rational men constituted "the ideal type of the race."[54] When he subordinated "the history of the fine arts" to "the general history of philosophy" in his social dynamics, similarly, he explained that it was because "the faculties of expression," which were "more intimately connected to the affective faculties" and shared the same part of the brain, had always "had to be subordinated . . . to the [more intellectual] faculties of direct conception."[55]

Although Comte acknowledged in the final volume of the *Cours* that the "philosophic or scientific spirit" and the "aesthetic or poetic spirit" were both parts of one "common contemplative spirit," he nevertheless continued to privilege the first category over the second by virtue of his claim that philosophy and science had a closer connection to "the fundamental conceptions [that were] destined to direct the universal exercise of human reason." Putting philosophers ahead of artists, he even went so far as to describe them as more fully human by virtue of their more extensive rational abilities: "In the philosophical class, the human type necessarily approaches its characteristic perfection by a superior expansion of the faculties of abstraction, generalization, and coordination, [the faculties] that certainly constitute the principal preeminence of humanity over animality."[56] As he summed up his project in the final chapter of the last volume of the *Cours* by looking at the results of positive philosophy for "two essential modes of human life," "two very distinct points of view," and "four classes of general considerations," reason consistently took first place in his presentation yet again: the "mental" modes came before the "social" ones; the "scientific" views came before the "aesthetic" ones; and the "scientific, or rather, rational" considerations were first on his list, while the "aesthetic" considerations, "destined [not to create, but only] to spontaneously reflect the ensemble of the diverse human aspects, social as well as intellectual," were last of all.[57]

In the *Discours sur l'ensemble du positivisme* and the *Système de politique positive*, by contrast, reason started to take second place to emotion most of the time. So, for example, in Comte's introductory summary of "the fundamental spirit of positivism," he stressed that "neither reason, nor even activity, could constitute the true human unity, whose individual, and above all, collective economy could never rest on anything but sentiment."[58] As he looked to the future, similarly, he predicted that "the new [positive] philosophy" would gradually "become more moral than intellectual," and that it would "place the affective life in the center of its . . . systematization."[59] Once the "positive principle" had achieved its "full maturity," he asserted, its "fundamental dogma" would focus on "the continued preponderance of the heart over the mind."[60] The final pages of his "general conclusions" proposed that true positivists should no longer speak "in the name of developed reason alone," but should also now advance the worthy claims of "sentiment."[61]

As Comte moved to discuss the Religion of Humanity itself later in the *Système*, similarly, he consistently subordinated reason to emotion there as well. So, for example, he predicted that the "reign of faith," which had been based on "the mind," would give way to a "reign of love" that would be based on "the heart."[62] The six-stage model of science that he had described in the *Cours* as culminating in the study of sociology would give way to a new seven-stage model that would culminate instead in the new study of ethics.[63] The positivist "sociocracy" would finally end the French revolutionary cycle by eliminating what Comte had identified as its chief characteristic, "the increasing insurrection of intellect," while "positive ethics" would put feelings first instead, instituting "the final preponderance of sentiment."[64] In Comte's ideal positivist future, then, love would come to be seen not only as the guiding principle but also as the ultimate goal of both order and progress.

Even as both the introductory *Discours* and the complete *Système* privileged the role of emotions and the affective life, however, Comte also devoted shorter sections of each work to the ways in which reason would remain critically important for the new religious, scientific, social, and sexual order. On the one hand, for example, he seemed to subordinate reason to emotion when he stressed that "the mind should address only the questions posed by the heart for our just needs." On the other hand, however, he also suggested a more complicated relationship between the two powers of mind and heart when he stated in both texts that "the mind should always be the minister of the heart, and never its slave" and highlighted on the title page of each publication that "the mind is the minister of the heart, but not its slave."[65] Michel Bourdeau, who has devoted particular attention to the significance of these evocative slogans, has concluded that the new approach that Comte elaborated in the *Discours* and the *Système* required not only "the necessity of a moral impulsion for the intelligence" but also the necessity of "an intellectual discipline for the emotions."[66]

The potential insufficiency of emotion alone in the *Discours* and the *Système* is especially clear in the sections of the shared chapter on "the feminine influence of positivism" that Comte devoted in each work to the relationship between women and men.[67] On the one hand, for example, he seemed to stress the new subordination of reason to emotion when he dramatically reversed the significance of the link between women and the affective life that he had asserted in the *Cours* by proclaiming that a superior "sociability" actually made women, not men, "the purest and the most direct type of humanity." On the other hand, however, even as he now admitted that "this [female] sex is certainly superior to ours [the male sex] as regards the most fundamental attribute of the human species," he also instantly asserted that such a "natural preeminence" could never be enough to constitute a "social ascendancy." Women might have the "affective superiority," but "the ordinary laws of the animal kingdom" still ensured that men would retain certain other key sorts of superiority by virtue of the fact that they were stronger "in every kind of force, not only [in the strength of] the body, but also [in that] of mind and character."[68] In the end, then, despite the superior claims of emotion in Comte's second system, men would maintain their social authority and political control by virtue of the superior claims of their strength, their energy, and, perhaps most important of all, their reason.

Rational Men, Emotional Women: Gender in Comte's Philosophy and Politics

As the passages above already suggest, Comte frequently linked the distinction between reason and emotion to the distinction between men and women. While he was certainly not alone in this set of gendered associations, which was common to a wide range of nineteenth-century French thinkers from conservatives such as novelist René de Chateaubriand to radicals such as historian Jules Michelet, Comte's analysis of the implications of this pair of associations was particular to him alone, especially in his discussions of the positive polity and the Religion of Humanity. This becomes particularly clear if one considers his ideas in comparison to those of the utopian socialist Saint-Simonian school whose work had inspired some of his most extensive criticism ever since his break with Saint-Simon himself around the publication of the *Plan des travaux scientifiques nécessaires pour réorganiser la société* in 1824.

While Saint-Simon himself died in 1825, his surviving students continued to elaborate on his projects for social renewal and reform as they founded what historian Claire Goldberg Moses has characterized as "first a 'school,' then a 'society,' and finally a 'religion.'"[69] As figures such as Saint-Amand Bazard and Prosper Enfantin elaborated increasingly controversial proposals for the reorganization of marriage and family life, Comte responded to these Saint-Simonians' new guidelines for the social, sexual, and religious renovation of society by using his *Cours de philosophie de positive* to criticize them as the work of "audacious sophists who [had] . . . tried to raise a metaphysical axe against the elementary roots of the social order."[70] Even when he transformed his own emphasis on philosophy, reason, sociology, and science into a new emphasis on art, emotion, ethics, and religion in the *Discours sur l'ensemble du positivisme* and the four volumes of the *Système de politique positive* that he produced from 1848 to 1854, he continued to attack the Saint-Simonians' earlier ideas on gender, sex, and society as "vain retrograde utopias" and dangerously "subversive dreams."[71]

Saint-Simon's own rare, but potentially significant, comments on women's rights and roles in a renovated society had included a proposal to include women in the Conseil de Newton that he hoped to assemble as a guide for European reform, a meditation on the need for the moral regeneration of society that he began but did not live long enough to complete in his posthumously published *New Christianity*, and the famous statement that "man and woman together create the social individual," which he supposedly made on his deathbed in 1825.[72] As Bazard and Enfantin took up the direction of the school that Saint-Simon's work had inspired, they not only placed increasing importance on the exploration of proper relations between men and women but also attracted an increasing number of women to their work. By 1831, one-third to one-half of the participants in the Saint-Simonian school were women of the middle and working classes who acted with the men of the movement in mixed couples to establish "common houses" in the poorest neighborhoods of Paris, to recruit sympathetic members from across the city, and to set up programs for worker education and assistance. Among the members of the unemployed and working poor that male-female pairs such as Henri Fournel and Claire Bazard recruited to the cause, the gender breakdown was roughly half and half, with men mostly artisans and women mostly textile workers.[73]

From 1825 to 1831, Bazard and Enfantin's Saint-Simonian movement stressed not only the complementarity of men and women but also their potential equality in difference.[74] Each section of the group had both a male and a female director, for example, and each educational meeting had both a male and a female discussion leader.[75] Over the course of 1831, however, Enfantin started to challenge this approach as he centralized the organization, established male control, and elaborated the two especially controversial notions that became known as "the rehabilitation of the flesh" and "the wait for the woman." In the first of these doctrines, Enfantin argued that just as men had been falsely seen as superior to women, so the spirit had been falsely seen as superior to the flesh. Speaking in the name of both sexual equality and sexual liberation, he proposed that women as well as men should have the right not only to marry but also to divorce, to remarry, and even, under certain circumstances, to take multiple sexual partners without ever marrying at all.[76] In the second of these doctrines, however, he seemed to contradict the radical egalitarianism of the first one when he not only established himself as the male pope of a new Saint-Simonian church but also argued that the movement should reorganize under exclusively male leadership until he had found the right "feminine Messiah" to serve as the female pope at his side.[77]

Enfantin's innovations inspired a rupture with Saint-Amand Bazard, Claire Bazard, and other major figures in the movement, which split the Saint-Simonian school in two. While men as well as women rejected Enfantin's ideas, women were particularly likely to decide that they needed a new movement of their own in reaction to their unexpected exclusion from his hierarchy. In the hands of some of the most active Saint-Simonian women, then, utopian socialism turned into socialist feminism.[78] Seamstresses Désirée Veret and Reine Guindorf, for example, founded and edited the pioneering periodical known successively as *La Femme libre* (The free woman), *L'Apostolat des femmes* (The apostolate of women), *La Femme nouvelle: l'affranchissement des femmes* (The new woman: the enfranchisement of women), and, for most of its two-year run from 1832 to 1834, *La Tribune des femmes* (The tribune of women), the first French publication in which women wrote all the articles.[79] Writer Eugénie Niboyet responded to the Revolution of 1848 by founding *La Voix des femmes* (The voice of women), a daily newspaper that inspired the creation of a political club by the same name in the early days of the new Second Republic.[80] Jeanne Deroin, who had written for all of these earlier publications, not only took advantage of the new political climate of the Second Republic to found her own weekly, *La Politique des femmes* (The politics of women), and monthly, *L'Opinion des femmes* (The opinion of women), but also defied the Constituent Assembly's rejection of women's suffrage in April 1848 by presenting a public platform and claiming the right to run for political office in the new Legislative Assembly's first set of national elections in April 1849.[81]

While Comte always shared the Saint-Simonians' sense that men and women could unite to form perfectly complementary couples, he consistently opposed the controversial defenses of sexual freedom that emerged out of the Saint-Simonian movement. Furthermore, although he had once read and appreciated both Mary Wollstonecraft's famous *Vindication of the Rights of Women* (1792) and the Marquis de Condorcet's equally significant defense of women in the *Esquisse d'un tableau historique des progrès de l'esprit humain* (1794), his thoughts had started turning away from such defenses of sexual equality as early as 1825, when he wrote his friend Pierre Valat that "men of merit" should choose wives who combined "a certain intellectual mediocrity" with "attachment, devotion of heart, and sweetness of character."[82] Comte's increasingly unhappy marriage to the intelligent and independent Caroline Massin combined with his political opposition to the liberal individualism of the July Monarchy to convert him into an opponent of women's emancipation and equal rights.[83]

Although the *Plan des travaux scientifiques nécessaires pour réorganiser la société* had said nothing on the topic of men's relationships to women, then, the three volumes of the *Cours de philosophie positive* that Comte devoted to social statics, social dynamics, and general conclusions featured a variety of striking statements about the real and ideal relationships between men and women in the family and society that served to highlight Comte's utter rejection of the Saint-Simonians' positions on these same topics. Indeed, even though Comte's own split with Saint-Simon had been acrimonious in the extreme, he felt so strongly about the error of the Saint-Simonians' dual support for sexual liberation and sexual equality that he devoted a portion of his "personal preface" to the final volume of the *Cours de philosophie positive* in 1842 to clearing Saint-Simon himself of any and all responsibility for both the "shameful ephemeral aberrations that [Saint-Simon's followers] had dared to introduce under his name after his death" and the "profoundly subversive ... maxims that were subsequently impudently attributed to him [Saint-Simon] by tricksters whom he had never known."[84] In the conclusion to this same volume, Comte reacted to these unnamed Saint-Simonians' "dangerous aberrations" by claiming that "as far as domestic morality, a decisive comparison will soon no doubt produce an appreciation of the spontaneous superiority of positive philosophy."[85]

Comte based his own view of "domestic morality" in the *Cours* on the analysis of the legitimate family, which he characterized as "the true social unit," and especially on the analysis of husband and wife, "the elementary couple" that formed the family's "principal base." While Comte initially characterized this fundamental couple in complementary and potentially egalitarian terms as "a sort of complete fusion of two natures in one," he went on to insist upon the importance of familial, sexual, and social hierarchy instead: "No matter what vain notions one forms today about social equality, every society, even the most restricted one, requires ... not only diversities, but also inequalities." "From this point of view," he asserted, "the sociological theory of the family can be reduced to the rational examination of two fundamental orders of necessary relations, that is: [first] the subordination of the sexes, and then that of the ages." Starting with "healthy biological philosophy," especially Gall's "important theory" about the "anatomical," "physiological," "physical," and "moral" differences between the sexes, Comte used the insights of his new "positive biology" to reject the whole range of what he characterized as "chimerical revolutionary declamations on the supposed equality of the two sexes." "Completing this indispensable scientific assessment," he concluded, "sociology will demonstrate after its own fashion . . . the radical incompatibility between all social existence and this chimerical equality of the sexes by characterizing the special and permanent functions that each of them is supposed to fulfill in the natural economy of the human family."[86]

Just as the *Cours* clearly subordinated art to science and emotion to reason, then, so also it subordinated women to men, and for many of the same reasons. Looking at "the feminine organism" from "the social point of view," for example, Comte asserted that one could easily see both women's "fundamental inferiority" and their "secondary superiority." His list of "fundamental" categories in which women were inferior was long, including not only the all-important, reason-related "intellectual faculties," "speculative activity," and capacity for "mental work," but also additional categories such as "the fine arts, even under the most favorable circumstances," and "the functions of government, even if they were reduced to the most elementary state, and were only relative to the general conduct of the family alone." The corresponding list of "secondary" categories in which women were superior, while significant, was nevertheless quite short by comparison: "the affective life," "sympathy," and "sociability."[87] Coupling his biologically based sociological conclusions with his philosophically inspired study of gender relations from primitive times to the present, Comte asserted that marriage should be monogamous, that men should work to support women, that women should not work themselves, and that the best societies would always be those in which marriage was permanent, divorce was illegal, and women took no role in public life, politics, or religion.[88]

Just as the *Discours sur l'ensemble du positivisme* and *Système de politique positive* initially seemed to demonstrate a fundamental break in Comte's thinking when they turned from the early focus on science, reason, sociology, and positive philosophy to a new focus on art, emotion, ethics, and positivist religion, so also the two works initially seemed to demonstrate another fundamental break when they turned from the related early focus on a rational male standard to the interconnected new focus on a new emotional female standard. Where the first volume of the *Cours* featured a dedication to Comte's "illustrious [male] friends" Baron Joseph Fourier and Professor Henri Marie Ducrotay de Blainville at the Académie Royale des Sciences, for example,[89] the first volume of the *Système* featured a dedication to "the sacred memory" of his "eternal [female] friend, Madame Clotilde de Vaux (née Marie)."[90] Where Comte's first "author's announcement" to the *Cours* opened with his thanks to a series of public male scientists that included Alexander von Humboldt, Blainville, Louis Poinsot, Fourier, Henri Navier, François Broussais, Jean-Étienne Esquirol, and Jacques Binet,[91] his opening "preface" to the *Système* focused on a trinity of private female associates that featured his "Saint Clotilde [de Vaux]"; his "noble and tender mother," Rosalie Boyer; and, most strikingly of all for its apparent challenge to any system based on intellectual training, scientific understanding, or masculine philosophical investigation, "his daughter of choice," Sophie Bliot, the domestic servant whom he both characterized as "the eminent proletarian woman who deigns to devote herself to my material service without suspecting that she also offers me an admirable moral example" and praised for "the happy inability to read that . . . brings out not only her affective superiority, but also the rectitude and the penetration of her mind, which has spontaneously used all the advantages of a wise feminine experience."[92]

In his new chapter for the *Discours* and the *Système* on "the feminine influence of positivism," Comte promised that the positive polity would be especially good for women, offering them not only "noble social career[s]" but also "just personal satisfactions."[93] Challenging the Civil Code that had served as the basis of French family law without interruption since 1804 as a "coarse evaluation" of women that had been "brutally formulated" by the "retrograde hero" Napoleon, Comte rejected the code's identification of marriage with fertility and family inheritance in favor of his own description of marriage between complementary male and female opposites as "the best type of true friendship," "the principal source of moral improvement," and "the essential basis of true human happiness."[94] Furthermore, unlike the code, which stressed not only a husband's absolute marital authority over his wife but also a father's absolute paternal authority over his legitimate children, Comte hoped that his own "positive theory of marriage and the family" would inspire a new arrangement in which the direction of children's education would shift from the hands of husbands and fathers into the hands of wives and mothers, women whose superior aptitude for "tenderness" best qualified them both to hold the "spiritual power" in the family and to provide the "education of the sentiments" that constituted the true source of all "moral education."[95]

While Comte stressed that emotional women's primary role would be within the family, he also imagined several key extensions of this tender domestic role that would enable them to extend their moral influence through the whole of society as well. As the only leaders of a new series of "positivist salons," for example, they would have the chance to use their "sweet moral influence" to discourage philosophers from any "misguided ambition" or "prideful ramblings" and to discourage proletarians from any "reappearing seeds of violence or envy."[96] Similarly, as the holders of six special seats for "ladies of the elite" on the Positive Occidental Committee that was to serve as the "permanent council" of the new Religion of Humanity, they would be able to participate in advisory conversations with the positivist representatives of countries across Europe and, eventually, around the world.[97]

Most strikingly of all, as Comte described the advent of the new Religion of Humanity itself, he not only projected that "each woman would become for each man the best personification of Humanity," but also asserted that "only the worship of Woman, first private and then public, could prepare men for the real worship of Humanity."[98] Indicative of women's crucial new position in the positive polity of the future, Comte's initial plan for its new international flag in 1848 featured the face of a woman as the "symbol of Humanity" over the "fundamental" slogan of "universal love."[99] When he replaced this single flag with two separate flags for religion and politics in 1851, similarly, the new "religious flag" featured "the symbol of Humanity personified by a thirty-year-old woman holding her son in her arms," while the new "political flag" floated from a standard that had been adorned at its top with a "statuette of Humanity," presumably also female in form.[100]

While the *Discours* and the *Système*, with their valorization of art and emotion, offered more scope for women in the family, the positive polity, and the Religion of Humanity than the *Cours*'s valorization of science and reason, however, women still remained subordinate to men at almost every turn even in this second system. In radical contradistinction to the utopian socialists and socialist feminists in the Saint-Simonian school, who had continued to challenge traditional French marriage and family practices by campaigning for the legalization of divorce even after the split between Enfantin and Bazard, for example, Comte privileged indissoluble monogamous marriage so strongly that he not only rejected divorce but also recommended perpetual widowhood after the death of either spouse and eventual burial for husband and wife together in a common casket.[101] While he wanted to free women from the pressure to marry by eliminating dowries, he also planned to isolate them from the potential for economic independence by eliminating female inheritance.[102] While he wanted to ensure their freedom from work by requiring their husbands to support them, he also wanted to eliminate their freedom to work by refusing to let professional schools accept them.[103] While he wanted to give mothers more influence over their children's initial education in "aesthetic exercises" and "spontaneous morality," he never imagined women in the corps of philosopher-teachers who would be in charge of children's more advanced education in topics such as "healthy historical theory," "scientific preparation," and "systematic morality."[104] While he opened new intellectual vistas for girls when he stressed that their education should be "almost uniform" with that of boys, he nevertheless imagined that even educated women would remain subordinate to educated men.[105]

As was the case in the family, so was it also in the larger society, in the positive polity, and even in the Religion of Humanity. While Comte provided for six women on the Positive Occidental Committee, for example, these women would always constitute a minority in a body whose ideal membership was projected to grow from thirty-six to sixty.[106] While he imagined that women would be "spontaneous priestesses" with a special access to the emotions, similarly, these female priests would serve only as "worthy" auxiliaries (and potential wives) for the fully trained "systematic" male priests who would retain special responsibilities for both science and art, both fields in which only men could excel.[107] Even on the symbolic level, as Annie Petit and Bernadette Bensaude-Vincent have pointed out, "women enter[ed] the [positivist] calendar [only] as exceptions," serving as the subject of only 21 days out of 365 in the initial "concrete calendar" that was supposed to usher in the "final transition" to the positive "occidental republic" and as the subject of only one month out of thirteen in the final "sociolatric table" that was to serve as the ultimate "summary of the universal adoration of Humanity in eighty-one annual festivals."[108] In this light, then, the crucial point about Comte's idealized feminine representation of Humanity is not so much that she is herself a woman as that she is a woman who is mothering a son who will grow up to become a man.

Science, Reason, and the Positive Priesthood: Continuity and Change in Auguste Comte's Philosophy and Politics

Where earlier generations of Comte scholars once stressed the discontinuity between Comte's initial *Cours de philosophie positive* and his subsequent *Système de politique positive*, contemporary Comte scholars are more likely to stress the continuity between the first system and the second system instead.[109] Perhaps because the Religion of Humanity has been more controversial than the science of society, the defense of Comtean unity has typically relied on explorations of the ways in which the subject matter, the philosophical approach, and the political concerns of Comte's eventual vision for the positive polity are already apparent in his earlier and more acceptable manifestoes for positive philosophy.[110]

While this approach to continuity rightly focuses on the ways in which Comte's later plans for religion and ethics were already present in his early work on science and sociology, however, I would like to conclude by highlighting the merits of an alternate approach to continuity that focuses on the ways in which Comte's early ideas about the positive value of the sciences, philosophy, the intellectual life, and the social reforming role of rational men remain surprisingly present even in the later works in which he supposedly turned so decisively toward the defense of the arts, poetry, the affective life, and the social reforming role of emotional women. On the one hand, for example, Comte's acknowledgment of the social role of artists is present as early as the *Plan des travaux*. On the other hand, however, Comte's refusal ever to allow the practice of art for the sake of art is present even in the *Système de politique positive*. Even in the final volume of the *Système*, where Comte concluded both that the "positive priesthood" at the head of the Religion of Humanity would represent a "necessary fusion between philosophy and poetry" and that its members would be "interchangeably labeled as philosopher[s] or as poet[s]," he nevertheless presented this new dual combination as resulting from the work of "healthy philosophy" alone.[111]

The pattern of continuity and change is more complicated in the case of the relationship between Comte's earlier and later work on the relative importance of reason and emotion. Even as the *Cours de philosophie positive* criticizes the "scientific pedantocracy" for its overspecialization, stresses the necessity of sciences that are socially useful, and codifies that usefulness in the slogan "Science, from whence comes foresight; foresight from which comes action," it also holds open a possibility for the practice of science for the sake of science when it reminds readers that "we would form a very imperfect idea of the sciences if we conceived of them only as the basis for the arts."[112] This latter possibility is one that the *Discours sur l'ensemble du positivisme* and the *Système de politique positive* decisively eliminate from the early pages of their shared opening chapter on "the fundamental spirit of positivism," where both works stress that "the mind is not destined to reign but to serve," that "the satisfaction attached to the discovery of truth alone never has enough intensity to direct [our] habitual conduct [without] the impulsion of a passion," that any "mental impulsion" that might result from "a sort of exceptional passion for pure truth" would remain "profoundly egotistical" if it had no "social destination," and concludes, as it began, by asserting "the fundamental philosophical and political dogma of positivism: the continued preponderance of the heart over the mind."[113]

Even in the positive polity and the Religion of Humanity, however, science, reason, and the intellectual life continue in highly visible and influential roles. The positivist calendar that Comte included in the *Système de politique positive*, for example, begins under the auspices of Moses for the opening month on "the initial theocracy," but closes under the auspices of Bichat as it designates the thirteenth month for "modern science."[114] The calendar for the "abstract worship of Humanity" that Comte outlined in the *Catéchisme positiviste* in 1852 and reprinted as "the sociolatric table" for the "universal adoration of Humanity" in the fourth volume of the *Système de politique positive* in 1854, similarly, may highlight the importance of the affective life when it allocates six months to the celebration of the "fundamental ties" of humanity, marriage, paternity, filiation, fraternity, and domesticity, but even as it describes the month of the woman as the month of "moral providence," it still highlights the importance of the intellectual life when it describes the month of the implicitly male priesthood only as the month of "intellectual providence."[115]

The pattern of significant continuity in spite of dramatic change is particularly apparent in the case of Comte's earlier and later work on the proper relationship of women to men in the family, society, politics, and religion. While emotional woman's absolute social status rises as the status of both art and the affective life also rises from the first system of the *Plan des travaux scientifiques nécessaires pour réorganiser la société* and the *Cours de la philosophie positive* to the second system of the *Discours sur l'ensemble du positivisme* and the *Système de politique positive*, her relative social status remains largely secondary to the still-significant social status of rational man as science and the intellectual life continue to be central to the positive polity. Women may play key roles in the positive family of the future as mothers, sisters, and daughters, but the "sociolatric table" honors each of these positions with only one day each, while it highlights men's bonds of affection as fathers, brothers, and sons with an entire month each instead.[116] Women may serve as "worthy representations" of Humanity by virtue of their superior access to emotion, but the positive polity and the Religion of Humanity continue to underscore the insufficiency of emotion alone when they reserve most of their most significant economic, intellectual, political, and religious roles for men by virtue of their superior strength and, especially, their supposedly superior reason.

This dynamic is particularly apparent in Comte's discussion of the relative places of women and men in the intricately interconnected practices of philosophy, science, and art that appears in the *Discours sur l'ensemble du positivisme* and the *Système de politique positive*. While Comte reserved the practice of science for men alone when he noted that "scientific successes are incompatible with [women's] true nature," he never reserved the practice of art to women alone in the same way.[117] Even as he described "poetic culture" as something that "naturally belongs" to women and argued that "most poetic, and perhaps also [most] musical works" should be seen as the special province of "the loving sex," for example, he also limited women's poetic ability and expertise to the forms of "private poetry," "the genres of poetry that concern personal existence and domestic life," and those "poetic compositions that do not demand intense and prolonged application."[118] Because "great epic or dramatic poems" are "beyond the strength" of women, similarly, the best and only poets for the great "public festivals" of the positive polity and the Religion of Humanity will all still be men, "new philosophers" who will turn their attention from "scientific activity" to "aesthetic activity" as new "social needs" require.[119]

Comte's conclusions to the *Discours sur l'ensemble du positivisme*, the pivotal work that became the link between his first system in the *Cours de philosophie positive* and his second system in the *Système de politique positive*, feature a vivid prescription for the philosophers who are to become priests of his new positive religion: "The priest of Humanity will not develop his necessary superiority over the priest of God until his systematic reason combines [as] worthily with the enthusiasm of the poet as with feminine sympathy and proletarian energy.[120] Unlike the "savants" of the *Plan* or the sociologists of the *Cours*, whose authority derives from their ability to apply the tools of science and reason to the improvement of society and politics, the philosopher-priests of the *Discours* and the *Système* must be not only rational but also enthusiastic, sympathetic, and energetic. Even as Comte's second system clearly accords new importance to poetry, the affective life, and the "tender souls" of women, who serve as representatives of Humanity by virtue of their superior access to the supreme emotion of love, however, a closer examination of his trajectory from the *Plan des travaux scientifiques nécessaires pour réorganiser la société* and the *Cours de philosophie positive* to the *Discours sur l'ensemble du positivisme* and the *Système de politique positive* reveals the persistence of key areas where art still serves to further the ends of science, emotion still requires the crucial assistance of reason, and women still appear chiefly as auxiliaries to men.

8

The Religion of Humanity and Positive Morality

Andrew Wernick

There is no part of Comte's oeuvre that is harder to digest, or indeed take seriously, than the writings associated with what he called, in the preface to the first volume of the *System of Positive Polity*, his second career.[1] As the systematizer of positive philosophy, he had aspired to be the Aristotle of the scientific-industrial age. As the founder of positive religion, he was now to be its Saint Paul.

The pivotal moment, he tells us, was 1845–1846: the brief time of his "purified" relationship with Madame Clotilde de Vaux, of his "moral renovation" in her company, of her untimely illness and death, and of the private cult he immediately made of her inspiring memory, its ritualized effusions already the prototype for the elaborate system of worship that he feverishly began to invent. Personally transfigured, he devoted himself henceforth to establishing the new church, organizing its priesthood (with himself as *le grand-prêtre*, that is, high priest), and expounding its doctrine.

The fullest exposition of Comte's religious project came with the *Système de politique positive*, whose four volumes were published between 1851 and 1854. At one level, this second system was a natural follow-up to the *Cours de philosophie positive*. In the first, we get the positive synthesis of knowledge understood as the historically updated mental basis of social unity; in the second, an account of the reorganized polity that this synthesis and the laws of social development implied. But the prayerful dedication to Clotilde indicates clearly enough that there was a shift in register, as does the work's subtitle: a "treatise of sociology instituting the Religion of Humanity." Not only had sociology, as the abstract science of the social totality, been reframed as "a replacement for theology," but the discourse on "the final religion," which the *Système* unfolds, is interior, as doctrine, to that religion. At the same time, there was a conceptual change, in which the "subjective synthesis" supersedes the "objective" one and a binary schema of theory and practice was expanded into a ternary one through the addition of the sentiments. The synthesis to be aimed at entailed the union of thought, action, and love.[2]

The *Système* is a complex text. It was meant, in the first instance, for adepts who had already labored their way through the encyclopedic scale of knowledge presented in Comte's previous synthesis. It proceeded, moreover, from first principles, and while there was a helpful preamble,[3] the more casual reader would not have a full picture of the religion being instituted until the final volume "on the human future." For the laity, something more straightforward was required. Thus, between the publication of volume 2, on the static laws of sociology, and volume 3 on the dynamic ones, he penned *The Catechism of Positive Religion*, a popular work intended to convey the main points to a wider audience, especially women and workers.[4] It is a remarkable document, a tender instructional dialogue between a (positivist) priest and a younger woman, modeled on a conversation that Comte might himself have had with the saintly Clotilde.[5] Helpfully appended to it were a positivist calendar of festivals and memorials, a schema for sociolatry (the *culte public*), a modified hierarchy of the positive sciences, and a table showing the "Positive Classification of the internal functions of the brain."

To read these texts today with the kind of sympathetic but critical detachment needed to see past all that is risible, symptomatic, and hopelessly dated in them to the epochal questions about religion and its future with which Comte is wrestling, and to interrogate his thinking in these terms, is no easy task. Yet only in this way will we succeed in engaging with his thought at all.

This is not to suggest that his formulations about the future of religion after God and Christianity are cogent, let alone "scientific." It is to suggest, though, that the dismissive marginalization that his work suffered after, and largely because of, his religious turn, has obscured what might not be so easily dismissed. His Religion of Humanity was a plausible move on the contemporary intellectual board; and the very singlemindedness and mania for systems with which he attempted to think it through illuminate not only that position but the constellation of problems in which it was embedded. To read his work is to place oneself in the midst of that constellation, if from a perspective that is entirely eccentric to the dominant, which is to say, German, philosophical idiom in which the religion question ("what after theism?") came to be formulated.

But to get to the question of positive religion's larger significance, we must first follow as faithfully as we can the labyrinthine thread of Comte's own thought. What exactly, then, was the religion that Comte had in mind? And what was its rationale?

Positive Religion

Let us note first that Comte's Religion of Humanity was designed to be much more than a personal faith. Like its fetishist, polytheist, and monotheist predecessors in the genealogy it gave itself,[6] it was to be an all-encompassing social institution. Through its punctiliously worked-out system of *dogme*, *culte*, and *régime*, it would reach into every corner and aspect of social life. This extension was necessary, given what positive sociology took religion itself to be.

In each of its historically prevailing forms, religion had articulated a system of worship together with an overarching complex of beliefs and an apparatus of moral regulation. Anchored in veneration for the powers imagined to rule the world, it was an affectively ramified orientational system, in fact a system of systems, which, for both individual and society, brought thinking, feeling, and acting into harmony with one another, and in so doing secured the unity of both. "Religion," Comte stated at the beginning of volume 2 of the *Système de politique positive*, "will always be characterized by the state of complete harmony proper to human existence, whether collective or individual, when all its parts are properly coordinated." The dual function of religion, as suggested by its etymology, was to "bind together [*lier*] the within by love and to bind it again [*relier*] to the without by faith."[7] As such, it was an essential organ of any functioning society, and the absence of a new religious institution in France and indeed throughout the advanced nations of the "Western Republic" to replace the one that had become defunct was abnormal and pathological.[8]

Of course, in the maintenance of social order, there had always been coercive mechanisms, too. But across the arc of human development, military rule had gradually given way to industrial rule, and forced obedience had been increasingly replaced by higher levels of voluntary sociability. With the full emergence of industrial-positive society, the coercive role of the state could be expected to decrease, and the role of religion as a purely moral force would become correspondingly more important. Both in the scale of the religious institution and in the progress of the heart, "humanity becomes more and more religious."[9]

But why religion at all?

In a positivist optic, the perennial human problem that religion in every form was constituted to address was how to reconcile the refractory egoism of the individual with the love for others—general as well as specific, toward superiors and inferiors as well as toward equals—needed for sustainable social cooperation. In effect, these were two interlinked problems: on the one hand, the mental-moral one posed for the *individual* by the greater strength of lower egoistic impulses over higher and altruistic ones, and on the other hand the problematic character of the *collective* entity that human beings constituted in the social state. The first exacerbated the second and made it chronic. However, the second, which concerned the contradictory character of social being, was an ongoing problem in itself.

In the great distinction drawn by positivism within the natural order between the inorganic and the organic,[10] sociology, as the science which treats of the most dependent, complex, modifiable, and particular order of being, was to be accounted a life science. This was not simply because the constituents of (a) society were living organisms. The human collectivity, "man in the social state," was to be considered a life form in itself, indeed the highest known to us, and the most deserving of our affections. In addition, then, to those sociological laws that apply specifically to it, the same general "laws of life" applied to society as throughout the organic domain, chief among which was that a living entity must be regarded as a "unified whole"[11] and that the constant renewal of its own substance was the necessary condition of its existence.[12]

At the same time, however, the social organism was not a biological entity like others. It was a composite being. That is, its constituent elements were themselves independent life forms, each subject to the laws of life proper to such an entity, each an independent and spontaneous center of action, and each pushed and pulled by its own interests and instincts. In consequence, the consensus of parts necessary for society to cohere and function as the complex entity it was, and to maintain that consensus through time, could not at all be guaranteed.[13]

The risk to vital unity was particularly acute during periods of transition from one settled stage of social development to another, and the *grande crise* that climaxed the transition to industrialism and the positive state was the riskiest of all.[14] The chronic and conflictual instability of post-Revolutionary France bore witness to the danger. The Revolution had swept away the corrupt and retrograde institutions of the Ancien Régime, including its church, without putting anything durable in their place. Not only had an intricate system of solidarities been dismantled, there had been a great rupture to *continuité*, with the past rejected as hateful by one party, and the future by another.[15] Meanwhile, the metaphysical individualism powering the rhetoric of revolution had become virulently negative, with outright anarchism proclaiming itself.[16]

But the present crisis was not only one of unresolved transition. Industrialism came with its own disintegrative risks. To be sure, through large-scale production, commerce, and trade, *l'industrie* was drawing larger and larger populations into collective activity. That was its great moral as well as material promise. But the increasing division of labor that made this possible—and that was indeed a law of progress for all organic life[17]—also narrowed individual perspective, so that task-divided workers within the collective tended to lose their sense of connection with the whole production process. The same dispersive effects of specialization were evident in the intellectual world, with its *esprit de détail* and tendency toward *pedantocratie*.[18] In addition, escalating class conflict between industrial workers and *patrons* (business owners), pitting envy against greed in a clash of particular material interests, indicated the urgent need to remoralize relationships of property and power.

Thus the rallying and unifying function of religion, which had been indispensable for social unity in previous epochs of social development, remained essential. And it remained so, even after the old belief in fictive gods had faded before the spread of a scientific worldview. Nor was the spread of that worldview sufficient in itself unless it, too, was ramified into an englobing institution able to orient the heart as well as the head and able to reproduce, individually and collectively, the requisite synthesis of intellectual and moral consciousness.

As for the other side of the problem, concerning individual instincts and selfishness, all previous religions had devised a panoply of devices for moral education and for making individuals more selflessly loving toward others. And none more so in the West than Christianity in its medieval heyday, the age of "Catholic feudalism." But for all the civilizing results, for example in the ending of the ancient world's slavery, and in elevating (in its way) the position of women,[19] Christianity had misconstrued the process in which it was engaged and indeed the very nature of the soul. In Christian doctrine, human nature was at root fallen; in response, it posited Grace from above to save the sinner from sin and the Judgment to come as a self-interested motive for good behavior. It meanwhile urged a war against the lower instincts, while holding that the highest form of love was of emphatically divine origin. By contrast, for positive religion, in which this schema was scientifically corrected and brought down to earth, there is no "original sin." As Gall had purportedly demonstrated in his "cerebral physiology," altruistic sentiments, like the other more animalistic ones, had their physical source in a locatable organ within the brain.[20]

The moral problem posed by the unruliness of human instincts could thus be redefined in more tractable terms. It was a question of energetics. According to positivism's "science of the soul," different sentiments have different degrees of strength. As we ascend the scale from the seven lower ones—the nutritive (self-preservation), sexual, and maternal (preservation of the race) impulses; the two instincts of improvement (i.e., the urge to destroy [military] and the urge to construct [industrial]); and the two motors of ambition (the thirst for power [pride] and the need for esteem [vanity])—to the fully social ones at the peak with their three species of love—attachment, veneration and *bienveillance* (or *humanité*)—the relative force of each succeeding sentiment declines. *Bienveillance* (benevolence or kindness) is the highest and purest, but its spontaneous impulse is exceedingly weak. It nevertheless remains the case that the sentiments of love are no less innate than those of hunger and sexual desire. If dormant, they can be awakened. And cerebral organs, like other parts of the body, can be strengthened through exercise.[21]

Moral improvement (*perfectionnement*) was possible, then, through human means alone. And it was so without recourse either to a disguised egoism (the snakes-and-ladders game of salvation[22]) or to the extremes of an asceticism that would mortify the flesh and extinguish individual desire altogether. It was enough to moderate the lowest impulses through habits of restraint while strengthening as much as possible the highest ones through the regular exercise provided, for example, by festivals, ceremonies, and prayers. At the limit, with suitable technologies of the self, the biologically spontaneous organization of the personality around the egoistic instincts, and the peripheralization of the altruistic ones, could even be reversed.

Of course, the weight of moral muscle training had never fallen solely on the shoulders of religion. Religion is reduplicative. It rebinds and recombines. In its practices and beliefs, it reinforces what already exists as a spontaneous process of collective self-moralization. Both historically and for each individual, the cultivation of the higher sentiments begins in the family and extends outward toward the wider group, engendering successively love for the city, the nation, and ultimately humanity as a whole.[23] Lamarck's principle of the inheritance of acquired characteristics suggested too that the strengthening of social sentiments in individuals could be passed on from one generation to the next.[24] But as a check on this ascending movement from particular to universal love, the extension of the group also brought fractiousness, interfamilial strife, and the constant drag of the lower instincts on public and private morality. And the larger the group, the more indirect and attenuated the social ties among its members, and the greater the need for supplemental structures to sustain and strengthen the sentiments underlying them.

However, the human nature problem that religion was called upon to address was not only that posed by the biologically preponderant strength of egoistic over altruistic instincts. There was a further problem concerning the harmonious operation of the "soul." Taking his cues again from a modified version of Gall's phrenology, Comte conceived the problem in terms of the relations shaped by the physiology of the brain between its three major departments of action, thought, and feeling. For individual as well as collective subjectivity, it was axiomatic, Comte thought, that overall harmony required a coordinating center.[25] Since "the organism acts usually under the stimulus of some desire and it thinks only for the purpose of acting," this coordination could not be provided by the intellect.[26] (That it could was a rationalist error that Comte himself had flirted with in the *Cours de philosophie positive*, before firmly recognizing the moral primacy of the sentiments and the social viewpoint.[27]) But if the sentiments, as the actual motor force, were to play the coordinating role, they themselves, with their two modes and ten impulses, would have to be coordinated.

That could be accomplished in one of two ways: *either* through the dominance of the lower seven instincts and an overall orienting egoism; *or* through the dominance of the three higher social sentiments and an orientation to the Other.²⁸ The former sufficed for many in the mundane world, and a certain measure of egoism was in any case required (especially in the task-specialized society of industrialism) to motivate socially necessary activity. However, a self organized around its egoistic instincts was not robustly stable, both because of the strength of the lowest ones and because of their volatile, clashing, and episodic character. It was a further axiom that mental equilibrium required the dominance of external impressions over internal ones.²⁹ An egoism closed in on itself risked a reversal and the confounding of internal and external reality. It would also distort or undercut the perceptual orientation to the outside that was essential if the processing intellect were to provide realistic guidance to action. For all these reasons, an altruistic orientation—over and above the social benefits—was the only route to a full harmony of the soul.

Such harmony, however, could only become complete if, in step with the inner subordination of egoism to altruism, the mental impressions that came from without could be sufficiently detached from self-interest (the world as field of play for egoism) as to be themselves suffused with altruistic love.³⁰ This begins to happen when contact with the world engenders an appreciation of its objective reality and order, and the limits of its modifiability. Recognition of necessity and eventually of the actual laws in which it consists leads not only to enhanced human power; it also provides an antidote to willful pride and limitless desire and is thus the beginning, for the infant as well as for the species, of moral training. However, even in the highest form of *amor fati*, the orienting attachment to the outside that this can provide is indirect and falls short of being cathectic. The genius of religion—which, when positivized, becomes fully aware of its own instrumentality—has been to provide a directly emotive mechanism through various forms of anthropomorphic projection. In this way, external reality was imagined to be ruled by "fictive beings" endowed with the highest attributes of life, with whom, though they be mightier than us, communication was possible and an emotional bond could be formed.

The functions of these beings have been multiple. In addition to serving as an external and transcendent rallying point for the highest sentiments, both for individuals and for the community they unite, the spirit world has supplied material for the imagination to fill gaps in positive knowledge (at first enormous) and to supply answers for the great unknowns. At any given stage in the uneven development of knowledge from theology through metaphysics to positivity this has enabled the intellect, in sync with the current religious imaginary, to develop a coherent if false map of reality, able to serve as a unifier of individual and collective mentalities.[31]

In the history of religion, then, a procession of fictive beings has been associated with a procession of "philosophies," each with its own manner of systematically connecting the highest sentiments to some version of an absolute Other outside. In the animism of our ancient ancestors, external reality was immediately charged with feeling via the spirits imagined to inhabit the entities that composed it. In polytheism, these spirits were substantialized in human or animal form, and although they still influenced the course of the world, they had their separate home in the distant heights or depths. In monotheism, the one true God—exalted as Lord of Creation but also, and still more, as our only salve for sin, death, and despair—became still further remote from a cosmos increasingly understood in terms of its impersonal laws until, in its Deist sunset, it was no more than their initial author.

At the end of the line, when the human-moral domain of knowledge had been wrested from the church and itself positivized, the fictive gods were destined to disappear altogether. They would be replaced, however, not by nothing but by a commanding center for altruistic feeling, thought, and action that would finally be real—wherewith we come to the very heart of positivist doctrine and to the focal point of its worship, that is, to Humanity itself: the true Great Being that we make up together and in which we actually, not mystically, are, in Saint Paul's phrase, "everyone members of one another."[32]

Humanity as Great Being

It should be said at once that although Comte's Great Being is designed to play a similar role in the psychic and mental economy of positive religion as the old God did in monotheism, it has a quite different ontological status.

In the first place, while its vastness, along with the implication of the knowing subject in it, puts Humanity likewise outside the range of immediate sense-perception, it is neither intangible nor ethereal. It is encountered daily through its effects. Moreover, through the synthesis of sociohistorical data and with the application of a suitable methodology, it has come into purview as the object of a positive science. Secondly, while the power and wisdom of Humanity compared with those of the individual are overwhelming, it is neither omniscient nor omnipotent. It is simply, in reality, the highest and most powerful being that human beings can come across. Nor, at the same time, is it in principle unique. It is alone on Earth, certainly, for on one planet, there can be only one Great Being. And that being can form around only one center—which, in the instance, is the advanced countries of Western Europe led by post-Revolutionary France.[33] But all human societies, indeed some animal ones too, have the potential to become part of the Great Being.[34] And there may be other Great Beings on other planets orbiting other stars in the universe. We just don't know, nor can we.

Nor, finally, is the Great Being eternal.[35] Enduring, yes, through numberless individual lifetimes. But its existence at every moment depends on the whole natural order and on the fate of the Earth that is its immediate, life-supporting milieu. As the highest, most dependent, and most complex form of being, it is, moreover, the most modifiable.[36] Not only is it, therefore, like lower organisms, vulnerable to injuries and disease that throw its equilibrium out of gear, it is also vulnerable to endogenous disturbances that come from its being an être composé, including those modern ones that emanate—in the constitution-making chaos of the Revolution, for instance—from ignorant meddling by those who are trapped in a metaphysical understanding of society.

What takes the place of God in positive religion, then, lacks the full indices of divinity. It is the transcending ground of individual existence and the highest form of being in the natural order. But it remains part of that order, is subject to its laws, and cannot breach the limits of its conditions of existence. What it gains in reality as a god-term, in short, it loses in absoluteness. This is just the point: the Great Being that serves as the *arche* and horizon of the positivist universe is *relative* in its greatness. In our tiny speck of the cosmos, it is simply the Great Being we are part of and the only one we know. And if that knowledge can be only ex parte, through a glass darkly, it is not because of its infinitude or supersensible mode of being, but because human beings can only ever know the how and not the why, the laws and not the causes, of anything in the world of its experience. In sum, Humanity is the "true Great Being" not absolutely, but relatively: for *l'homme* (man).

For that same reason the religious attitude comported toward it marks a modal shift. The old God may have demanded worshippers, but in its serene meta-ontic self-sufficiency, it did not strictly need them.[37] In worshipping, loving, and serving Humanity, in contrast, we each become more perfectly integrated into it and in so doing advance the order and progress of the Great Being itself. Thus its adherents are not just in a fideistic relation with it, but a practical one: both objectively and subjectively, Humanity is a human production.[38] The implications for religious practice are profound. But to see how the reconfigured relation between believer and deity translates into the baroque edifice of worship and moral regulation that Comte erects on its basis, we must take account of three further features of the Great Being that shape his design.

★ ★ ★

1. While its tissues and organs are all made up of individual human beings, the Great Being is first and foremost an integral whole. It does not follow, then, that all individuals are part of it. Those are members of the Great Being who participate in it, who execute its functions, and who serve it in whatever way. Excluded by definition are nonparticipants, parasites, and the actively antisocial. Indeed, says Comte without hyperbole, many animals have "a far higher claim" to be included "than many useless members of the human race."[39] Whole animal species that humanity has been able to tame and enter into a cooperative relationship with, have become incorporated into the Great Being as "our worthy auxiliaries." So if Humanity is less than the sum of all humans, it is more than them as well. At any one time, in other words, there is an empirical, and not only conceptual, nonidentity between humanity (small *h*) as the aggregate of all individual human beings, and the aggregate of human beings who make up Humanity (big *H*) as the Great Being.

The love for Humanity that positive religion seeks to elicit, accordingly, is not a warrant for a disguised form of collective narcissism. Nor is it indiscriminate. People's activity may be such that they are deluded in thinking they themselves are members. Under the aegis of its humbling and inspiring spirit, rather, the ideal set before the worshipper is to become, through an active life of useful service, Humanity's worthy servant—worthy, at best, of being permanently incorporated and remembered as one of its permanent organs.

★ ★ ★

2. The Great Being is, constitutively, a temporal being: not just in the sense that it perdures, and does so beyond the lifespan of its members, but in the sense that it develops, undergoes metamorphoses, and has a history. Of this, there are three important corollaries.

First, to love Humanity is not just to love Humanity in the present, in solidarity with living contemporaries. It is to embrace its past and future as well. Toward past generations, that love takes the form of veneration and gratitude for the long and ever-accumulating gifts they have bestowed; toward the future, that love is a solicitous benevolence as for one's children.

However, there is an asymmetry. What is not present *in* the present can only be conjured up in the imagination. In the case of the future, even for positivists, this is purely so. To be sure, in terms of the direction of its development, the future of Humanity can be predicted from the past on the basis of sociology's static and dynamic laws. The *Système de politique positive* describes the end point of positivist transition as the Great Being's "destiny." Comte's later writings are written as if posthumously, from *une tombe anticipé* (anticipated tomb).[40] Volume 1 of the *Synthèse subjective*, his unfinished final work, is confidently set in 1927. But to write in the future perfect is a performative gesture. For Comte, 1927 has not happened. Contingencies can interrupt. *Bienveillance* is admixed with hope and intention, and its object cannot be said to be real even at secondhand. Love for the human future is, in fact, a call to action, setting "progress" as its overriding goal. In Comte's summary formula: *l'amour pour principe, l'ordre pour base, le progrès pour but*.

The case of the past is more complicated. Though not now present, the image of what it was can be recomposed before the mind's eye. That indeed is a condition of its veneration, whence comes the prominence in positivist worship of the memorialization of past benefactors, who become a continuing inspiration as well as the means for inculcating a sense of the whole ancestral inheritance. Whence comes also the memorialization of the stages—agrarian-fetishist to industrial-positive—through which the Great Being has gone. All of this helps worshippers not only to appreciate their indebtedness but to conceive the Great Being in its temporal fullness and the place of their own time in the great procession of human becoming. At a collective level, it also serves to reinforce *continuité* as a crucial dimension—even more crucial than synchronic solidarity—of Humanity's vital unity.[41]

But this, *secondly*, is not the only effect of memorialization. To the extent that the past is remembered and subjectively incorporated, past Humanity comes to have a second existence in the subjectivity of the living. At any one time, then, the Great Being exists as a combination of the two modes:

objective life plus subjective life, in Comte's terminology.[42] Moreover, the older humanity becomes and the larger its accumulated inheritance of past achievements, the greater the proportion, within that composite, of subjective to objective life. "More and more the dead rule the living."[43] The older humanity becomes, the greater the place, too, of the collective memory and of the activity needed to keep it alive. That is why, at the culminating industrial-scientific stage of history, positive religion is marked by the overwhelming weight of venerative memorialization within its cultic practices. In addition to the daily memorials prescribed in the positivist calendar—culminating each year in a Day of All the Dead—this is spatially marked by the placement of the Temples of Humanity in the midst of the sacred groves in which the worthiest souls are buried and venerated.[44]

For those divested of belief in heaven, this same process affords the only possibility of personal immortality, i.e., through memorialization in turn. Positive religion systematizes this commonplace insight, taking its place within the wider nineteenth-century cemeteries movement.[45] But let us note here one further point. The transformation, for those worthy, from objective to subjective life is a purifying move. The memory that survives of the individual servant of Humanity is not that person warts and all, but those features and achievements that most make them worthy. "Thus in all cases the process subjects, as its natural result, the being incorporated to idealization."[46] Nor, indeed, is everyone to be remembered. In positive religion, the judgment as to whether the dear departeds were true servants of Humanity and thus worthy of passing into subjective life is made by the priesthood seven years after their death, when grieving has ended and a dispassionate assessment can be made.[47] The remains of those who pass the test are kept in the civic cemetery, with the most worthy being solemnly removed to the sacred grove for public commemoration. For reprobates, criminals, subversives, and so forth, there are unmarked graves in wasteland, and oblivion. And the same is true at the collective level. What the collective memory retains, under the aegis of positive religion, and converts into a representation of the Great Being itself, is a history from which evil, tragedy, and waste have been removed, or recuperated as necessary defects on the way of progress.[48] In passing from objective to subjective existence, then, Humanity itself is purified—with the purity of the whole increasing as the ratio of the dead to the living within it grows.

It is evident, *thirdly*, that the time dimension of the Great Being is teleological. Human history has a direction, which the dynamic side of sociology establishes as laws. However, the nature of that teleology is not as straightforward as may seem. At first sight, it is the natural entelechy shared by plants and animals, which is transposed, via the law of three stages, into a model of collective mental maturation. In this vision, Humanity, through the gradual and stepwise positivization of knowledge and worldview, advances at once toward enhanced production to meet material needs, toward optimized modification of the environment, and finally toward conscious self-direction via scientific self-knowledge. But if this is Comte's Saint-Simonian starting point and the quasi-materialist spine of his grand narrative, it is supplemented by a perfectionism that comes to overtake it. The telos of the Great Being is not only—if it survives the storms and stresses of its metaphysical adolescence—to come maturely into its own as what it is (essence become existence, and the riddle of history solved, one is tempted to say). For at that point, what steps to the fore is order itself. Or rather: order in its highest form as harmony— a harmony, that was being perfected all along, and whose ultimate perfection now becomes the conscious goal of progress.

Whereas in the *Cours de philosophie positive* progress and order simply went hand in hand, in the *Système de politique positive* the accent is firmly on the second term. Progress is the progess of order. This is to say not just that each advance in knowledge and material civilization has to be accompanied by a way of holding the social ensemble together, and that therefore the destructive work of 1789 needs to be followed by the construction of a new social order, but also that the whole dialectic of progress tends toward the perfection of order itself. In the case of society, the most perfect order would be one in which the voluntary adhesion of each to the good of the collectivity enables a perfect harmony of its organs and functions. At the limit, with feeling, thought, and action perfectly harmonized at both the individual and collective level, the Great Being becomes the very embodiment of the highest sentiment that binds it together. Order is love. "By a happy coincidence of language," Comte notes, "the same expression is used to designate the widest exercise of the highest affection and also the race in whom it exists in the highest degree."[49]

Besides providing a regulative ideal and a mirror in which to see the imperfections of the present, this omega point of human development offers a fixed direction for progress and a goal for action. Progress does not stop, however, once the final stage of social development is reached. After the full establishment of the positive polity, the progress to be aimed at would be that which comes from correcting, so to speak, natural imperfections in the human animal and its environment. An instance of the first would be the scientific realization of the virgin-mother, a parthogenic maternity unpolluted by men or sex.[50] An instance of the second, at the furthest imaginable limit of human powers, would be a correction of the Earth's orbit through a "series of explosions like those that gave rise to the comets."[51] Thus, with the scientific-industrial growth of its "modifying power," Humanity becomes that which, in the order of nature, rebounds back onto that order not merely for adaptive survival—all life does that—or even for the greater satisfaction of material needs (about which Comte has little to say), but to perfect it. To amend Aquinas: *Humanitas non tollit naturam, sed perficit* (Humanity does not abolish nature, but perfects it).

★ ★ ★

3. The third distinctive feature of the Great Being that shapes the practices of positive religion is the pivotal role of women in its movement toward the Good. If the moral ascent of *l'homme* (man) begins in the family, it does so not just through the attachment of each with all, which elevates the group above its members and prepares the way for wider association, but also through the angelic influence of *la femme* (woman), without whose *tendresse* the affective bonds of the family could scarcely be formed. In general terms: if men are the active and intellectual sex, women are *le sexe affectif*.[52] As such, they not only hold up half the sky; their role is salvific. "The most important duty of woman is to form and perfect man,"[53] and "Women are the intermediary between man and Humanity."[54]

Comte's essentializing of women (and correspondingly, men) intertwines familiar Victorian motifs: idealized medievalism—Dante's Beatrice is a guide star for Comte as it was for the pre-Raphaelites; the bourgeois nuclear family, projected back to the origins of hominization; and a masculinist reaction to the equal rights feminism that had first surfaced in the French Revolution.[55] There are resonances too with Comte's own life. One well understands why he rejected the loyal but free-spirited Caroline Massin and turned Clotilde into a saint. But being positivist, the schema is naturally underpinned by a positive theory of women's moral difference, a difference rooted, we are assured, in cerebral physiology. What distinguishes the instinctual apparatus of male and female is that, of the two components that make up the reproductive instinct (the "preservation of the race" is the second most powerful after that of self-preservation), the component of sexual desire is uppermost in the male, while care for the young—maternal love—is uppermost in the female.[56]

The consequence is strategic. If men are left to their own devices, their path to altruism is through a slow ascent up the egoistic instincts, whose successive cultivation leads only by degrees, via the desire for esteem and power, to social sentiments unalloyed with self-interest. But women have a shortcut through the strength of the maternal instinct, which disciplines the lower instincts from within and bridges directly to the social sentiments. This empowers them, in turn, to draw men lovingly upward and away from their more engrained egoism and animality. Moreover, the cradle-to-grave influence of women—as mothers, sisters, daughters, and above all as wives, a relation that blends the character of all three—not only tames and elevates men at home; it also radiates outward by imparting to men a measure of the altruism that alone can give their rambunctious, ego-driven, world of industrial, intellectual and political life a firm orientation to the social good.

The linchpin of positive morality, then, is the stable union of man and woman. It is this that, both domestically and society-wide, enables head and heart to be harmoniously combined, making of positivist marriage—which consecrates this model and is for life, even if a spouse dies—the nucleus and prototype of the highest level of social harmony.[57] To ensure this optimal result, positive religion prescribes a strict sexual division of labor. Women tend to the household, leaving wholly to men the world of work, money, and public life, whose materialism and competitive striving would otherwise degrade and corrupt them. In return, so that they may be secure in their exalted domestic role, "the man must [materially] support the woman."[58]

This is not to say, though, that women are narrowly confined to their familial roles. To women is entrusted the early education (to age fourteen) of their children, providing the rudiments of positivist principles and focusing especially on art, imagination, and the development of verbal and pictorial expression.[59] Women are entrusted, too, with an important role in the formation and deployment of public opinion as a moral force. To prepare for these roles, girls from fourteen to twenty-one are to receive from the priesthood the same public education as boys. En masse, in fact, women are to be considered an intrinsic element of the spiritual power, an affective and emollient complement to the intellectually based authority of the male priesthood. "All classes . . . must be brought under women's influence, for all require to be reminded constantly of the great truth that Reason and Activity are subordinate to Feeling."[60]

In accord with this status, finally, women, though more as a category than as concrete-historical individuals, figure prominently in positivist symbolism. The "three Guardian angels"—idealized representations of sister, mother, and daughter—are first among those to be invoked in daily private prayer. The tenth month is consecrated to women, and every leap year there is a festival honoring those *saintes femmes* (like Clotilde) held up as inspiring examples of their sex. The *Vierge-Mère* is a subject for sacred art, with a subcult of her own, as well as an inspiring goal for biological science. Specially chosen women acolytes—positivism's equivalent of Vestal Virgins—guard the temples' inner sanctum entrance. And in the paintings and statuary of positivist ceremonial, Humanity itself is to be depicted as a woman, forever thirty years old, cradling the future in her arms.[61]

Culte and *Régime*

It follows from the doctrine of the Great Being that positivist worship is neither propitiatory nor sacrificial nor petitioning. There is no magic or symbolic exchange here. It is designed, rather, to operate directly on the moral psyche of the worshippers by firing them with love for Humanity and strengthening their altruistic sentiments. "Worship is always a real exercise."[62] At the same time, the cycle of worship is ramified into an extensive system of memorialization, which builds up and keeps alive a picture of Humanity in its spatiotemporal totality and so sustains the subjective existence of the Great Being itself. What connects these two functions is that in both *le culte privé*—a do-it-yourself shrine plus half an hour or more of prayer, eyes closed,[63] three times a day—and the public cult organized around the annual cycle of social festivals, the Great Being is evoked metonymically, that is, through the recall of particular representatives and benefactors of Humanity, which condenses its essential features and catalyzes outpourings of love and gratitude.

In private worship, the prime intermediary is *la Femme*, beginning with those Guardian Angels closest at hand, though worshippers are encouraged to supplement these with the venerative recall of other personally significant figures as well as to apply to their devotions the utmost of their poetic powers. Each day, in addition, is dedicated to the memory of the great human benefactor specified for it in the positivist calendar. In the eighty-one festivals that make up the *culte public*, the intermediaries are more general, consisting of the modes of unity (religious, historical, civic, national), social ties (marital, paternal/filial, fraternal, master-servant), historical stages, and final components (priests, women, patricians, proletariat) into which the Great Being can be analytically decomposed, together making up a grand unfolding tableau of order and progress. Only on New Year's Day is Humanity to be worshipped "synthetically."[64]

Liturgically linking the private and public realms are the "social sacraments": the rites of passage in which each servant of humanity is publicly (re)dedicated to Humanity at each new stage of the life cycle.[65] For men, these sacraments are presentation (at birth), initiation (following successful home education), admission (following secondary education), destination (career choice), marriage, maturity, retirement, transformation (at point of death), and incorporation. For women, since they have no career other than marriage and motherhood, steps four (destination) and seven (retirement) are omitted. For both, as in Christianity, the climax is at the end. Approaching death is marked by the sacrament of transformation; at this point, the deceased begins the transition to the possibility of subjective life and then, if deemed worthy, to incorporation through which he or she enters into an enduring existence as an organ of the Great Being. For women, Comte adds, there is no need of a separate judgment. Their own incorporation, like that of servants and "auxiliary animals," is an automatic accompaniment to that of worthy husbands.[66]

The sacraments also link the system of worship to the *régime*, that component of positive religion that, through education, persuasion, and the moral pressure of "public opinion," orients and regulates action. The watchword is *Vivre pour autrui*, a principle that is drilled into individuals throughout what amounts to a lifelong process of moral education[67] and is reinforced by the disciplining effects of moral reputation in the court of public opinion. Channels for the latter were to be managed by the priesthood, including through control of the press.[68] For public opinion to operate on the basis of more than vague rumor, however, something further was required. Accountability, as they say these days, requires transparency. From this, in Comtean terms, emerges the complementary principle of *vivre au grand jour*. Living life in the open was particularly important in the case of *les patriciens*, the industrial and financial elite that, in the positive polity, makes up the temporal power. Here, too, Comte placed much importance on the judgment of the auxiliaries to the priesthood:

women—at home and also in the regular salons they were encouraged to run—and workers, speaking through assemblies charged with evaluating the worthiness of the bosses.[69]

Finally, sitting atop the whole triple system of *dogme*, *culte*, and *régime* and forming the leadership of the new spiritual power, was the priesthood itself. There was to be one priest per six thousand people. Priests were organized in a hierarchical structure headed by the *Grand Prêtre*, who chose his own successor and also presided over the French section. Under him were four superiors from each of the other advanced countries of the Western Republic (to be augmented as more countries were added), and under them four degrees of ordinary priests. All were to be married and were precluded from a full assumption of their role before the age of forty-two.[70]

Besides officiating at festivals and services and running the Temples of Humanity—their "consecrating" role—their functions were multiple. First and foremost, they were the guardians, teachers, and developers of the positivist doctrine. This they transmitted directly through their monopolistic control of public education.[71] The curriculum would lead pupils through the encyclopedic scale of abstract sciences and up to the study of Humanity through sociology and the composite (sociological and biological) science of morality that crowned the system in its religiously corrected form. (A less summary and more abstract version of positivist systematics would be reserved for trainee priests, as part of their own seven-year novitiate.) In this form—that of the "subjective synthesis" sketched out and initiated in the *Système de politique positve*—the natural order and its laws were presented from the standpoint of the Humanity in which the "objective synthesis" of the *Cours de philosophie positive* culminated. It would thus be simultaneously cognitive and affective, drawing pupils toward a love first, via mathematics and astronomy, of the abstract order of the cosmos; then, via physics and chemistry, of *notre planète*; and finally, via biology, sociology, and the composite science of individual morality, to love of Humanity in its rise to self-perfection.[72]

There was no place in the prescribed curriculum, it should be noted, for the concrete, as opposed to abstract, sciences, such as mineralogy or engineering, nor for the practical applications associated with them. Special institutes would be set up for these under the direction of the state and industrialists. But they were outside the purview of positivist education proper. The sciences to be taught by the positivist priesthood were exclusively abstract and generalizing; their paramount aim was to provide a moral basis and framework for whatever uses were made of them in the profane sphere of material life.[73] The model for the new educational content, which had been a major preoccupation for Comte since the start, was provided by the corpus of his own writings, the principal ones of which were conceived and initially delivered in the form of lectures. Needless to say, his idea of pedagogy was relentlessly monological.

With respect to the system of moral regulation, the priesthood's role was to counsel the public and the governing *patriciens* on matters of policy and behavior; to mediate and arbitrate local conflicts, especially industrial disputes; and to guide the formation and deployment of public opinion. The latter, as concentrated "moral force," was the only means of coercion at the disposal of the spiritual power. In this, as in other respects, there was to be a strict division between state and church. The temporal power "orders acts," while the spiritual power counsels, consecrates, teaches, and so "modifies the will."[74] The limit of the priesthood's sanctioning power, accordingly, was to shame and denounce. If the culprits did not change, only then were they to be handed over to the temporal power and its colder system of justice.

The last but not least function of the priesthood was to organize the Church of Humanity beyond the boundaries of France and eventually on a planetary scale. Comte's plan for the expansion of positive religion, like the organization of its priesthood, was entirely top-down. Leadership branches in target countries would be established from the center, and direct appeals to political authorities would give the venture their blessing.[75] Yet the global spread of the religion did not have to await political unification. To the contrary, the unification of the world would be religious rather than political, through, for example, a confederacy or a world government. As political units, national societies had already become too big and anonymous and should be devolved. The lead would naturally be taken by France at its center, with Paris as the seat of the new *sacerdoce*.[76] From there, the religion would fan out to the advanced countries of Western Europe and their New World colonies, to Russia and the Ottoman Empire, and finally to the peoples of Asia and Africa.[77]

Catholicism without Christianity

There is more than a grain of truth in Thomas Huxley's celebrated jibe that what Comte had spun out of his brain was "Catholicism minus Christianity."[78] From saints and Maryology to sacraments, guardian angels, the new priesthood, and a systematic (socio)theology, the transcriptions are notable. As for Christ, the Crucifixion disappears together with the Evangel; but the god-man fusion reappears as Humanity, and the great mediator as *la Femme*.

Not that Comte himself would have considered Huxley's jibe much of a criticism. His employment of mutant Catholic forms was deliberate and undisguised. His sociology itself, with its deduction of industrialism's "normal" form of religion, was a synthesis of "progressive" and "retrograde" thought, each corrected in light of the other.[79] From Condorcet's *Esquisse d'un tableau historique des progrès de l'esprit humain* came a model of human history whose direction is determined by the logic of collective intellectual development; from de Maistre and Louis de Bonald he derived a protoscientific conception of society as an organism, together with a deep appreciation of the Catholicism and chivalric culture of the Middle Ages. Where Condorcet had erred was in having no understanding of social order and in thinking that the vanquishing of superstition by knowledge terminated in freedom of conscience, individual rights, and a secularized democracy. Against this view, the Catholic counterrevolutionaries had been right to see in the individualism and critical rationalism of the Revolution a dissolvent and anarchic force, whose antitraditionalism and refusal of authority threatened the very basis of social unity. They had also been right to see in Protestantism the seeds of the individualist revolt and of the autoliquidation of religion itself. Where de Maistre and the whole retrograde party had erred—besides in their stubborn clinging to theism—was in not understanding the laws of progress and the way in which the positivization of knowledge and the rise of towns and industry had completely undermined the conditions in which the Christianity brought to magnificent flower in the high Middle Ages had been viable as the ruling form of religion.

The point, then, was to extract from the religious institution of the Ancien Régime its rational kernel. To be sure, Catholicism had been a contradictory assemblage of scholastic metaphysics and Pauline faith, of the altruism of universal love, and of the egoism of personal salvation. But in its institutional form and micropractices, it had gotten many things right. As such, it was the *religion préparatoire*, paving the way for religion in its final form.

Catholicism had been right above all in its concept of a spiritual power that combined intellectual and moral leadership in an institution independent of the temporal power of the state. As Comte's 1826 essay "Considerations on the Spiritual Power" attests, the necessity of establishing a new spiritual power to replace that of the Catholic church had been at the center of his project from the start. That essay leaves open the form it was to take, but the general assumption was that the community of scientists would form its basis, as foreshadowed in Francis Bacon's *New Atlantis*. For that to happen, though, it was necessary that the sciences be organized and mentally harmonized through the kind of philosophical synthesis that Bacon himself had been the first, in the age of modern science, to attempt. Hence arose the *philosophie positive*, though on a sociologically transformed basis.

What Comte had come to understand by 1845, however, was that although science, correctly understood, was the foundation for a new spiritual power, scientists themselves, at least as based in the university, were not. Like others who had joined the Saint-Simonian circle in the 1820s, Comte had been a student at the École Polytechnique. As he had been expelled for political agitation against the Bourbon Restoration, his time there had been brief. But he always looked back with fondness to the survey courses in mathematics and the physical sciences that he had taken and that had fired a lifetime of intellectual enthusiasm.[80] He held a junior teaching post at the École for many years. But if he had entertained the hope that his beloved alma mater would be the center for training the new priesthood and for reforming the education system from within, these were dashed by his failed campaigns to obtain a professorship and, after many years of acrimonious disputes, eventually (in 1851) by his being cut off from teaching there altogether.[81] Academia, he concluded, could not become the church, even that part of it newly founded by the Convention to promote empirical science and useful knowledge.[82] From top to bottom, the education system must be built afresh. Moving closer to the Catholic model, then, industrial society's new spiritual power would have to be a wholly independent institution, its organizing cadres supported through subscription and voluntary contributions. Eventually, with the political establishment of a suitable transitional regime and with the wise support of its rulers, a public education system under the positivist priesthood would be established, and priests themselves would receive a regular payment from the public treasury, proportionate (Comte does not hesitate to specify salaries) to the needs of their office.[83]

For all the borrowed clothes and continuity of forms, however, the Religion of Humanity could fairly claim to mark a decisive break not only with Catholicism but with any species of religion that had gone before. In the first place, it was drained of all mysticism. Positive religion's love for and knowledge of Humanity was *une foi démontrable* (demonstrable faith). In the second place, and paradoxically for that same reason, it did not claim for its doctrine the status of absolute truth. Nor indeed, with due reference to Kant, did it claim the status of objectivity. Its knowledge of Humanity, as of everything else in nature, was conditioned and limited by its always being human knowledge, knowledge for us. At the core of the doctrine as propounded in the *Système de politique positive* and subsequent writings, accordingly, is not the *philosophie positive*—though its propaedeutic value is given due weight—but its transformation into the *synthèse subjective*.

In the preface to the first and only volume of what was intended to be, under that title, his crowning work, his recognition of the subjective, mentally constructed character of the object of positivist faith went further still. To close the gap of doubt between positive/relative knowledge and certainty, the Religion of Humanity would practice a kind of conscious fetishism, in which the object of faith, though apprehended as real and law-bound, was reimagined as a kind of spirit endowed with the attributes of life. Indeed, it was reimagined as a trinity.[84] The first member was *l'Espace* (combing the "ether" of astronomy with the space of Cartesian geometry) figured as *le Grand-Milieu* and endowed with love. The second was *la Terre* (our home and the site of the physical sciences), figured as *le Grand-Fétiche*, endowed also with activity. And the third, at the summit of the natural order, was *le Grand-Être*, endowed with will and intelligence as well. In this scenario, be it noted, the middle and transitional stage of history is no longer the metaphysical, as in Comte's earlier work. It is theism, both poly- and mono-. "Theologism is . . . but an immense transition from Fetishism to Positivism."[85] Thus, in its positivist culmination, the history of religion returns to its fetishist starting point, but at a higher level, one in which the fetishism that lends affective power to its condensed images of the real rests on an acknowledged fiction.

Moreover, it was not only from Catholicism and fetishism that the Religion of Humanity selectively borrowed. It took elements from polytheism as well—for example, the pieties of ancestor worship, its idealizing figurative art, and its advance from the "logic of sentiments" to "the logic of images." In its emphasis on unity and purity, there are elements, too, of Islam, and the proposed fetishism of space makes reference to the Chinese idea of heaven. Going further than Leibniz's ecumenical imagination dared to travel, Comte aimed, in fact, at a grand synthesis of all the major religions he knew of. This in part reflected the intended universalism of the project as an idea. But it also had a practical point in the ease with which he thought it would be possible to persuade the civil and religious leaders in every cultural region, with their different cultures and states of progress, to introduce the Religion of Humanity into their own domains. This was to be, in short, not only the Religion *of* Humanity, but the Religion *for* Humanity, the first one to truly embrace all the peoples of the world.

Comte and the Religion Question

It is hardly surprising that Comte's religious turn after 1845 split his followers and lost him the support of respectable figures like Mill and Littré. The trenchant judgment of Nietzsche, for whom Comte's earlier work had represented "the cock crow of positivism," expressed the same disappointment. The groundbreaking author of the *Cours de philosophie positive* had regressed to become the mere leader of a sect. "In the weariness of old age," Nietzsche noted, Comte was one of those thinkers "no longer able to bear that dreadful isolation in which every intellect that advances beyond the others is compelled to live. From this time forward he surrounds himself with objects of veneration, companionship and love; but he also wishes to enjoy the privileges of all religious people and to worship what he venerates in his little community—he will even go so far as to invent a religion for the purposes of having a community."[86]

Even this dichotomous assessment, though, did not last. In the context of twentieth-century totalitarianism, Comte's earlier work, too, with its confident assertion that a positive social science could form the basis of a scientific politics and thus of what came to be excoriated as social engineering, came under a cloud. For Karl Popper, he was one of the intellectual enemies of "the open society."[87] For Friedrich Hayek, Comte had been a leader in "the counter-revolution of science."[88] For Herbert Marcuse, he was an intellectual forerunner of the "totally administered society."[89]

What links these assessments, wherever they get off the Comtean boat, is that the core of his thinking is to be found in the *Cours de philosophie positive*, in relation to which the later work was either an aberration or an extreme and unwarranted extension. However, some more recent commentators, accepting in part Comte's own protestation in the *Système de politique positive*, have seen a thematic continuity such that the later writings made explicit what had always been their driving force.[90] This has not paved the way for a rehabilitation. But it does suggest that interpretations of his oeuvre that exclusively view it from the side of his philosophy of science miss the mark. Whether Comte, for whom scientific discovery was effectively a closed book with nothing fundamental remaining to be discovered, was even a real friend of positive science can be doubted. It can well be argued, indeed, that if we are to place Comte's whole systematizing project correctly in the context of nineteenth-century European thought, we need to see him less as the apotheosis of scientism and herald of technocratic politics and more—perhaps primarily—as a thinker of the religion question, such as it transpired in his day.

In these terms and if we strip away its idiosyncrasies, Comte's Religion of Humanity can be seen as a French variant of the same posttheistic teleological humanism that, via Immanuel Kant and Georg Hegel, surfaced more radically in Germany with Ludwig Feuerbach and the young Marx. Linked to a narrative of humanity's collective self-realization, Comte similarly posits and champions a final, demystifying stage of religious consciousness in which humanity shakes off false gods and comes to take itself as its highest source of value. However, in the markedly different circumstances of post-Revolutionary France, and inflected in a conservative (order and progress) direction, what distinguishes Comte's version of this figure is 1) its institutional elaboration as a full-fledged cult, church, and religion; 2) within a wider reconstructionist program, the strategic role this is assigned in completing the Revolution and overcoming the "grand crisis" of emergent industrial society; and 3) its link to neo-Baconian encyclopedism, the completion and correction of whose project through the founding of sociology at once made possible *both* the establishment of a postmetaphysical "new organon" (Bacon's term) as the basis of a reharmonized mental consensus *and* a scientific way to understand the history, social functions, and current travails of religion, and more broadly *la morale*, itself.

The German development in the 1830s and 1840s was doubtless influenced by the French. But the contrast with the way in which religious humanism surfaced (and then disappeared) in Germany is striking. In the land of Luther, where because of its political and economic backwardness, the Revolution, as Marx remarked in the *German Ideology*, only happened in speculative thought, the shift from faith in God to faith in humanity began as a movement not against but *within* Christian theology. Hegel's totalizing historicism—inspired as much by Joachimite eschatology as by the French Revolution and Napoleon's march to the east—continued along Kant's path of reconciling Enlightenment rationalism with the Gospel spirit and the God within. But it did so in a way intended to overcome the scientific objectivism that flourished in France and the *nihilismus* that Jacobi and other counter-Enlightenment philosophers had detected at work in Kant's "religion within the limits of reason alone." With David Strauss's *Life of Jesus* and Feuerbach's *Spirit of Christianity*, the theological framework was brought to a breaking point, issuing in utopian calls for a realization of the this-worldly community of love that official Christianity had ambiguously projected into the heavens.

As the successor to Reformed Christianity, then, the humanism distilled from it in German philosophy was borne on the wings of Protestant innerness toward the disappearance of organized religion as such. In Hegel's realm of absolute spirit and Feuerbach's anthropology, the disappearance of Christianity was its fulfillment. For Marx, after the Paris manuscripts, the whole problematic of consciousness and its posttheistic disalienation itself disappeared in the materialist immanence of the historical dialectic and proletarian revolution—though how completely was to remain a doctrinal issue for Marxism throughout its history. In France, on the other hand, where an unreconstructed Catholicism was a pillar of the Ancien Régime and an implacable foe of republicanism thereafter, the elevation of humanity into a foundational category was political from the start and, amid internecine conflict between clericals and anticlericals, emerged as the centerpiece of what became a conscious effort to replace Catholicism as one form of organized religion with another.

The first attempt at such replacement was during the Revolution itself. It had already been apparent to Rousseau that a democratic revamping of the social contract required a virtuous citizenry and that this required not only a reformed system of education to bring out man's natural goodness but also some institutional reinforcement. From this, following Locke, came his proposal for a civil religion and the principles he suggested for it in *Emile*. The lapsed Calvinist from Geneva, though, was an iconoclast and allergic to theater and ritual. This was not the case with the Jacobins, who, at the height of the Terror, sought to secure the "Republic of virtue" with a full-scale cult of Reason, Nature, and the Supreme Being. All manner of symbolic devices were employed, ranging from a new calendar, indexed to the seasons, to the extraordinary ceremonies, complete with statuary, backdrops, and living tableaux designed by David that were staged in the Champs de Mars and the (liberated) Notre Dame. After Napoleon's Concordat with Rome restored the Gallican Church and continuing on through the fuller restoration that followed his defeat, such nonsense was banished to the sidelines, together with the whole republican opposition.

It was in this context, swirling with political, intellectual, and ideological conflict, that Saint-Simon set up shop across the road from the École Polytechnique and that the elements came together that would eventuate in Saint-Simon's *Nouveau Christianisme*, its adventures under Bazard and "Père" Enfantin, and then Comte's own launch of his Religion of Humanity.

What gripped the milieu of young savants that Saint-Simon gathered around him was the idea, transplanted via Jean-Baptiste Say from Scottish political economy, that what was struggling to be born was a society based on industry and production, that completing the work of the Revolution entailed a shift in political power to those (bankers, engineers and capitalists) who commanded the industrial process; and that for a stable social order to be established, this shift in political power needed to be complemented by a corresponding shift in intellectual and moral power from the Catholic church to the scientists and sciences whose rise had been indispensable to the whole advance. For this configuration to take hold, it was also imperative to forge a new moral consensus based on a science of man that would complete the modern scientific revolution heralded, at its dawn, by Francis Bacon.

That the project moved in a religious direction, with scientific systematizing standing in for theology, is easily accounted for by the need to counter, in its own terms, the broken but still active hegemony of Catholicism. This also explains the aim of synthesizing progressive thought with the wisdom that could be extracted from conservative Catholics. But the importance of Francis Bacon to the Saint-Simonian, and Comtean, assemblage cannot be overstated. Bacon's "Great Instauration" had been a major reference point for Diderot, d'Alembert, and the *philosophes* who contributed to their *Encyclopedia*. And so too, though more *sotto voce*, had been Bacon's *New Atlantis* with its vision of a new spiritual power of scientist-priests, dedicated, in the House of Salomon, to the "achievement of all things possible" for "the relief of Man's estate." In all this those in the Saint-Simonian circle were the heirs not only of Bacon's general vision (with themselves at its gnoseoecclesial center) but also of his systematics, which, suitably corrected in the light of a positive science of man and morality, provided a starting point for their own.

There is no need here to go into the complexities of Comte's relations with Saint-Simon and his followers, which have been authoritatively documented and analyzed by Pickering.[91] Suffice it to say that regardless of who influenced whom and Comte's shocking inability to acknowledge, even to name, the "depraved juggler" whose circle he had joined and collaborated with between 1817 and the mid-1820s, it is evident that they operated within the same overall matrix. It is evident too that there are many parallels between the "new Christianity" that Saint-Simon launched just before his death in 1825 and Comte's own religious project. They shared, for example, a priesthood recruited from the scientific elite; a body of doctrine resting on a synthesis of positive knowledge and capped by a sociohistorical science of man; a set of moral principles centered on altruistic love; symbols and ceremonial designed to appeal to the heart as much as the head; and a *culte de femme* (which, in the case of the Enfantin-led commune led, after a phase of free love and "rehabilitation of the flesh," to a search in the Orient for the female Messiah).

At the same time, however, Comte's Religion of Humanity diverged in several fundamental ways from the religious mission that Saint-Simon bequeathed to his disciples. Most strikingly, the "Humanity" that is at once the epistemological, fideistic, and cultic crown of Comte's system was entirely of his invention. In New Christianity, the proposed community is not equated with the object of "social physics" and is not worshipped even as a fictive being. The vacated place of God is left empty. Secondly, although Saint Paul is credited by both as the founder of the Christian church, for Saint-Simon the essence of Christianity is to be found in Jesus's original message of peace and love, as embodied in the practices of the earliest Christian communities. For Comte, however, Jesus receives no mention, and the enduring value of Christianity lies in the church itself, brought to (relative) perfection in medieval Christendom. Thirdly, while love for the other is the central and unifying ethical imperative in both religious systems, in Saint-Simonianism it is a purely synchronic principle. To "love one another as brothers" corresponds, in Comte's vocabulary, to *attachement* and *solidarité*. But for Comte, attachment is the lowest of his three forms of altruism, and the solidarity based on it is complemented and even trumped by continuity, i.e., solidarity through time instantiated in love of ancestors on the one hand and of future generations on the other. For Comte, moreover, such an understanding is fully compatible with and even requires a hierarchical form of social order, whereas for Saint-Simon, the primacy given to brotherly love implies a community of equals and the moral obligation of the wealthy and powerful to pursue a social policy devoted to the interests of "the poorest class." In doctrinal terms, correspondingly, Saint-Simon's science of *morale* emphasizes the parallel between the law of gravity that he took to be the most general and universal principle in nature and love between living bodies as the unifying principle of society.[92] For Comte, besides his rejection of any such absolute objective synthesis, it is harmony itself that is elevated to the highest principle, in which direct solidarity among the living has only a subordinate role.

While for both, finally, a reformed religious institution was crucial for the post-Revolutionary establishment of a stable social order based on industry and science, and a love-based morality was important in overcoming the economic individualism, class conflict, and dispersive specialization inherent in the emerging industrial economy, Comte's estimation of the order problem presented both by this approach and still more by the unresolved crisis of transition was much more troubled. The specter that haunted his analysis was that of total social dissolution. This fear not only infused his program with a sense of practical urgency but also underpinned the intrusive, not to say totalitarian, cast of the Religion of Humanity he sought to establish. Theoretical warrant was provided by his organismic concept of society and its "vital unity," the flip side of which was the ongoing possibility of the social organism's death as a living being. One may speculate as well that there was an existential motive in Comte's (well-founded) fear of madness. Be that as it may, the biological model he adopted is absent from Saint-Simon's social physics, and the latter's *Nouveau Christianisme* is far closer to Bacon—and to socialism—in its attunement to philanthropy, social justice, and the sharing of the benefits of science-based production and material progress.

Overall, and explicitly so after the 1848 revolution when Comte's politics crystallized in a stridently conservative direction, one may say that Saint-Simon's (and Enfantin's) New Christianity and Comte's Religion of Humanity represented, respectively, left- and right-wing versions of Saint-Simonian religion. With respect to the religion question itself, however, Comte's excessive preoccupation with social order and the overcoming of *négativisme*—themes that he took over and reworked from de Maistre and de Bonald—also give him an important place in the history of another aspect of that question, over and above, that is, the thematics of nineteenth-century humanism. Four decades before the orphaned son of a Lutheran pastor from Saxony made the topic his own, Comte was wrestling with nihilism and the death of God.

This is not to say that Comte's position on the issues that Nietzsche was to formulate in the 1880s is at all satisfactory. Indeed, the Religion of Humanity would seem to present a paradigm case of "incomplete nihilism." Thus Comte insists on the irrevocable decline of "theologism" and on the necessity of forging a worldview free of all mysticism and metaphysics, but he then joins with religious conservatives in recoiling against active nihilism (qua *négativisme* and anarchism) and engages in a panicky effort to replace one foundational term by another, so guaranteeing a lease on life of a residually Christian moralism. The Comtean maneuver, however, is more complicated and ambiguous than this would suggest.

For one thing, the very distinction between "complete" and "incomplete" nihilism—*négativisme* in Comte's terminology—was first formulated by Comte.[93] Both are defined as transitional tendencies that, during the revolutionary period, were engaged in the destruction of "theologism." Incomplete negativism was that of "the metaphysicians" (like Rousseau) who had not completely torn free from Christianity and whose erroneous principles and chimerical projects of reform were "the main spring of Western disorders." Complete negativism was that of the "scientific thinkers" (like Condorcet) who eschewed the mishmash of deism but lacked a true science of society and history. Both, at the same time, are contrasted with the "organic" school of "completely emancipated thinkers" (like Comte), whose plans for reconstruction rested on a fully positive basis. Nietzsche envisaged a quite different outcome for the scientific destruction of theology and metaphysics. But the influence of the Comtean paradigm in his formulation of the problem (in *Twilight of the Idols*, for example) has been curiously overlooked.

Secondly, and for all its "demonstrability," Comte is quite aware that there is something fragile, even arbitrary in his faith. The Humanity (big *H*) that anchors the system is foundational in only a subjective and relative sense. Indeed, to sustain it as the ground of thought, feeling, and action requires an entire institutional and social-psychological effort. Its very existence depends on human will. "To complete the laws there must be wills. . . . Commandment must assist arrangement for order to be complete."[94] For Comte as for Nietzsche, moreover, will itself is neither an independent faculty nor an autonomous one. It is "nothing but the ultimate condition of desire when, after the deliberative mental process, the appropriate aim of some dominant impulse has been recognized."[95]

But here we come to a parting of the ways. Whereas Nietzsche was to formulate the problem of value-creation in terms of the biogenetically blessed heroic individual, for Comte the individual is an abstraction, and what matters is the action and subjectivity of the collectivity—of which the smallest unit, properly speaking, is not the individual but the family. Indeed, in contrast with all the formulations about European nihilism that came from Germany, the whole question of metaphysics and its overcoming is sociologized and doubly so: not only by the place that the transition to positivity is given in the laws of social progress, but epistemologically. For Comte, the metaphysical outlook is identified with the (absolutized) viewpoint of the individual, and the positive outlook with the (relativized) viewpoint of the collectivity. In this sense, Comte's positivism finds an echo in Martin Heidegger's account of the rising up of the (individual) subject, in whose terms Heidegger, another lapsed Catholic, was to accuse Nietzsche himself of not thinking nihilism nonnihilistically and so of being the "last of the metaphysicians."[96] At the same time, we may remark, not only does Nietzsche never get beyond individualism, neither does Heidegger. His attempt to surpass Nietzsche fixed on Being, and, his ethnocultural nationalism notwithstanding, he never got beyond an asocial concept of *Dasein* and an alienated understanding of the social and *Mitsein* in relation to the anonymous *Das Mann*.

Of course, Comte's move to the collective subject brings with it a host of problems of its own. The model of maturation and reality-testing autonomization that underlies the law of three stages—and thus Comte's whole thesis about religious progress—depends on a dubious analogy between individual and collective development. Moreover, on the face of it, adulthood does not imply recurrence, at however self-aware a level, to a form of fetishism. We may add, more generally, that the historical method he considered proper to positive sociology obscures the lack of comparativity needed for the induction that would establish its laws. If there is only one humanity we know of, the study of it as a totality must surely be ideographic. And this is not to mention all that flows from Comte's sociological employment of a biological metaphor and the unabashedly speculative nature of his "science of the soul."

There is, in any case, a hollowness in the whole construct—a hollowness that lies not so much in the pseudoreal status of Humanity as the Great Being, for this is a poetic fiction, but rather in the artificial character of Comte's very representation of the social, beginning with "the social tie" that supposedly holds it together. For Comte, it is love and the social sentiments that bind individuals into a group. But sentiments, as he conceives them, are individual and physiologically based, rather than anything shared, intersubjective, or relational. They are thus asocial. Moreover, once we move beyond primary groups and the face-to-face, and ultimately to the group of Humanity as a whole, the affective dimension of solidarity is mediated by a shared image of the whole—which itself, in a hall of mirrors, is predicated on the same abstractly "scientific" model of the social that provides the ground plan for the positive polity. In short, Comte's cornerstone conception of *société* is an example of what Baudrillard called a *simulacrum*: a copy without an original that is fashioned into a model for engineering Comte's finally normal and perfected society into being. It is a clear case of the map preceding the territory.

In spite of all his strenuous efforts, then, to repair the tear in the religious fabric that the Revolution had both caused and exposed, we might say that Comte succeeded only in revealing the depths of the nihilism that the posttheist has to overcome. Whether this is a necessary consequence of any attempt to escape the void of meaning and value by affirmatively situating oneself in the first-person plural and adopting a social viewpoint is, however, a more open question. Another open question, therefore, is whether the sacralization of humanity in any form is conceivable as more than a detour on the way to a more complete death of God.

On this score I will conclude with just one observation. Comte's grandiose attempt to found a new world religion ended up as no more than a marginal sect, if one with a rich history.[97] But his prognosis for the future of religion was not completely wrong. In the century and a half after his death the propagation of a cult of humanity can be discerned, though in a much weaker sense, in such ventures as the Olympics, world's fairs, and UNESCO world heritage, as well as in the more universalizable forms of national civil religion and—a long story in itself—in the ceremonials and ideological life of the variegated international left. After the disasters of the twentieth century, but in a paradoxically negative and disillusioned form, it can also be traced in humbling sites of memorialization. Perhaps today when Man too is dead, together with all the Enlightenment shibboleths of reason and progress, these have become our most important sites of the human-sacred. A present-day Comtean, recognizing that the metaphysical stage of religious consciousness has not at all passed and indeed that the theological has staged a comeback, would have to ponder the meaning and future of such developments in the light of all that rendered Comte's own religious project nugatory. The field of questions onto which that now opens is of course vast.

Conclusion
The Legacy of Auguste Comte

Mary Pickering

When Auguste Comte died in 1857, thirty-seven disciples belonged to his club, the Positivist Society, which he had established in 1848. Yet his impact on both academia and politics throughout the world was far larger than this small number of disciples suggests. In 1893 one journalist proclaimed, "No French thinker since Descartes has influenced other minds as profoundly as Auguste Comte."[1]

However, it is difficult to gauge Comte's legacy. Some people were loath to acknowledge his influence because he seemed mentally unstable if not ridiculous thanks to his self-representation as the savior of humanity. His bad style was also notorious, and French scholars who prided themselves on their writing, such as Hippolyte Taine, did not wish to be associated with someone so aesthetically abhorrent, so antithetical to the French ideals of classicism.[2]

Another problem in evaluating his impact is intellectual. People were not sure whether to call themselves positivists because positivism itself seemed hard to pin down and increasingly difficult to defend. Indeed, coming to grips with his legacy leads to the intractable issue of how to define positivism. In the mid-nineteenth century, despite Comte's frequent warnings, positivism became synonymous with hardcore empiricism, which admitted evidence based only on experience, observation, and the senses. It was equated with scientism and a cult of facts.[3] Yet how to define a "fact" was problematic, especially because Comte's explanation of what constituted a fact was unclear and his delineation of legitimate knowledge seemed arbitrary; as early as 1875 Charles Renouvier cast doubt on whether Comte's historical laws were little more than hypotheses reflecting his own mindset.[4] This scientific version of positivism increasingly fell into discredit. Beginning in the late nineteenth century, as different forms of German philosophy (e.g., neo-Idealism, Marxism, and Nietzscheanism) grew more appealing, epistemology became more developed, philosophers and psychologists turned their attention to intuition and the unconscious, and there was a severe questioning of the sciences (often called the "revolt against positivism"), calling someone a positivist became an insult and a convenient way to eliminate that person from serious debate. Given the danger, few people admitted to having been influenced by Comte.[5]

To make matters more confusing, in the late nineteenth century positivism morphed into almost its seeming opposite. Antirealists welcomed different facets of Comte's thought that had been neglected: his arguments in favor of constructing hypotheses, his many attacks on pure empiricism, his principle that science had to describe phenomena and not explain causes, and his insistence that induction had to be complemented by deduction in scientific research. These ideas seemed to give more of a role to the scientist and underscored Comte's own conviction that knowledge was relative. Scholars who adopted some of these concepts, such as Ernst Mach, could be called positivists.[6]

Another problem to keep in mind is the nonintellectual side of positivism. Is positivism also a movement for political, social, and moral reform, one that necessarily includes a religion that privileges emotions? To describe the Comtean movement is very difficult. If one sees positivism as a political undertaking, it is unclear whether it is left-wing or right-wing. Positivism was often used for purposes of nationalism and imperialism, as in France, despite the fact that Comte did not support either. It was also embraced by workers in some countries, such as France and England, to support their trade unionism, although Comte found the notion of individual "rights" too murky for his taste. Positivism as a religious movement is also confounding. Comte's antitheological *Cours de philosophie positive* was put on the Papal Index of Prohibited Books and was considered in 2005 by a conservative American journal to be the eighth most dangerous book published in the nineteenth and twentieth centuries—even more dangerous than Friedrich Nietzsche's *Beyond Good and Evil*.[7] Comte's infamy derives from his association with atheism and "secular humanism." And yet he clearly understood humans' spiritual longings and the role of religion in society. Paradoxically, with its rites, sacraments, and ecclesiastical hierarchy headed by himself as pope, the Religion of Humanity was not completely in tune with secular trends.[8] Although we may laugh at it today, this religion was popular for several decades after his death.[9] Its erasure of God and its austere morality, based on scientific evidence and founded on "altruism" (a word he coined around 1850), appealed to religious skeptics who were eager to believe in something universal and to work for the improvement of themselves, the human race, and the environment. Comte's religious views, as much as his political ideas, resist simple explanations and categories, a problem that makes charting his impact all that more difficult.

It is evident that Comte's philosophy is so rich and complex that its various components could influence individuals without their subscribing to the whole.[10] They were able to cherry-pick the elements they liked. A scholar might embrace Comte's synthesis of scientific knowledge, consider his classification of the sciences and law of three stages effective explanatory devices, and praise his scientific view of history and sociology, and yet this same individual might dismiss his political views, regard the Religion of Humanity as a throwback to theology, and object to his mania for systematization. A person in the throes of spiritual agony might be indifferent to Comte's scientific system but enthusiastic about the Religion of Humanity with its rituals, dogmas, and altruistic morality. Even devout Christians and Muslims could embrace Comte's humanitarianism and emphasis on the community. Politically, positivism posed challenges as well. A person on the right could approve of Comte's authoritarianism but reject his secular doctrine and his republicanism based on social justice. A person on the left might applaud Comte's concern for workers and use his secular system to attack traditional churches, but decry his praise of dictatorship and his proposals for an extremely regulated society. Yet could all these individuals who adopted aspects of his doctrine be labeled positivists? Not according to Comte. He demanded that every "orthodox" positivist accept his entire doctrine. Yet must we agree with his point of view? Is it not true that there are many kinds of positivisms and thus positivists of different stripes?

Another difficulty with untangling Comte's chains of influence is contextual. Comte's scientific approach to knowledge was echoed by other thinkers in the late nineteenth century. Many philosophers, historians, literary critics, and other scholars after 1860 were eager to claim that their field was scientific in order to raise its prestige in a period when science was all the rage.[11] Yet simply because a person had a scientific approach, did that mean he or she was a follower of Comte? Could Comte's influence be indirect so that an individual was unconsciously a positivist? Could a person be influenced by reacting against Comte's ideas? After all, John Stuart Mill seemed to fine-tune his defense of individualism after realizing how much he disliked the dogmatic aspects of Comte's philosophy that had initially attracted him. Finally, it is important to recognize that Comte's ideas changed after his death as they were misinterpreted or as they were mingled in the minds of posterity with those of other political and scientific thinkers, such as Mill, Herbert Spencer, and Charles Darwin, or philosophers, such as Immanuel Kant and Georg Hegel. If a person held a mutated version of Comte's doctrine, would his or her worldview constitute part of his legacy?[12] These are intriguing questions that make tracing his influence a challenge.

Comte's Legacy in France

Despite the mixed nature of Comte's reputation, he had a significant impact in France. His most influential French disciple was Émile Littré (1801–1881), author of the famous *Dictionnaire de la langue française* (1863–1873). Littré broke with Comte after the latter supported Napoleon III, introduced too many regulations into the Religion of Humanity, and argued with him about money and marital matters. Nevertheless, in articles for *Le National* and in numerous books, Littré worked for forty years as a popularizer of positivism and its scientific method. In his "Préface d'un disciple," published in *Principes de philosophie positive* (1868), he explained positivism's attraction: "Positive philosophy is the only [philosophy] that makes known how these three things are connected: the order of immanent properties, the order of the successive constitution of the sciences, and the order of their hierarchical teaching." Comte succeeded in making philosophy more scientific, while giving the sciences a philosophic turn. In relying on "the new instincts created by science and industry" to repel "the old habits created by theology and metaphysics," his new philosophy put "man in his place in the intellectual and moral world." It preached to people "resignation to what is unchangeable," and gave them the "knowledge to discern what can be changed, and the moral force to make use of the property of things to improve their material condition and themselves." Littré presented positivism as an exciting new worldview, one that would render obsolete the old philosophical and literary movements and traditional institutions. Because he was a respected intellectual and senator, he helped make positivism a major movement in the latter half of the nineteenth century. It gained favor because he showed that it need not lead to authoritarianism, could be compatible with republicanism, and would ensure order and stability. Attaching "mental and social stability to the stability of science," it represented "the remedy for a troubled era."[13] Above all, no one had to fear that the spread of positivism would lead to another revolution. Thanks partly to his effective propagation skills, positivism became an important doctrine upholding the Third Republic.[14] However, Littré, along with Mill, contributed to making it also a more one-dimensional philosophy, one completely dedicated to the cult of science and objectivity.

Another important figure was the mathematician and philosopher Pierre Laffitte (1823–1903), who was Comte's closest disciple in the last ten years of his life. Laffitte directed the positivist movement for almost fifty years. After Littré died in 1881, he became the "quasi-official philosopher" of the Third Republic.[15] Unlike Littré, he supported both Comte's scientific doctrine and the Religion of Humanity, although the latter was not a priority for him.[16] He gave lectures, wrote articles, and guided the Positivists, who were headquartered in Comte's apartment at 10, rue Monsieur le Prince. In 1892 the Collège de France awarded Laffitte the first chair in the history of science, the discipline that Comte had launched.

Comte's philosophy became well known in France thanks also to journals published by different groups of positivists. Grégoire Wyrouboff and Littré's *Philosophie positive* ran from 1867 to 1883. One interesting contributor with positivist roots was the feminist scientist Clémence Royer (1830–1902), who translated Darwin's *Origin of Species* into French in 1862.[17] Dr. Eugène Sémérie (1832–1884) published the *Politique positive* from 1872 to 1873. He belonged to a group of positivists that was more enthusiastic about Comte's religion and later broke with Pierre Laffitte, causing a schism in 1878. That same year Laffitte launched the highly successful *Revue Occidentale*. Charles Jeannolle (1822–1914) took it over at Laffitte's death in 1903 and ran it until he himself died in 1914. Jeannolle's control over the positivist movement was challenged, however, as he had a difficult character and no enthusiasm for the Religion of Humanity. Another schism occurred in 1906. Most positivists followed the more dynamic republican politician Émile Corra (1848–1934), who set up a new headquarters for the movement at 54, rue de Seine. He founded the International Positivist Society which resumed some of Comte's religious observances, such as feasts honoring the Dead and Humanity. In 1906 Corra started the *Revue Positiviste Internationale*, which stopped publication in 1940. The president of the International Positivist Society during World War II was André Haarbleicher (1873–1944), but as a Jew and Freemason, he was deported in 1944 and died on the way to Auschwitz. At the end of World War II, there were only five adepts of religious positivism left in Paris; after almost a hundred years, the organized positivist movement was dying out.[18]

Thanks in part to Littré's and Laffitte's high profiles and to these journals, which applied positivism to current social and political developments, Comte's impact remained strong during the Third Republic—an impact that resonates still today. As the writer Anatole France (1844–1924) noted, the "consciousness" of many French people was "profoundly" marked by Comte's "great ideas," including his critique of metaphysics, faith in the scientific method, history of progress, and "happy idea of a morality founded on human solidarity."[19] Many of these ideas found fertile ground in France, where they had been adumbrated during the Enlightenment, which underwent a revival in the second half of the nineteenth century.[20]

The problem of whether to regard people who shared in this consciousness as positivist is particularly evident in the case of Hippolyte Taine (1828–1893) and Ernest Renan (1823–1892), two important men of letters in the generation after Comte. The former created a "scientific method for metaphysics," while the latter sought a "scientific religion." Both were affected not only by Comte but by Hegel; such an intermingling of disparate philosophies makes it even more difficult to discern exact influence.[21] What also complicates matters is that neither called himself a positivist.

At the time of his death in 1893, Taine was regarded as the foremost French exponent of the principle that the past should be subjected to universal laws. He wanted not only history but the study of politics and morality to gain scientific status by relying on facts. At the same time, he insisted that the subject matter and methods of the moral and political sciences were complex and irreducible to the phenomena and methods of the natural sciences. All of these ideas were similar to those of Comte. Indeed, Taine praised Comte's theory of science, which he saw as the basis of modern logic. Taking up the study of milieu, which was strongly advocated by Comte in his *Système de politique positive*, Taine also sought to make literary criticism a science; it would focus on the influence of the milieu on the artist, a topic he took up in his *Histoire de la littérature anglaise* (*Introduction to the History of English Literature*), published in 1863–1864. (Taine's disciple Émile Hennequin [1858–1888] continued his efforts to make literary criticism a science—efforts that culminated in *La Critique Scientifique* [1888], which offered a three-part analysis of literature—aesthetic, psychological, and sociological—and focused on the effects of a text on the audience, foreshadowing twentieth-century reception theory.[22])

Noting the similarities between Taine and Comte, many past scholars, such as Edmond Scherer, Émile Faguet, Alfred Fouillée, Harald Höffding, Alfred Weber, and Émile Meyerson, considered him a positivist. However, pointing to Taine's search for causes, his acceptance of metaphysics, and his effort to base the study of man on psychology, a discipline rejected by Comte, and arguing that Taine discovered Comte rather late, in the early 1860s, after he had already formed his worldview, more recent scholars, such as Jean-Paul Cointet, do not find him to be a positivist.[23] Nathalie Richard argues that Taine was more of an "admirer of Stuart Mill than a disciple of Auguste Comte."[24] Yet if Taine did pull some of his ideas from Mill, one must remember that Mill obtained many of his own concepts from Comte. Again, tracing influences presents many challenges. Did Taine embrace Comte's ideas that were taken directly from his works or indirectly from secondhand sources, such as Mill or Dr. Charles Robin (1813–1878), whose course he audited? (Robin, an important histologist, had been a close disciple of Comte and propagator of his views.) Or did Taine simply absorb scientific ideas that were in the air? It is impossible to say conclusively.

Besides Taine, the historian and philosopher Renan is often considered a positivist.[25] Just as Taine preferred Mill to Comte, Renan admired Littré, with whom he was very friendly because of their common interest in philology; he frankly did not like Comte. Nevertheless, there are undeniable parallels in the two men's doctrines. Like Comte, Renan rejected supernatural and metaphysical concepts as not sufficiently based on observation. His most famous book, *La Vie de Jésus* (*The Life of Jesus*), which was published in 1863, treated Jesus as a historical figure and Christianity as a product of mythmaking. Like Comte, he sought to create a philosophy based on science. Indeed, he believed that science was a religion and could solve humankind's problems, leading the way to progress. This religion of science would not only cultivate the mind but fulfill people's need for faith. He set out many of these "positivist" ideas in his main work, *L'Avenir de la science* (*The Future of Science*), published in 1890. Yet in this work, he also took his distance from positivism, lashing out at Comte for his narrow, dry conception of human nature. It seemed to Renan that Comte did not recognize the value of psychology, traditional religion, the arts, and works of imagination in general. Comte's *Cours de philosophie*, with its overriding law of three stages, showed a singular lack of appreciation for the histories of diverse cultures, whose characteristics were best grasped through philology, the true science of humanity.

Comtean positivism was simply too intellectual, reductionist, and analytical for Renan's taste. In arguing the case for the reconciliation of religion and science, Renan, who had been a seminarian, presented them in a different fashion than did Comte. His religion, unlike Comte's Religion of Humanity, embraced the unobservable, the transcendent, and the infinite. It seemed at times even to acknowledge God. A historian of different creeds, Renan was more fully entranced by religion than was Comte, who saw it chiefly as an instrument to reorganize society. Renan also created more of a cult of science, though his definition of science was not precise. It seemed sometimes to mean simply the orderly use of the mind, which could include intuition. His science could also seek causes, something anathema to Comte. In addition, Renan wanted scientific research to be pursued for its own sake, not for a practical or social purpose as Comte demanded.[26] One could argue that Renan deserves to be called a positivist even less than Taine, who had at least a more favorable opinion of Comte.

Besides Taine and Renan, scholars in a variety of disciplines grappled for decades with the legacy of positivism, especially its key principle that science was a model for knowledge. According to Anastasios Brenner, positivism dominated French philosophy in particular until World War I. "Comte set the agenda" by obliging philosophers, especially philosophers of science, to address important issues such as "the classification of scientific disciplines, the role of hypotheses and the empirical criteria of meaning."[27]

The role of hypotheses was particularly controversial. As Warren Schmaus has demonstrated, the theory that hypotheses could guide scientific research was upheld by Comte and a dozen or so other prominent European scholars.[28] Yet Comte did a great deal to make this relatively new concept acceptable. It countered the traditional, empirical approach that tended to reduce science to a collection of data. The idea that one should begin research with a provisional hypothesis—a hypothesis that could later be disproven—also challenged the absoluteness of knowledge. It was hard for many scholars to accept the paradox that one commentator has found to be at the heart of positivism, that is, "the belief that the key to the advancement of knowledge is the awareness of the limits of knowledge."[29] Positivism's more tentative attitude toward knowledge ran counter to the tenets of the reigning academic school of philosophy in France, the spiritualist or eclectic school of Victor Cousin (1792–1867), which embraced Thomas Reid's commonsense and induction-oriented approach. As Reid and Cousin believed in absolute truths, they rejected the employment of hypotheses in science.

Their views were attacked by the neocritical idealist philosopher Charles Renouvier (1815–1903), who had been Comte's student at the École Polytechnique in the mid-1830s. He adopted Comte's approach to hypotheses and the relativity of knowledge and his principle that knowledge should be limited to the laws of phenomena. However, Renouvier disparaged Comte's illiberal social philosophy, pointed out his vague definitions of law and phenomena, and criticized the law of three stages, especially for inadequately describing the development of the sciences. Renouvier in the end decided to develop Kant's legacy.[30]

Renouvier's contemporary Paul Janet (1823–1899) was very familiar with Comte's ideas, covering them in depth in an article for the *Revue des Deux Mondes* in 1887 and in his overviews of philosophy. Like Comte, he maintained that hypotheses could direct experimentation, challenging the approach of Cousin, his mentor. Janet attempted to find a middle ground between Cousin's older, more literary tradition and the newer, more cutting-edge positivist movement.

Janet's ally in the fight against materialism was another philosopher, Émile Boutroux (1845–1921). Though ultimately a defender of spiritualism like Janet, Boutroux likewise upheld the "hypothetical, contingent nature of science."[31] He learned a great deal about Comte from his friend Paul Tannery (1843–1904), a graduate of the École Polytechnique and an enthusiast of the history of science. Like Renouvier and Janet, Boutroux incorporated elements of both Kant and Comte into his own theories, which challenged simple scientism.[32] Some scholars claim that Boutroux espoused a "metaphysical positivism."[33]

Janet and then Boutroux held the chair of History of Modern Philosophy at the Sorbonne. The philosopher who succeeded Boutroux in 1908 was another scholar influenced by Comte, Lucien Lévy-Bruhl (1857–1939). In 1900 he published a hefty, still much-esteemed book on Comte—*La Philosophie d'Auguste Comte*—that points to the rejuvenation of interest in positivism around this time. Indeed, in 1902 Lévy-Bruhl and Boutroux argued in favor of looking again at positivism at a meeting of the Société Française de Philosophie; they hoped to encourage its development along Kantian lines by examining the conditions of knowledge, a subject glossed over by Comte. Lévy-Bruhl also pushed a positivist agenda in his attempt to develop morality as a science and to import Comte's scientific approach into the new field of anthropology. With his respect for fetishism, Comte inspired his theory of the existence of a "primitive mentality," which stood in opposition to the "modern" mindset. Furthermore, Lévy-Bruhl, like Comte, believed that ideas—whether "primitive" or "modern"—should be considered reflections of their historical context. Intellectual and social developments were particularly interconnected.[34] Lévy-Bruhl's theory was an important contribution to the sociology of knowledge.

Besides philosophers and philosophers of science, French scientists were intrigued by Comte's legacy, focusing often on questions of methodology. According to D. G. Charlton, the physiologist Claude Bernard (1813–1878) fully adhered to the "positivist theory of knowledge," especially its principles that "the scientific method is our only means of knowing reality" and that such knowledge "will always be relative and provisional."[35] Bernard was also influenced by Comte in his approach to experimental medicine, where he focused on understanding the internal environment of organisms.[36] In addition, Bernard's idea that the pathological was similar to the normal could be found in Comte's works.[37] Annie Petit points out that Bernard "apparently advanced several Comtean orthodoxies," including "the wisdom of abandoning first causes, absolute knowledge, and realist pretensions; [and] the need to limit positive knowledge to phenomenological relationships," but she argues that he was not a positivist in his dismissal of history and philosophy.[38]

Bernard's contemporary the chemist Marcelin Berthelot (1827–1907) adhered to the principle that scientific knowledge must limit itself to observable phenomena, which led him to deny atomism as too speculative. After all, no one had seen an atom. Many scholars, such as Maurice Crosland, who tend to reduce positivism to empiricism and scientism, see this position as a typical positivist one.[39] Crosland and others point out that Berthelot, an anticlerical politician during the Third Republic, did much to advance positivism as a social philosophy, for he believed the scientific method could renovate morality and lead to material progress.[40] Yet Bernadette Bensaude-Vincent downplays the similarity between Berthelot and Comte, pointing out that Comte himself was not opposed to atomism, nor were two of his disciples who promoted it in the nineteenth century: Alexander Williamson (1842–1904) and Jacobus H. van't Hoff (1852–1911).[41] Petit tends to agree with her, making the interesting argument that scholars seeking to explain the dominance of positivism in the nineteenth century have mistakenly latched onto Berthelot and his friend Renan as exponents of the philosophy because they were so famous and influential. Laffitte, who was in truth the more orthodox positivist, was not prominent, and thus no one can imagine that he could be the source of Comte's large impact.[42]

Around the turn of the twentieth century, the two physicists Jules Henri Poincaré (1854–1912) and Pierre Duhem (1861–1916) elaborated on the use of the scientific method as the way to achieve legitimate knowledge.[43] They did not proclaim themselves disciples of Comte. Duhem referred to Comte only once in his works. (He mentioned Mill more often.) Poincaré rarely spoke of him. Nevertheless, to some scholars, they seem to be in the positivist tradition and owe a great deal to Comte.[44] Indeed, Édouard Le Roy (1870–1954) called the intellectual movement that they helped launch "a new positivism."[45] Bensaude-Vincent and Jonathan Simon regard Duhem as a positivist: "Duhem, like other positivists, regarded scientific theories as . . . simply instruments for the organization and classification of data, which, while they could be better or worse, did not denote any underlying reality."[46] With his rich notion of scientific theories as simply helpful tools, Duhem helped spread Comte's idea that science can give only an approximation of reality.[47] Poincaré, the brother-in-law of Boutroux, was a conventionalist who also denied that scientific theories exactly represented reality. Theories consisted of conventions; they reflected a consensus among scientists who determined that they were convenient in terms of the situation under consideration at the moment.[48] According to María de Paz, "conventionalism considered the role of the scientist in the theoretical constitution of science, instead of being a simple data-collector or an event-descriptor. This meant that conventionalism inquired into the creative role of the scientist."[49] Comte pointed out the role of the scientist's imagination in the construction of useful provisional hypotheses and underscored the importance of the social context of science, but it is not clear whether he was the source of the two men's theories. Nevertheless, it is indisputable that his thoughts on the scientific method inspired the continued development of the philosophy of science.[50] Philosophers of science Georges Canguilhem (1904–1995) and Michel Serres (1930–), though not positivists, acknowledged his influence.[51]

Comte's imprint, according to Anastasios Brenner, is also evident in the tendency of French philosophers of science to focus on the history of science instead of scientific logic.[52] Duhem, Poincaré, and Émile Meyerson (1859–1933) certainly approached science through its development in the past. Léon Brunschvicg (1869–1944) added a strong neo-Kantian, less empirical feature to this historical approach. All four men provided at first an inspiration but then a departure point for the great philosopher of science Gaston Bachelard (1884–1962), who came to oppose positivism.[53]

Paul Tannery (1843–1904), Abel Rey (1873–1940), and George Sarton (1884–1956) did much to organize Comte's history of science as a new discipline. Sarton (who was Belgian and later worked in the United States), was particularly important. In 1912 he founded *Isis*, the journal devoted to the history of science that still operates today. Sarton maintained Comte's view that the sciences should be studied together by epoch; that the history of science should be a narrative of both intellectual and moral advances; and this history was the story of humankind's liberation from superstition, servitude, and intolerance. Sarton was even attracted to Comte's religious ideas but rejected his dogmatism, which he found repugnant in all realms, including the sciences.[54]

The history of the "sciences de l'homme"—sciences of man—in France also attests to the pervasiveness of Comte's ideas. We have seen that Lévy-Bruhl enriched anthropology, which had already been much influenced by positivism. The Société d'Anthropologie de Paris was founded in 1859 and, according to Joy Harvey, adopted a positivist program in that it attempted to found this new science of human races on observation, measurement, induction, description (as opposed to explanation of first causes), facts, and experimentation, where possible. One important member was Comte's disciple Dr. Charles Robin (1813–1878), who had helped establish the Société de Biologie on a similar basis ten years before. Paul Broca (1824–1880), another founder of the Société d'Anthropologie, was also a positivist.[55] The only female member, Clémence Royer, had a positivist background as well. Contesting Comte's influence, Claude Blanckaert has recently pointed out, however, that there were few real positivists in this organization and that the ones who were members, including Robin, hardly participated. He claims that Broca was a free thinker who eschewed all philosophies, including Comte's. Nevertheless, the société seemed eager to associate itself with the movement, as shown by its making Littré and Renan honorary members.[56]

Two close friends at the École Normale—Théodule Ribot (1839–1916) and Alfred Espinas (1844–1922)—were attracted to positivism in their fight against spiritualist philosophy and brought its outlook into the two sciences that they helped shape: psychology and sociology, respectively. They pulled these studies of the mind and society away from traditional philosophy, which resisted positivist incursions.[57] Ribot, who became professor of psychology at the Collège de France in 1888, attempted to rid psychology of the introspective method that Comte condemned.[58] (In the 1870s and 1880s, Jean-Martin Charcot [1825–1893] similarly attempted to put his studies of hysteria at the Saltpêtrière on a positivist foundation, insisting on observing the external behavior of his ill patients [hence the importance of photography in the study of the mentally ill], dismissing unseen causes, and creating universal laws relating to mental pathology.[59]) Appointed to the newly created chair in "social economy" at the Sorbonne in 1893, Espinas went further than Ribot in arguing that the social sciences had beneficial political and moral effects and could increase social harmony.[60]

Espinas, along with Littré, Renouvier, and Boutroux, had a large impact on Émile Durkheim (1858–1917). Durkheim occupied the first chair in sociology at the Sorbonne in 1913. Comte's influence is evident in Durkheim's belief that society could be studied in a scientific manner; his desire to shape sociology around the search for consensus, social solidarity, and moral values; and his stress on the specificity of social phenomena and distinctiveness of sociology.[61] Durkheim also referred to Claude Bernard in his philosophy course and adopted his positivist definition of the experimental method as one rooted in the testing of hypotheses, not in the simplistic manipulation of data. Comte's indirect influence is obvious.[62]

Evidence of Comte's impact can be seen in French novels as well. Although he never read any of Comte's works, the realist novelist Émile Zola (1840–1902) knew Littré, Wyrouboff, and Laffitte and felt indebted to Taine and Claude Bernard. Inspired at least indirectly by Comte, Zola claimed to be a positivist, practicing sociology. Positivist influences can be discerned in his commitment to empirical data, that is, to observations of everyday life. In addition, he looked upon his novels as scientific experiments; characters set in motion a series of events that were determined by the phenomena being studied. At the end of his life, Zola even expressed approval for Comte's attempt to found a new religion.[63]

Maurice Barrès (1862–1923) called Comte one of his three "gods."[64] With an eye to strengthening nationalist feelings, he praised Comte's "cult of the dead," whose "heroes" allowed us to bypass traditional religions while still giving us the social connections fostered by religion.[65] He also appreciated Comte's struggle to make morality the leading science, one that made sure all scientific research served human interests.[66]

The contemporary novelist Michel Houellebecq (1956–) is an avowed fan of Auguste Comte. Like Comte, he is adamant about the importance of understanding scientific methods and principles in the modern age, critical of the liberal discourse of individualism and rights, insistent on the social nature of humans, and convinced of the end of traditional religion—an end that he finds troubling in terms of maintaining the cohesiveness of the collectivity. Houellebecq's pessimistic novels depict individuals motivated primarily by material values with no faith in anything. To him, the metaphysical stage of history is one that we are still struggling to exit.[67] But Houellebecq does not approve of the Religion of Humanity, Comte's solution to modern skepticism and despair, because it rejects eternal life.[68] Houellebecq has recently become notorious for his denunciation of Islam, a position that would not have been upheld by Comte, who respected this religion.[69]

In addition to leaving its imprint on academia and literature, positivism had a significant political influence. Its impact was most apparent during the Third Republic. The workers' movement during this period was affected by positivism. In 1880 Auguste Keufer (1851–1924), the head of the Printers' Union, took over the Cercle des Prolétaires Positivistes, which had been founded by a carpenter and devoted follower of Comte's courses, Fabien Magnin (1810–1884). Keufer supported positivist doctrines, including the Religion of Humanity; pushed for a moderate, that is, reformist form of syndicalism; and helped establish the French national labor union, the Confédération Générale du Travail (CGT) in the 1890s. Another advocate for the working class was Victoire Tinayre (1831–1895), a friend of the revolutionary Louise Michel (1830–1905). Comte's disciple

Henry Edger was her neighbor in Paris and converted her to positivism in 1880. A socialist and feminist, she campaigned for secular primary education, especially for girls, an idea advanced by Comte in the *Système de politique positive*.[70]

The leading politicians of the early years of the Third Republic, Jules Ferry (1832–1893) and Léon Gambetta (1838–1882), were also directly influenced by positivism, which shaped their anticlerical, educational, and imperialist policies. Reflecting the law of three stages, they were convinced that religions had accomplished their role but were now obsolete and should disappear. This argument formed part of the French doctrine of laïcité, which claimed that religious dogmas prevented the proper use of reason.[71] Positivism attracted attention especially during the Third Republic because after 1871 many people blamed French backwardness for the victory of the Prussians, whose universities and research in the sciences seemed superior; the French increasingly looked at the sciences as an effective tool with which to rejuvenate the country.[72] It was not a coincidence that a chair in the history of science was created and given to Laffitte in 1892—a chair Comte had requested without success from François Guizot in 1832.

Gambetta seemed close to Laffitte and knew Littré, whose works on positivism he read with care. Calling Comte the most powerful nineteenth-century thinker, Gambetta used his ideas about freedom of thought and the importance of science and reason to support his anticlerical campaign. He was convinced that if the influence of the church remained strong, the Third Republic would die. Positivism was his weapon of choice to ensure its survival.

Ferry, unlike Gambetta, read Comte's books. He knew not only Laffitte and Robin but Comte's friend and disciple Hippolyte Philémon Deroisin, who initiated him into positivism in 1857 by having him read the *Discours sur l'ensemble du positivisme*. Ferry was also friends with Littré, whom he met at the Freemason lodge of Clémente Amitié.[73] Hoping to diminish the influence of the Catholic church on the educational system, Ferry sought to use schools to inculcate the "laic" values of the Third Republic in a positivist fashion. Girls were to be given a solid education as well—an idea espoused by Comte. In general, Ferry believed that positivism could assure political and social unity on the basis of the sciences and a new morality centered on love for humanity.

Reflecting Comte's insistence on honoring contributors to civilization and thereby ensuring them immortality, the Third Republic made respect for the dead part of its civic instruction, as seen in the establishing of the Panthéon as a republican site of worship in 1885. In the late nineteenth century, there was a kind of "statue mania" and a predilection for big state funerals, all of which bolstered national solidarity.[74] As Philippe Ariès remarked, "the cult of the dead became the great popular religion of France."[75]

This cult extended to Comte himself. In 1902 a statue of Comte flanked by a worker and a woman was erected on the Place de la Sorbonne, attesting to his influence.[76] According to Catherine LeGouis, the Sorbonne was "the bastion of the positivists" during the fin-de-siècle, thanks especially to the dominance of Gabriel Monod and Ernest Lavisse, historians who advocated the scientific method.[77] As we have seen, other social scientists and philosophers at the Sorbonne from the 1880s to 1919 were also influenced by Comte even if they were not positivists per se, most notably Espinas, Durkheim, Abel Rey (the historian of science and philosopher), and Lévy-Bruhl. The scientists Bernard and Poincaré did a stint there as well. In addition, from 1880 to 1888, Laffitte had been allowed by Jules Ferry (then minister of public instruction) to give public lectures at the Sorbonne, which attracted a large audience.[78] One person at the inauguration ceremony of 1902 who objected to those lauding Comte exclusively for his antireligious stance was Pierre de Coubertin (1863–1937). He preferred to dwell on Comte's global vision, which had inspired him to found the modern Olympics movement in 1894.[79]

A philosopher who paid his respects to this statue but was unassociated with the Sorbonne was Émile Chartier (1868–1951) a *lycée* teacher who went by the name of Alain. He was the spokesperson for the Radical Party, which espoused radical republicanism, a political movement that was socially conservative but anticlerical and loyal to the principles of the French Revolution. Alain considered Comte to be the founder of radical republicanism because it echoed the positivist philosopher's wariness of large states and his position that political institutions should be based on small social units, such as the family.[80] Georges Clemenceau (1841–1929), who was the hardcore anticlerical leader of the Radical Party and the prime minister of France during World War I, was influenced by positivism when he studied medicine under Charles Robin; indeed, he translated Mill's famous book on Comte into French in 1868.

Although Comte opposed large, self-aggrandizing states and colonization, positivism was nevertheless used to bolster the civilizing mission at the heart of French imperialism and to strengthen French nationalism. Imperialists claimed to spread the benefits of science and industry and guide so-called backward nations so that they could evolve from primitivism to the level of the more secular, hygienic Western Europe.[81] (Paradoxically, during the same period, British followers of Comte were using his ideas to campaign *against* imperialism.[82])

The monarchist Charles Maurras (1868–1952) was a leading exponent of nationalism in the early twentieth century. He embraced positivism and its insistence on the need for submission as a means of self-improvement. Along with other members of the Action Française, he used positivism's authoritarian principles and scientific aura in his fight against parliamentary democracy, the rights of man, and egalitarianism. Like Comte, he privileged the collectivity over the individual. Because the Action Française was associated with fascism, its embrace of positivism hurt Comte's reputation.[83]

Positivism in Great Britain

Comte's influence was widespread outside of France. Because of space constraints, we cannot examine it in depth nor introduce readers to all the controversies surrounding the relationship of positivism to each thinker and movement. However, we will give readers an overview that will hopefully stimulate more study on their part.

Comte had a significant impact in Victorian Britain with its long-standing empirical tradition.[84] Having corresponded with Comte from 1841 to 1847, John Stuart Mill (1806–1873) was his most influential British follower. Early on, Mill persuaded leading intellectuals such as John Herschel (1792–1871), William Whewell (1774–1866), and Edward Bulwer-Lytton (1803–1873), to read his works.[85] But their correspondence ended after Mill disagreed with Comte's handling of finances and his dismissive remarks about women. Nevertheless, although he was embarrassed by his early enthusiasm, which had led Comte to consider him a disciple, Mill continued to support the Frenchman's philosophy of the sciences and extensive use of the scientific method. He frequently referred to Comte's idea of the need to assure both order and progress and his notion of a guiding educated elite. In addition, he maintained Comte's interest in replacing traditional creeds by a religion of humanity based on feelings of altruism. Yet he deliberately distanced himself from Comte's authoritarianism and misogynist tendencies in his second edition of the *System of Logic* and in other books, notably *On Liberty*, *The Subjection of Women*, and *Auguste Comte and Positivism* (1865). In the end, Mill helped create a different, arguably more influential version of positivism, one that advocated a formal epistemic system inspired by the sciences and permanent rules for directing scientific practices.[86] As mentioned previously, he contributed to making positivism a scientific manifesto for the industrial age.

Many of Mill's contemporaries were taken with Comte. Harriet Martineau (1802–1876) freely translated and condensed the *Cours de philosophie positive* in 1853. George Henry Lewes (1817–1878) wrote on Comte in the *British Foreign Review*, *Westminster Review*, and the *Leader* and included a large section on him in his *Biographical History of Philosophy* (1845–1846). The Scottish logician and psychologist Alexander Bain (1818–1903) gave up his evangelical beliefs and embraced Comte's stress on induction and sensory knowledge as key aspects of proving the truth of an idea or law. Henry Thomas Buckle (1821–1862), whose multivolume *History of Civilization in England* began to appear in 1857, was called a positivist for presenting history as a science based on laws; he wanted to demonstrate that physical causes, such as climate, shaped societies in certain ways.[87] None of these persons was excited about the positivist religion.

Other contemporaries, however, did understand the attraction. Richard Congreve (1818–1899) was the dominant orthodox positivist. He translated the *Catéchisme positiviste* into English in 1858 and wrote books to propagate the Religion of Humanity. At Wadham College, Oxford, he gathered around him Frederic Harrison (1831–1923), Edward Spencer Beesly (1831–1915), and J. H. Bridges (1832–1906), Oxford students who became eager acolytes of positivism. (They translated the *Système de politique positive* into English in 1875–1877.) In 1867, Congreve established the London Positivist Society, which began meeting at 19 Chapel Street in London in 1870 and had forty-seven contributors by 1878. That same year, Congreve broke with Pierre Laffitte, whom he did not view as sufficiently interested in founding churches of Humanity. The London Positivist Society dissolved due to internal squabbles about Congreve's authoritarian, cold personality and his dogmatic approach to religion, which seemed to hew to empty Catholic and Anglican formulas.

While Congreve maintained a London Church of Humanity on Chapel Street, Harrison, Beesly, and Bridges led the dissident positivists in a new London Positivist Committee (later known as the English Positivist

Committee), which met at Newton Hall and remained loyal to Laffitte. (An affiliate was established in Manchester.) It experimented with different approaches to the cult of Humanity—adding music, for example—but it also devoted itself to social activism. Maintaining Comte's interest in the working class, Beesly and Harrison promoted trade unions. To help his causes, Beesly founded the *Positivist Review*, which ran from 1893 to 1925. Bridges contributed over a hundred articles. In 1901, Newton Hall joined for financial reasons with Chapel Street. The joint organization finally closed its services in 1934. Some people who had been worshippers included the sociologist Patrick Geddes (1854–1932) and the Greek scholar Gilbert Murray (1866–1957). Churches of Humanity that were active in Liverpool (founded in 1883) and Newcastle upon Tyne (founded in 1885) survived until the 1940s.[88]

Comte's religion was popular in England outside of strictly positivist circles.[89] The social reformer Annie Besant (1847–1933) wrote her own *Secular Song and Hymn Book* because of her enthusiasm for Comte's religion in the 1870s.[90] The Fabian socialists Sidney Webb (1859–1947) and Beatrice Webb (1858–1943) considered themselves quasi-positivist priests and in 1895 founded the London School of Economics, where Britain's first chair of sociology was established in 1903.[91] William Morris (1834–1896), Walter Pater (1839–1894), Thomas Hardy (1840–1928), and George Gissing (1857–1903) applauded the worship of Humanity. Although she refused to be labeled a Comtist, George Eliot (1819–1880), Lewes's partner, approved of Comte's making Humanity the center of thought and activity; her novels alluded to positivist ideas regarding the altruism inherent in human nature.[92]

One important social theorist who resented being viewed as a disciple of Comte was Herbert Spencer (1820–1903), who was friends with Eliot and Lewes and briefly met Comte in 1856. When Spencer published his *Social Statics* in 1851, people assumed he was elaborating on one of the chief divisions of Comtean sociology: social statics. Although he denied any influence, he had absorbed Comte's ideas about the need to replace religion by an all-embracing system of philosophy based on the sciences, adopted from him key terms such as "sociology" and "altruism," and followed his principle that society was an organism that developed throughout history according to laws that could be discovered by scientists. However, Spencer differed from Comte in his support of laissez-faire, freedom of the individual, and social Darwinism. More of a proponent of empiricism than Comte, Spencer dismissed the law of three stages, classification of the sciences, and Religion of Humanity.[93]

The Spread of Positivism on the Continent

Positivism penetrated many places in continental Europe. In Sweden, a Positivist Society was begun in 1879 in Stockholm by Dr. Anton Nystrom (1842–1931), who was close to Laffitte and translated several of Comte's works into Swedish. Like Comte, Nystrom was disgusted by the greed and hypocrisy of the bourgeoisie and put his faith in workers and in popular education. Indeed, workers made up the majority of members of the Positivist Society. In 1881, Nystrom founded the Workers' Institute, which offered scientific, practical, and cultural courses to workers. He also supported the trade union movement and established the Stockholm Liberal Voters' Association, which demanded the improvement and secularization of the school system, universal suffrage, and freedom of religion. The astronomer Karl Hjalmar Branting (1860–1925), who edited the radical journal *Times* in Stockholm, also was a positivist until he switched to socialism and became a politician. (He became the first socialist prime minister of Sweden.) In general, as Madeleine Hurd has written, "The scientific teachings of Positivism were lumped with other, subversive workers' ideologies."[94] Positivism was instrumental in pushing for the separation of church and state, the secularization of society, and scientific inquiry, which challenged those in power.[95]

In the late nineteenth century, Comte won adherents in the Netherlands and Belgium. Many used his ideas to critique religion and faced accusations of atheism, materialism, and socialism. Nevertheless, some enthusiasts were very prominent. One of Comte's early disciples was Menno David, Count van Limburg Stirum (1807–1891), who became a Dutch minister of war. Concerned about workers, Ernest Solvay, a Belgian chemical manufacturer, claimed to be influenced by Comte and created the Solvay Institute of Sociology in 1902 to promote his new doctrine of energetics, designed to lift productivity in a humanitarian fashion.[96]

In Italy, the strength of the Catholic church and Comte's hostility to the Risorgimento limited positivism's influence, especially in politics. Nevertheless, positivism proved attractive to Carlo Cattaneo (1801–1869), a public intellectual, editor, and republican who supported liberalism and science and helped lead the revolt of Milan against Austria in 1848. Cattaneo is usually considered the source of positivism in Italy. Comte's ideas (as well as those of Darwin and Spencer) entered more widely into Italian intellectual life after the unification of Italy, when intellectuals sought ways to make it a strong modern state. The main centers of positivist debate were Bologna, Florence, Turin, and Naples. Roberto Ardigò (1828–1920), a former priest, was perhaps the most important positivist. He was a philosopher devoted to the cause of Italy and the spread of scientific enlightenment. He sought to separate philosophy from theology and make it more scientific. However, unlike Comte, he believed that psychology could be a science. Indeed, he claimed to have read Mill and Spencer, not Comte, although he called himself a positivist. Positivism also influenced to some degree the philosopher Andrea Angiulli (1837–1890), the anthropologist and novelist Paolo Mantegazza (1831–1910), the doctor Salvatore Tommasi (1813–1888), and the historians Giuseppe Ferrari (1811–1876), Pasquale Villari (1827–1917), and Niccola Marselli (1832–1899).[97] Cesare Lombroso (1835–1909), his disciple Raffaele Garofalo (1821–1934), and his student Enrico Ferri (1856–1929) attempted to apply the scientific method to the study of criminals. Hoping to eliminate crime by figuring out the laws of criminal behavior, they established the influential Positivist School of Criminology.[98] Vilfredo Pareto (1848–1923) endeavored to create a scientific sociology and is often associated with positivism, but he critiqued Comte for being too dogmatic, metaphysical, and optimistic about the capabilities of science.[99]

As for Germany, in 1825 Comte had some initial contact with the publicist Friedrich Buchholz (1768–1843) thanks to Gustave d'Eichthal, but the relationship went nowhere. In Paris he had two German disciples, Adolphe von Ribbentrop, whose aristocratic airs annoyed him, and August Hermann Ewerbeck (1816–1860), who was friends with Marx, Feuerbach, and Engels. Apparently, Comte's positivism was well known in Germany by the 1850s. However, there was not much interest in positivism because of the dominance of the hermeneutic "life-philosophy" of Wilhelm Dilthey (1833–1911), which was generally considered incompatible with positivism, and the existence of another scientific philosophy, that of Hermann von Helmholtz (1821–1894) and Emil du Bois-Reymond (1818–1896). Comte's criticisms of Germany and Lutheranism for blocking progress to the positive stage of history and his Catholic-infused Religion of Humanity also made him distasteful, as seen in the derogatory remarks of Nietzsche. Not surprisingly, Comte's work was therefore not translated into German until the 1880s and 1890s. Finally, in the early twentieth century there was a greater organizational push for positivism. A teacher of French in a gymnasium, Heinrich Molenaar (1870–1965) founded a German positivist society in Munich that supported anticlericalism, European cooperation, and pacifism. He also published *Die Religion der Menschheit* (1901–1902) and the *Positive Weltanschauung* (1904–1906). The Society for Positivist Philosophy was established a few years later in Berlin in 1912; it published the journal *Zeitschrift für Positivistische Philosophie* from 1913 to 1914.[100]

As for individual Germans, recent scholars suggest that despite his disparaging remarks about "this shit-Positivism," Karl Marx (1818–1883) incorporated elements of the philosophy in his theory of history.[101] Marx was a friend of Comte's English disciple Beesly, who in 1864 chaired the meeting that led to the creation of the International Workingmen's Association. Marx also became familiar with Comte's ideas through his son-in-law, Paul Lafargue (1842–1911), who had been a fan of Comte and the positivist method when he was a medical student in Paris in the 1860s.[102] Marx finally read Comte's work in 1866 because, he said, "the English and the French make such a fuss over the fellow."[103] Marx shared Comte's interest in discerning the laws and stages of human development and in applying scientific reasoning to the study of history.[104] But he found Hegel's synthesis of knowledge to be superior.

The scholar Frederick Beiser has recently called the anti-Marxist philosopher Eugen Dühring (1833–1921) the "founder of German positivism." Dühring thought Comte and Ludwig Feuerbach (1804–1872) were the most prominent thinkers ushering in the modern age. He approved of positivism's stress on facts as the basis of reality especially because that approach provided a good argument against metaphysics. He believed that because we can know only facts of experience, we should find the meaning of life now on this Earth and seek to reform, not overturn, institutions, which we should accept as part of reality. Dühring was, however, disappointed that Comte made the suggestion that there were realms that were unknowable—a suggestion that opened the door to religion. In his eyes, Comte had not gone far enough in his positivism.[105]

The chemist Wilhelm Ostwald (1853–1932) was a great admirer of Comte and even translated his early essay *Plan des travaux scientifiques nécessaires pour réorganiser la société* into German in 1914. Comte's impact can be seen in Ostwald's effort to find the natural laws of the history of scientific thought and his antimetaphysical animus. Ostwald also approved of his hierarchy of the sciences. But Ostwald's attack on hypotheses would not have been condoned by Comte.[106] Another German enthusiast was the literary critic Wilhelm Scherer (1873–1936). Like Taine and Hennequin, Scherer attempted to study the development of literature in a scientific manner, tying texts systematically to history, philology, society, and national character. Yet as reflected in his preoccupation with whether literature shed light on the German people and fostered German unity, his nationalist fervor was decidedly nonpositivist.[107]

David Lindenfeld maintains that the German philosopher Franz Brentano (1838–1917) was a positivist in his devotion to the scientific method in philosophy and to Comte's idea that science advanced by studying more complex matter.[108] Barry Smith appears to agree, pointing out that Brentano was particularly taken with Comte's insistence that science avoid any kind of association with metaphysics and focus only on "phenomena."[109] This idea would be more fully developed in the phenomenological movement of Brentano's student Edmund Husserl (1859–1939).[110]

Brentano helped spread positivism to Austria. In the 1870s, while a professor at the University of Vienna, he introduced his student Alexius Meinong (1853–1920) to the scientific way of philosophizing, which included a strong respect for observation and experience. Like Brentano, Meinong contributed to the development of scientific psychology, going beyond Comte by focusing on mental phenomena or "acts."[111] Brentano was also friendly with the positivist philosopher Theodor Gomperz (1832–1912), who promoted the academic career of the physicist and liberal reformer Ernst Mach (1838–1916).[112] Thanks to the growing circle of sympathizers, Vienna became the center of Austrian positivism, attracting many liberals who were worried about the dominance of the Catholic church in politics and nationalist movements threatening the unity of the empire. According to Deborah Coen, they clung to the empirical sciences as models of "independent thinking and consensus building."[113]

Mach reinforced Vienna's importance in the development of modern philosophy. Positivism nourished his skepticism regarding speculative knowledge and his belief that scientific laws were models for organizing observed facts in a rational fashion. Thanks in part to Mach, the Vienna Circle of thinkers, which included Moritz Schlick (1882–1936), Rudolf Carnap (1891–1970), Philipp Frank (1884–1966), and Otto Neurath (1882–1945), turned Comte's doctrine into "logical positivism," which became influential beginning in the interwar period in both Vienna and Berlin. Logical positivists echoed Comte's critical attitude toward metaphysics, especially in social and political discourse, and his Enlightenment faith in the liberating effects of the sciences. Like him, they sought to make philosophy scientific and to unify knowledge on the basis of the sciences, which privileged observation.

One old-fashioned Comtean positivist who approved of the agenda of the Vienna Circle was the Frenchman Marcel Boll (1886–1971), whose sister had a positivist wedding in 1908. A physicist, he sat on the editorial board of the *Revue Positiviste Internationale*, wrote dozens of articles for the journal, and was a member of the Société Positiviste. Yet in the interwar period in other, more mainstream journals, such *Les Nouvelles littéraires*, and in the association that fought for *laïcité*, the Union Rationaliste (to which Marie Curie and Lucien Lévy-Bruhl belonged), he popularized the ideas of the Vienna Circle, which, according to Peter Schöttler, he considered "a renewed form of positivism, rejuvenated and opened to the present."[114]

However, logical positivists differed from Comte in their respect for pure empiricism, which they separated from theory, and in their search for an epistemology that would fit all the sciences. In addition, they dismissed metaphysics not only because it dealt with issues that could not be known, which was Comte's position, but because it asked questions about God and freedom and so forth that were meaningless. While Mach stressed the importance of mathematical equations in representing the world, other logical positivists looked at language; they deemed statements that could not be verified, especially those involving values, as intellectually hollow. In general, logical positivists spent a great deal of effort examining logic and formal language, which were not the concerns of Comte, who did not think one could come up with universal rules of scientific investigation. Finally, logical positivists differed from Comte in not seeking to offer moral guidance.[115] Thus when the Frankfurt School famously attacked positivism in the 1960s for being too scientistic and detached from normative engagement, it was a different kind of philosophy than the one Comte had advocated.[116] However, Boll's case demonstrates that there were solid links between the old and new versions.

Positivism made an impact elsewhere in Eastern Europe. Another student of Brentano's was the philosopher Tomás Masaryk (1850–1937), who championed the Slavic cause in Austria-Hungary and became the first president of Czechoslovakia after World War I. He was deeply influenced by Comte's idea of experts, scientists of society who in an ideal world would play an instrumental role in politics and promote the common good and morality.[117]

After the failure of their 1863 rebellion against Russia, young Poles embraced positivism as a replacement for romanticism and idealism; with its emphasis on reason, it seemed to offer a more sober, realistic approach to social renovation and progress than direct confrontation or appeals to Poles as a chosen people. As the Polish writer Czeslaw Milosz (1911–2004) explained, "Without the possibility of expressing political aspirations, they had to place all their hopes in science and in economic progress."[118] Many journalists and men of letters, especially in Warsaw, debated positivist ideas, including those of Spencer, Mill, and Buckle, and their potential help in modernizing and liberating Poland or at least creating a sense of Polish community. Polish positivists coined the slogan "organic work," referring to their positivist belief that society was an organism that everyone should develop by working together, especially on economic and educational improvements. Marginalized groups, such as Jews, women, and peasants, should be given equal rights so that they too could contribute to Polish society and culture. Downtrodden groups and all of Poland would be uplifted if everyone were given an education, one grounded in the sciences and reason. Schools could help spread Polish culture as well.

Notable Polish positivists include the philosopher and economist Józef Supiński (1804–1896), Father Franciszek Krupiński (1836–1938), the sociologist Ludwig Gumplowicz (1838–1909), the literary historian Piotr Chmielowski (1848–1904), and the philosophers Adam Mahrburg (1860–1913), Julian Ochorowicz (1850–1917), and Aleksander Świętochowski (1849–1938). The Polish novelists Bolesław Prus (1845–1912), author of *The Doll*, and Eliza Orzeszkowa (1841–1910), author of *Nad Niemnem* (On the banks of the Niemen), are also usually considered positivists.[119] One of Orzeszkowa's admirers was Marie Curie (1867–1934); she was deeply influenced by positivist ideas as a teenager growing up in Warsaw and, according to her biographer, "would continue to live by parts of the positivist credo for the rest of her life"—especially its emphasis on the importance of education and "empirical proofs."[120]

Positivism also deeply affected alienated members of the intelligentsia in Russia. Comte's ideas first arrived in Russia in the late 1840s. The sciences at the time were admired at the universities, which were considered almost the only progressive and modern institutions in Russia. Reformers studying or teaching at the universities hailed the sciences as full of promise.[121] Positivism found an immediate audience in the Petrashevsky Circle, a small group of "utopian socialists" who were committed to democracy and who believed the sciences could be applied to the study of the mind and society. (Fyodor Dostoyevsky was a member.) The literary critic Valerian Maikov (1823–1847), who had studied at Saint Petersburg University and belonged to this reformist group, was one of the first Russians to study Comte's works. He first referred to Comte in 1845 in a publication titled "The Social Sciences in Russia," arguing in favor of removing metaphysics from social theory. His friend, the economist Vladimir Milyutin (1826–1855), who also studied at Saint Petersburg University, praised Comte soon afterward in a series of articles in 1847, urging the use of the scientific method in the study of social facts and criticizing socialist theories for not having attained the positive stage of history.[122]

Comte became better known in the 1860s and 1870s, when censorship laws became less strict after Russia's defeat in the Crimean War. The radical intelligentsia increasingly considered science the key to weaning the masses from traditional religious and political beliefs and the chief means to effect Russian progress. Coming from the West, positivism was a new, exciting doctrine, one that replaced German idealism as a source of inspiration for intellectuals. Ivan Turgenev (1818–1883) made Bazarov, a central character in *Fathers and Sons* (1862), a spokesman for "militant positivism." Bazarov's freedom from religion and other illusions was lauded by the influential nihilist thinker Dmitrii Pisarev (1840–1868), who learned of positivism at Saint Petersburg University, where intellectuals embraced it for personal certainty in a time of upheaval and as a means to improve society.

In 1865 Pisarev wrote a long article on positivism that caught the attention of more members of the intelligentsia. Pisarev used positivism to argue against the vitalist notion of a mysterious life principle, and he embraced the law of three stages, which confirmed the importance of intellectual developments as the basis of progress and the role of the sciences in liberating the mind. Like Hennequin and Scherer, he sought to make his own field, literary criticism, into a science by coming up with laws governing the relationship between literature and the environment. Like the other sciences, literature could stimulate thought, and he hoped it would eventually inspire social change. He was very critical of two other radical literary critics, who have been called left-wing positivists: Nikolai Chernyshevsky (1828–1889), author of *What Is to Be Done?*, and Nikolai Dobroliubov (1836–1861). Pisarev found them too idealistic and Fourierist—that is, not sufficiently scientific.

Another Russian leftist who admired Comte was the anarchist Mikhail Bakunin (1814–1876). He commended Comte for using science to go beyond theology and metaphysics and for creating a systematic philosophy. The populist social critic Nikolai Mikhailovskii (1842–1904) liked Comte's moral and historical theories and the Religion of Humanity, which he advanced in radical circles. Both he and his fellow populist the philosopher Petr Lavrov (1823–1900) stressed the originality of Comte's subjective method, which viewed man as an active agent, one who integrated all the sciences because he created them. Thus all sciences should be considered in their relation to man, especially his social needs, as a way of organizing and unifying them.

The academic field in which Russian positivists made their greatest contribution was sociology. Comtean sociologists include Eugène de Roberty (1843–1915), who wrote *Sociology* (1880), which introduced Russians to Comte's variety of the science of society; Maksim Kovalevsky (1851–1916), president of the International Institute of Sociology in 1907 and supporter of the Comtean idea that progress involves increasing social solidarity, not class struggle; and Nikolai Kareev (1850–1931), a liberal historian, who embraced the "subjective sociology" of Mikhailovskii and Lavrov.[123]

The Russian who did the most to spread Comte's theories was Grégoire Wyrouboff (or Vyrubov; 1843–1913). When he attended Lycée Alexandre in Saint Petersburg in the late 1850s, his literature teacher was Louis Edmond Pommier, a French disciple of Comte, who introduced him to positivism and praised Littré's version in particular. After studying medicine and the natural sciences in Saint Petersburg and Berlin, Wyrouboff settled in Paris, where in 1862 he became friendly with Littré. Five years later the two men founded *La Philosophie positive*, where Wyrouboff propagated a dogmatic, nonreligious version of positivism as a philosophy that believed in absolute truths verified by science alone. After Laffitte died in 1903, Wyrouboff replaced him as professor of the history of science at the Collège de France. Yet some positivists did not like Wyrouboff's approach, which seemed to them excessively scientific. Russian historians and philosophers who debated whether positivism was a science or religion include Vladimir Lesevich (1837–1905), Konstantin Kavelin (1818–1885), and Vladimir Solovyov (1853–1900).[124]

In general, Russian radicals appreciated Comte's scientific method, his argument that a scientific study of society could solve social ills in a rational manner and stimulate progress, and his religion with its strict code of ethics. Because of their belief that the positivist era was imminent and their fears of censorship, they often withdrew from politics to focus on intellectual and moral renewal. Yet many facets of their doctrine would be kept alive by the Bolsheviks, who maintained a comparable strict morality and a similar faith in science, the laws and stages of history, and the elites' role in guiding the masses.[125]

Further east, Comte's views were taken up around the turn of the twentieth century by the Young Turks, who opposed the autocratic sultan of the Ottoman Empire and eventually drove the movement to make Turkey a modern secular nation. As in France, Italy, Austria, and Russia, the scientific thrust of positivism could be used against the dominant religious institution. One of the early spokesmen of the Young Turk movement was Ahmed Rıza (1859–1930), who led the movement off and on from 1895 to 1908. He came to Paris in 1889 to wage war against the sultanate and became a disciple of Pierre Laffitte and a regular contributor to the *Revue Occidentale*. He was attracted to positivism partly because it was one of the few anti-imperialist doctrines in Europe. After all, Comte had proclaimed that in the future, Constantinople would replace Paris as the capital of the world, uniting the Occident and Orient. In the 1890s, Rıza became the leader of the militant exiled Turks in Europe. Helping to make positivism the ideology of the Young Turk movement, he pictured a new state founded on scientific laws and directed by a strong central government in keeping with Comte's principles.

The Young Turks were enthusiastic about Comte's separation of church and state as a way to challenge the Ottoman Empire, where the two were intermingled. In considering these issues, Rıza and the reformers insisted that experts should guide the masses, who needed to be educated—again, a Comte-like position. (Even girls should go to school.) Besides French positivist ideas of laicization, Rıza called on all Young Turks to accept Comte's slogan "Order and Progress," which was the motto of *Mechveret*, his Young Turk journal. Founded in 1895, the journal used the positivist calendar for its dating system and was smuggled into the Ottoman Empire. Eventually, the Young Turks adopted the motto but substituted "Union" for "Order." They formally called their association the Ottoman Committee of Union and Progress (CUP), which later became a political party.

Union referred to their Muslim and positivist belief that people were responsible for other members of the community. Like Comte, the Young Turks stressed the collectivity and duties, not individual rights; similarly, they emphasized the need for maintaining order and regulating private life. A key party ideologue and Young Turk intellectual was the poet and sociologist Ziya Gökalp (1876–1924), who was also influenced by Comte.

The Young Turks had a big impact on Mustafa Kemal Atatürk (1881–1938), who founded modern Turkey. Thanks to Rıza and Gökalp, positivism had been fashionable when Atatürk was a young officer, and he was deeply struck by Comte's insistence on science, solidarity, and the law of three stages, which suggested that Turkish society could modify its Islamic practices and resemble more closely contemporary secular European societies without losing its identity. Atatürk joined the CUP in 1907. A year later, in 1908, the Young Turk Revolution occurred, which was encouraged by the French Positivists. Shortly thereafter Ahmed Rıza returned to Istanbul and became the first president of the Chamber of Deputies, where he pushed a positivist agenda. Encouraging industrialization and education for boys and girls in keeping with positivism, the CUP dominated politics until 1918. Five years later, Atatürk became the first president of Turkey, which he ruled until his death in 1938. Atatürk's key principles of secularism, statism, republicanism, and reform from above as keys to the modernization of Turkey were nourished at least to some extent by positivism.[126]

Comtean Influence in the United States

In the United States, Comte's philosophy seemed to be an impractical, foreign import and displayed an authoritarianism and secularism that had little appeal to a people proud of their democratic tradition and in the throes of a religious revival. His Religion of Humanity was excessively theological for free-thinkers, too Catholic for Protestants, and insultingly critical for Catholics.[127] Nevertheless, British writings on Comte and a number of American writers helped spread positivist ideas.

John Henry Young translated Émile Littré's *De la philosophie positive*, an overview of positivism, in the *United States Magazine and Democratic Review* in 1847. Edgar Allan Poe (1809–1849) mentioned Comte in his poem "Eureka" of 1848. Between 1851 and 1854 George Frederick Holmes (1820–1897), a lawyer and professor, wrote influential articles on Comte for the *North British Review* and the *Methodist Quarterly Review*, which was edited by John McClintock (1814–1870), who appreciated Comte's logic and application of the scientific method. One of McClintock's friends, James O'Connell, wrote *Vestiges of Civilization* (1851), whose attacks on religion were heavily influenced by Comte's arguments. William Mitchell Gillespie (1816–1868), an engineering professor, translated the first volume of the *Cours de philosophie positive* in 1851.[128] The term sociology was used in 1854 in the title of two books by southerners, George Fitzhugh (1806–1881) and Henry Hughes (1829–1862), who employed this new science to defend slavery.[129] In 1855 the anarchist Calvin Blanchard (1808–1868) published an American edition of Harriet Martineau's translation of the *Cours*. In 1873 William Graham Sumner (1840–1910), a social Darwinist at Yale University, taught the first course on sociology in the United States. In 1889, the University of Kansas instituted the first department of sociology. By the end of the century, many college textbooks referred to Comte as a worthwhile thinker. The first president of Cornell University, Andrew Dickson White (1832–1918), was partly inspired by positivism when he decided to help found this nonsectarian university.[130]

Paradoxically, one group interested in positivism consisted of Protestant clergymen from New England, who were impressed by Comte's erudition and emphasis on social solidarity, altruism, and the limits of human knowledge. Yet his reduction of religion to ethics, rejection of the supernatural, and insistence on scientific demonstrations for every belief constituted a challenge. Indeed, Protestant theologians in both Britain and the United States, such as Benjamin Jowett (1873–1893) and James McCosh (1811–1894), grappled with positivism for fifty years before turning to the next modern "materialistic" doctrine, evolution. In the process of coming to terms with positivism, which was often implicitly equated with modern thought, Protestant thinkers deepened their faith, increasingly prioritizing the group, not the individual, and reiterating their commitment to serve humanity.[131]

With their openness to new ideas and their social values, liberal Unitarians were especially interested in positivism. The Reverend Joseph Henry Allen (1820–1898) wrote the first overview of Comte's ideas outside of Europe in 1851 in the leading Unitarian periodical, *Christian Examiner*, where he praised Comte's ethical stance. John Fiske (1842–1901), a young philosopher who called himself a positivist, gave lectures on Comte at Unitarian Harvard University in 1869. Many Unitarian clergymen were transcendentalists and were interested in Comte because, like him, they sought to reconcile science and religion. Considering that Comte had argued that one could not use reason to prove God's existence, they aimed to demonstrate that there was a God by pointing to social and historical evidence.

The transcendentalist clergymen William Henry Channing (1810–1884) and Theodore Parker (1810–1860) helped spread Comte's ideas in their writings and sermons. Parker was particularly inspired by Comte's dislike of supernaturalism and his rejection of an anthropomorphic conception of God. He bought several volumes of the *Cours* from the bookstore owned by Elizabeth Peabody (1804–1894), a Unitarian educator with transcendentalist sympathies, who appreciated the Frenchman's critique of traditional natural theology. In 1858, she wrote a review of the *Catéchisme positiviste* for the *Christian Examiner*. Parker influenced Lydia Maria Child (1802–1880), the abolitionist and women's rights advocate, who wrote in 1855 *The Progress of Religious Ideas through Successive Ages*, which seemed to incorporate Comte's ideas about the history of religion without mentioning him. And one member of Parker's congregation was Julia Ward Howe (1819–1910), the social activist and composer of "The Battle Hymn of the Republic." She was so interested in Comte's "bird's-eye" approach to the sciences that she gave lectures on his work in the 1860s. However, she, like the transcendentalists, did not take to the Religion of Humanity. Indeed, another transcendentalist, Orestes Brownson (1803–1876), was influenced in an unlikely manner by Comte: he converted to Roman Catholicism after learning of Comte's appreciation of its constructive role in earlier ages. As is clear, very few Americans accepted all of Comte's religious, scientific, and social ideas.[132]

There was in truth only one important orthodox positivist, the lawyer Henry Edger (1820–1888), who had a long, intimate correspondence with Comte. As Comte's chief disciple in the United States, he endeavored to create a positivist community (Modern Times) on Long Island, but by 1859 it had only six positivists. Yet a series of lectures by Edger in the mid-1860s in New York City led to the founding of the New York Positivist Society, which developed into the Positive Society of North America in 1869. Members tended to be upper middle-class journalists, editors, writers, and professionals. Many of these early American positivists, such as Edger, liked the ritualized Religion of Humanity. They were in search of a new faith that did not rely on a traditional God but would not threaten social norms; positivism had just the right mix of atheism and piety. They believed it could help common people without challenging the social hierarchy and corporate capitalism.

One member of the Positive Society was David Croly (1829–1889), editor of the *New York World*. He had learned about positivism from the British and had read Comte's works. It was he who invited Edger to lecture in New York. An enthusiast of the Religion of Humanity, Croly later launched a Comtean journal called *Modern Thinker*. Another member was Thaddeus Wakeman (1834–1913), a radical New York lawyer. In 1876 he organized the Society of Humanity, which kept its distance from the more orthodox positivist groups. He and other positivists of the late 1870s and 1880s considered positivism a scientific surrogate for religion and sought to Americanize it by eliminating its Roman Catholic rituals and papal hierarchy, narrow dogmatism, and ultraconservativism.

Pushed to the left by Wakeman, positivism became more acceptable to other American liberals and radicals. One such leftist affected by Comte was Edward Bellamy (1850–1898), the author of *Looking Backward* (1888). He appreciated Comte's notion of serving Humanity, stress on social solidarity, and insistence on a government managed by a disinterested elite to promote the common good.[133] Other leftists, members of the Progressive Movement, were especially influenced by Comte. One of its leading ideologists was Herbert Croly (1869–1930), who was brought up in positivism by his father, David, and was eager to inject progressivism with Comte's ideas about the need for interdependence in an organic society, the importance of cultural consensus and social harmony, and the role of a meritocratic bureaucracy in managing government. Croly spread these principles in his important periodical, the *New Republic*, which he cofounded in 1914, and an influential book called *The Promise of American Life* (1909).

Another progressive influenced by Comte was Lester Ward (1841–1913), who wrote the influential *Dynamic Sociology* in 1883. He used the scientific authority represented by positivism to bolster his campaign for state-led reforms in the interest of the common people. Eventually serving as the first president of the American Sociological Association in 1906, he helped bring positivist ideas into progressivism and anchor sociology in matters of social reform. Ward's *Dynamic Sociology* influenced two progressive sociologists: Albion Small (1854–1926) and Edward Alsworth Ross (1866–1951). In 1895, Small founded the *American Journal of Sociology* and in 1905 helped establish the American Sociological Association. Besides Ward, Small and Ross became presidents of this Association. All three men incorporated into American sociology Comte's stress on order and social planning by means of a controlling elite of experts, such as sociologists.[134]

Comte's blend of progressivism and conservatism and his fusion of science and religion appealed to reformers in the Gilded Age and Progressive Era. Their positivist ideas eventually transformed liberalism itself. Reformers who were dubious about Christianity approved of his emphasis on altruism, which they regarded as a promising nonreligious foundation of society. By calling attention to the common good and highlighting the collectivity as Comte did, they demonstrated the abuses of laissez-faire capitalism with its unrestrained individualism. Comte's ideas helped liberals construct the case for active government intervention in the economy in order to help the working class and to secure social harmony and social justice. The liberal reformers liked Comte's stress on the important role of positive philosophers, for they saw themselves as a similar neutral, educated professional elite—that is, experts who could use findings from the scientific study of society to direct the government and create a harmonious, organic community. Like Comte, they saw society's building blocks in terms of interdependent groups, not isolated individuals. Thanks partly to positivism's influence, American liberalism became in the twentieth century favorable to interventionism, statism (or corporatism), and the notion of a technocratic elite in charge of managing the government, economy, and society, positions that are still controversial today. Yet Comte's imprint can also be found in American liberalism's elitism, paternalism, and unwillingness to question the private accumulation of capital.[135]

The Spread of Positivism throughout Latin America

In many ways, positivism had its greatest impact in areas outside of Europe and the United States despite the fact that these regions of the globe were not industrialized, that is, not close to the positive stage. With its law of three stages, positivism gave middle-class intellectuals from such disparate places as India and Brazil a historical explanation for their countries' backwardness, thus absolving them of culpability, and a scientific basis for their campaign to reform conditions. It gave them a blueprint for action, which could be used against elites who monopolized power and held back technological and organizational advances.[136] Young Turks, Poles, and Russians who were also influenced by positivism had similar motivations. Like Marxism, positivism was seen as holding the key to modernization in many parts of the world.

Countries in Latin America embraced positivism with great enthusiasm. In the nineteenth century, most of them battled Spain and Portugal for independence and then underwent a period of instability as each tried to establish a new, separate identity. Reformers turned their backs on their Iberian heritage and the Catholic church, which they blamed for the backward, chaotic state of their countries. They sought to transform their culture and establish republics with the help of French and Anglo-Saxon models, which seemed to guarantee political, material, and intellectual progress. There were several reasons why the positivist model became so popular. In the mid-nineteenth century, many doctors, engineers, army officers, and other professionals and intellectuals studied in Europe, read or heard of Comte, and perused *El Eco Hispano-Americano*, published from 1853 to 1872, by José Segundo Flórez (1813–1900), Comte's main Spanish disciple. They were receptive to positivism because they had studied Cartesianism; the Enlightenment philosophers, including those in Scotland; the Idéologues' doctrine; French conservatism; utilitarianism; and utopian socialism (including Saint-Simonianism), all of which paved the way for positivism because it amalgamated ideas from all of these movements. These Latin Americans went back to their countries and joined other educated Latin Americans in presenting positivism as the latest intellectual fashion, one capable of using the sciences to guide political and social reforms and economic development.

At this time, France was seen by many Latin Americans as the most admirable country in the world. With its law of three stages and popularity in respected French and English circles, positivism seemed to offer an exemplary ideological basis for change, a philosophy capable of eliminating the traditionalist mindset that was allegedly holding back Latin America. After all, Comte had pointed out that intellectual transformations were the foundation of political, moral, and economic change. Indeed, positivism suggested that given the right circumstances, a nation could skip the metaphysical stage of history and enter the last, positive one. In short, Comte's ideas offered Latin Americans an explanation of their messy, humiliating colonial past, which they sought to overcome; a secular, disciplined morality; and an optimistic vision of a progressive future based on science and industry. Latin Americans were hopeful that thanks to positivism with its accent on the tangible and useful, science could guide change and reforms, industry could enrich the middle class, the new political order could be marked by stable republics, and the influence of the Catholic church could be diminished. No other doctrine seemed as broad in its scope or as capable of being adapted to each country's needs. No other seemed as supportive of an educated elite as the key agent of change. Indeed, a profusion of positivisms erupted within Latin America.[137] By 1900 positivism was the most dominant philosophy in Latin America, having served the liberal elites well since around 1850. It would retain its influence through World War I.[138]

Brazil

Positivism had the most impact in Brazil, where the transition to independence was the least violent. There were several direct connections with Comte. Antônio Machado Dias studied with Comte in the 1830s and became a mathematics professor in Brazil. Nísia Floresta Brasileira, an abolitionist and feminist who opened a girls' school in Rio de Janeiro, became a good friend of Comte's when she was in Paris in the 1850s. She, Machado Dias, and other teachers helped introduce the Frenchman's ideas to Brazilians. Marie de Ribbentrop, the daughter of Comte's leading German disciple, was a tutor for a Brazilian family in Brussels, where she taught positivist ideas to Luis Pereira Barreto (1840–1923), Francisco Antonio Brandão, and Joaquim Alberto Ribeiro de Mendonça, all of whom became spokespersons for positivism upon their return to Brazil. The most important of the three men was Pereira Barreto, who in 1874 published the first volume of *As Três Filosofias* (The three philosophies), the first Brazilian book that outlined positivism in detail, presenting it as the philosophy of the future. It could help move the country beyond Catholicism, especially by promulgating education, which could change the mentality of the masses.

Pereira Barreto's book boosted the positivist movement, which had taken off in the 1850s, when many young, middle-class Brazilians of Rio de Janeiro had become attracted to positivism at their military, engineering, medical, and law schools; shaped by their education, they appreciated positivism's emphasis on practicality and expertise. They were dismayed by their monarchical government, which was supported by a small oligarchy of wealthy conservative landowners and the traditional Catholic church, and they hoped for a better future under a positivist-type republic. An increasing number of Brazilian intellectuals and members of the commercial and bureaucratic middle classes became taken with positivism's dismissal of the supernatural and promises to eliminate anarchy and promote progress in a scientific and efficient manner. They thought of themselves as constituting a Comte-like sociocracy, a new elite of educated, cultured men who could direct government, advance industrialization, and maintain order. Even positivism's subtle authoritarianism appealed to them. In sum, their ideas for change coalesced in positivism. Eventually, at least four hundred people would declare their allegiance to positivism.[139]

One of the most famous Brazilian positivists was Benjamin Constant Botelho de Magalhães (1833–1891), a professor of mathematics at the Military Academy (Escola Militar) in Rio de Janeiro, which was attended chiefly by middle-class men eager for a free education. In 1857, he joined the budding positivist movement and made the school the center of positivist education for future elites. One of his students was Euclides da Cunha (1866–1909), who became a positivist and wrote the Brazilian masterpiece *Os Sertões*. In 1876, Constant helped found the first Sociedade Positivista (Positivist Society) in Rio to propagate Comte's doctrine.[140] Pereira Barreto became involved in its reorganization two years later.

Yet divergences soon emerged. Miguel Lemos (1854–1917) and Raimundo Teixeira Mendes (1855–1927) were drawn to positivism while studying engineering. They joined the Positivist Society but left Brazil in 1877 to pursue their education in Paris. Initially they supported Littré. But Lemos was charmed by Laffitte, who converted him to the Religion of Humanity and considered him an aspiring positivist priest. Lemos persuaded his friend Teixeira Mendes to adopt the religion too. As Lemos became more fanatical and authoritarian, Constant resigned from the Society, and Pereira Barreto kept his distance. In 1881 Lemos became the new president of the Positivist Society, which he made into the Positivist Church of Brazil. He proclaimed himself the head and made Teixeira Mendes his deputy. Two years later he switched his allegiance from Laffitte to Congreve, who more aggressively pushed the Religion of Humanity. Afterward, the number of Brazilian adherents fell from a high of over fifty-nine in 1882 to thirty-four in 1884. Nevertheless, thanks to the urging of Lemos, the Templo da Humanidade (Temple of Humanity), modeled on the Panthéon, was built in 1891 in Rio de Janeiro. This is the one of the only places in the world where the Religion of Humanity is still practiced.[141]

Whereas Lemos and Teixeira Mendes dedicated themselves to religious ceremonies, going so far as to commemorate Clotilde de Vaux, Pereira Barreto and Constant played a leading role in Brazilian academic and political life. They used positivism to fight the Catholic church and the landed aristocracy, which represented the colonial legacy, and to take the lead in assuring their country's orderly, just development. Positivists and other republicans began to focus on the evils of slave labor. Comte's ideas of miscegenation, his praise of fetishism and blacks' emotional superiority, and his denunciation of slavery influenced the abolitionist movement in the 1880s. Positivists were among the first to demand the complete abolition of slavery. Thanks partly to the positivist campaign, slavery, which was supported by the conservatives and their allies, the clergy, was finally abolished in 1888.[142]

Constant and the military officers whom he taught played a big role in the fall of the monarchy the following year. They spread the powerful positivist ideology that insisted on the historical inevitability of republicanism. Bolstered by this doctrine, republicans, with the assistance of the army, established the Brazilian Republic in 1889.[143] Constant helped organize the military coup. As the "soul of the republican movement," he became minister of war and then minister of education in the provisional government, assuring positivism's impact in these early years.[144] The positivist engineer Demétrio Ribeiro (1853–1933) became minister of agriculture, commerce, and public works. With allies in the army, many positivists tended to support a dictatorial republic, but a few other positivists and numerous republicans preferred a democratic, federal republic. In 1891 a federal republic with a constitution and president was finally instituted. The small vocal group of positivists in the Assembly was relatively pleased with this republic, though there were complaints that too few people could vote in a meaningful manner and the military retained too much influence. Both Teixeira Mendes and Constant wanted to follow Comte's idea of turning the army into a police force, but they had no success. Nevertheless, Teixeira Mendes designed a new flag showing the country's debt to Comte, whose motto, "Order and Progress," was displayed in the middle.[145] The motto is still on the flag today. Constant, who died in 1891, was officially proclaimed "Founder of the Republic."

Positivism's emphasis on the separation of church and state and its concern with social welfare helped shape politics in the late 1880s and early 1890s. Positivists of all stripes, even the religious ones, adopted a progressive agenda; besides insisting on the abolition of slavery, which they obtained in 1888, they pushed for civil marriages, secular cemeteries, the elimination of corporal punishment, an increase in the number of lay public schools, and labor reforms, such as higher wages, to improve workers' lives. However, the Brazilian positivists also favored conservative economic policies, typical of the middle class in all Latin American countries. The mixture of progressive and conservative policies was attractive. The number of members of the Positivist Society went from 59 in 1889 to 159 in 1890 and 174 in 1891. Lemos and Teixeira Mendes continued to promote a positivist agenda until 1910.[146]

The constitution of Rio Grande do Sul, a southern state, which was intact from 1891 to 1930, was modeled expressly on positivist principles. A republican journalist, Júlio de Castilhos (1860–1903), was a disciple of Comte, whose influence can be seen in Castilhos's call for a strong executive and a legislature that would supervise only the budget. He obtained the support of the positivist military officers in the region, wrote the constitution of the state in 1891, and took over as governor. He was often criticized for being a positivist dictator.[147] A Chapel of Humanity—the Templo Positivista—was constructed in 1928 in Pôrto Alegre, the capital of Rio Grande do Sul. Like the Templo da Humanidade in Rio de Janeiro, it is still open. (A third Brazilian Chapel exists in Curitiba in Paraná in the south.)

Another dictator took over the entire country of Brazil in 1930: Getúlio Vargas (1882–1954). Vargas had attended a military academy and began his political career in Rio Grande do Sul, where he learned about positivism. He seized power in Brazil in 1930 and claimed to follow Comte's model of a dictatorial republic. Like the earlier Brazilian republicans, he relied on the military for support because he did not wish to depend on politicians, whom he distrusted as much as Comte did. As a result, his government became known for being militaristic, corrupt, repressive, and close to wealthy elites. Yet, thanks partly to positivism, he was eager for economic and social changes. He pushed industrial development in a technocratic, authoritarian manner and embraced social reforms for workers with the goal of incorporating them better into society.[148]

For his own political purposes, Vargas protected the Brazilian Indians with the help of Cândido Mariano da Silva Rondon (1865–1858). A student of Constant at the Military Academy in Rio de Janeiro in the mid-1880s, Rondon was a fervent positivist. Unlike his professor, however, he also embraced Comte's Religion of Humanity. Eager to facilitate scientific progress and serve Humanity, he dedicated himself to economic modernization and social justice. An army officer and military engineer, Rondon headed several commissions that set up state telegraph lines in northwest

Brazil and sought to incorporate the indigenous peoples in this region into the Brazilian nation in a peaceful manner. He became head of the Indian Protection Agency. He, Teixeira Mendes, and his fellow army officers, most of whom were positivists, believed native peoples were not racially inferior. In close proximity to the fetishist stage of history, they would slowly evolve to a higher stage of history. He thus allowed the indigenous peoples of Brazil to continue following their religious, linguistic, and cultural practices; defended their land holdings; and insisted that whites return property they had stolen. His Indian pacification techniques became standard from 1910 until his death in 1957. Some scholars commend Rondon for upholding the humanitarianism at the heart of positivism—the humanitarianism that is also evident in its campaign to abolish slavery.[149] Others maintain that his work made it easier for politicians to control the Brazilian Indians.[150] It does seem that positivism's reputation suffered because of its association with the military and dictatorship.[151] Perhaps, as Paul Arbousse-Bastide suggested, positivism was more successful in Brazil as a religion with principles of social justice than as a political movement.[152]

Mexico

As in Brazil, positivism spread initially in Mexico thanks to personal connections. Gabino Barreda (1818–1881) is regarded as the founder of Mexican positivism. Born in Puebla in 1818, he took part in the Mexican-American War in 1847 and then several months later went to Paris to study medicine. There he befriended Pedro Contreras Elizalde (1823–1875), a Mexican medical student who was Comte's disciple and the first Mexican positivist. In 1849 Contreras Elizalde persuaded Barreda to attend Comte's course on the history of humanity, which had a big impact on him. Returning to Mexico in 1851, he practiced medicine in Guanajuato, fought for liberal reforms, and began to popularize positivism at a moment when Mexicans were looking for a new guiding program after their defeat to the Americans in 1848. At the time of the Mexican victory over the French in 1867, he gave his famous "Civic Oration," calling for Mexico to enter the positive stage of history by becoming a secular republic devoted to scientific progress. According to him, it needed positivism to free itself from servitude to the church, the army, and old colonial ways of thinking and behaving. Education was the key to this process of mental and political liberation, which he emphasized would lead to material prosperity and thus progress.

Shortly after this momentous speech, the republican Benito Juárez (1806–1872), who took over as president weeks later, persuaded Barreda to work for him. Juárez knew Barreda's two brothers-in-law, Francisco and José Diaz Covarrubias, both of whom were also positivists. In addition, Juárez was close to Contreras Elizalde, who had returned from Paris to Mexico in 1855, had worked as his political associate, and would marry his daughter in 1868. Thanks to this network, positivism became part of the triumphant liberal, anticlerical, antimilitary ideology. It legitimized the new republic's campaign to overcome both the anarchy that had periodically plagued Mexico since its independence from Spain and the conservative colonial mindset that continued to hinder its progress. Above all, positivism seemed practical and scientific, two characteristics that appealed to the rising Mexican middle class, which wanted to create an independent, modern, prosperous country.[153]

To take the first step and eliminate the clergy's control over education, Juárez appointed Barreda to a five-man commission charged with organizing a national school system. Contreras Elizalde and Francisco Diaz Covarrubias also belonged to the commission, but Barreda dominated it. He was enthusiastic about the chance to rejuvenate Mexico by giving it a new scientific culture and an elite of positivist intellectuals who could direct the country. The commission contributed to a new law, promulgated in December 1867, which reformed the entire system of education according to positivist principles. Schools were to stress the importance of fact-based, scientific learning, thereby challenging Catholicism. In addition, they would educate people to become altruistic. It was hoped that wealthy people, who would be allowed to accumulate riches in accordance with economic freedom, would learn to help other members of society. Comte's emphasis on society as an organism with the interests of the poor at heart was evident here. Yet as has been often pointed out, the new Mexican middle class embraced positivism chiefly because it upheld their economic and social status and created order—order that maintained their dominance and guaranteed further progress, especially economic progress.[154]

As minister of education, a post he obtained in 1867, Barreda oversaw until 1878 the new model high school, the Escuela Nacional Preparatoria, which was established in 1868 by his commission. As a blow to the church, it replaced the Jesuit Colegio de San Ildefonso. The new preparatory school, which taught the children of the intellectual and political elite, was instrumental in making positivism, according to one historian, "the official philosophy of Mexico" until 1910.[155] Faculty chosen on the basis of their academic merit, not politics, taught courses in mathematics, the natural sciences, languages, and logic. The school not only raised children according to positivist principles, especially the hierarchical classification of the sciences, but disseminated such ideas to the public. Thanks to this model, public education did expand, but very few poor indigenous people in the provinces had the chance to attend any of the new schools. There was still no universal primary education, despite the law stipulating that such education be made free and compulsory.[156] The main beneficiaries of this positivist curriculum were middle- and upper-class students in Mexico City.

To further his goal of shaping a positivist elite, in 1877 Barreda founded the Asociación Metodófila, which published works on positivism and Darwinism. Deliberately meeting on Sundays, it attracted medical and law students and tried to devise solutions to Mexico's problems. In this way, positivist ideas gained a larger audience and showed its practical applications [157]

The engineer Augustín Aragón (1870–1954) was one person converted by Barreda. Unlike Barreda, who highlighted scientific positivism, Aragón promoted Comte's Religion of Humanity. He worked to secularize Mexican culture, spreading positivist principles from 1888 to 1914 through his positivist journal, the *Revista Positiva* (1901–1914), and the *Sociedad Positivist*, which he helped found. Meeting from 1900 until 1911, the *Sociedad Positivist* was chiefly directed by his friend, the logician Porfirio Parra (1854–1912). Reflecting the strength of positivism in Mexico, Aragón raised a quarter of the cost of the statue of Comte that was erected in front of the Sorbonne. However, Mexico never had a Positivist Church as Brazil did. [158]

It was primarily positivism's scientific and practical approach that continued to be most compelling during the presidency of Porfirio Díaz (1830–1915), who ruled from 1876 to 1880 and from 1884 to 1911. Positivism was embraced by the "Científicos," the technocratic elite that supported him. These men included professionals, businessmen, bankers, and intellectuals. Many of them had been educated by Barreda and his Escuela Nacional Preparatoria. Just as Comte had at one point looked favorably on Napoleon III, Científicos, such as the positivist politician Francisco G. Cosmes (1850–1907), supported Díaz as an "honest" dictator who could bring order to the country. They hoped he would advance economic freedom, hygiene, medicine, and education. In its technocratic, practical guise, their positivist philosophy seemed to strengthen Díaz's own reform program and legitimize his authoritarian government.[159] Díaz thus gave the Científicos a prominent role in his administration. Following Comte's representation of positive priests as advisers to statesmen, they fancied themselves the counselors to a strong regime that would give Mexico the stability it required to break away from the past and achieve progress. Positivists established a journal, *La Libertad*, and their own political party, the Unión Liberal (known as the Partido Científico). They also carried a banner with Comte's motto, "Order and Progress."

Eventually positivists appeared to monopolize positions of power, leaving their mark on legislation, education, and politics especially between 1892 and 1911. These positivists included Justo Sierra (1848–1912), who was editor of *La Libertad*, minister of education, and first rector of the national university; José Limantour (1854–1935), who was minister of finance; Manuel Romero Rubio (1828–1895), who was minister of the interior and Díaz's father-in-law; and Miguel Macedo (1856–1929), the last subsecretary of Díaz's government. Positivists Pablo Macedo (1851–1919), Rosendo Pineda (1855–1914), and Francisco Bulnes (1847–1924) became members of the Chamber of Deputies. Díaz and these positivists continued Juárez's campaign against the traditional army and the Catholic habits and mindset associated with the Spanish colonial legacy, and promoted the benefits of social progress and modernization. They even pushed a kind of civil religion, getting Mexicans to worship national heroes instead of saints.[160]

But under Díaz, these men, like Barreda before them, increasingly downplayed the importance of rights and liberty. To them, the Mexican people needed to be educated before they could have a real democracy. Society had to evolve in an orderly fashion under a strong state and an educated bourgeois elite; progress for the many would come later. One way they justified the domination of the new bourgeoisie that was emerging thanks to the development of railroads and industry was by mixing positivism with the utilitarianism of John Stuart Mill and the social Darwinism of Herbert Spencer.[161] Due to their faith in the survival of the fittest, they promoted the immigration of white foreigners to improve the allegedly degenerate indigenous stock that was holding back progress.

Díaz rewarded the Científicos, giving more economic freedom to them and to industrialists, landowners, and foreign capitalists in general. This elite enriched themselves as Mexico became industrialized. They were pleased that positivism justified in scientific terms their political and social interests and their sense of superiority. Eventually, positivism was linked to "Porfirism" and was attacked by both conservatives and liberals. Catholic conservatives denounced positivism as atheistic, materialistic, and leftist. Liberals associated it with not only the social and political injustices of the dictator but the greedy capitalism and racism of the wealthy landowners, which left many people impoverished. By the early twentieth century, even positivists, such as Justo Sierra and Aragón, were criticizing Díaz and the Científicos for misusing power and mistreating indigenous peoples. Although they tried to disassociate positivism from Porfirism entirely, its reputation was irreparably damaged.[162] The wealthy middle class whose mentality was dominated by positivism had created a huge gap between themselves and the rest of society, repeating the problems of Mexico under Spain. Instead of dedicating themselves to advancing intellectual life and working for social harmony and justice, as Comte had wanted, positivists promoted a scientific rhetoric of progress that was embraced chiefly by politicians and the new bourgeoisie to justify their own power.[163]

Chile

In Chile, where there were many struggles between the church and state, positivism led the way to liberalism and shaped the educational system and the economy. Three positivists considered Comte's ideas to be important for liberalism: Francisco Bilbao (1823–1865), José Victorino Lastarria (1817–1888), and Valentín Letelier (1852–1919). Bilbao was a radical who blamed Chilean backwardness on its Catholicism and colonial past and argued for republicanism as a solution for its woes. His former teacher was Lastarria, who was a lawyer, political theorist, man of letters, and dean at the University of Chile. Having read Comte's work in 1868, Lastarria was the first avowed positivist in Chile. As a Liberal Party member, who served as a cabinet member, deputy, and senator, he retained his faith in individualism and embraced positivism chiefly to weaken the Catholic church and defend people's freedom and progress. He believed that with social progress, liberty would increase. He was thus far more adamant about the limits of state power than were Mexican positivists. Yet his younger positivist colleague and friend Letelier believed in "responsible" authoritarianism, arguing that the state should be devoted not chiefly to liberty but to social progress—that is, the improvement and development of its members.

The preeminent intellectual in Chile for thirty years, Letelier was a law professor and rector of the University of Chile (1906–1911), where positivism was the dominant philosophy, thanks in part to the Instituto Pedagógico that he had established there in 1889. This institution helped spread positivist ideas to teachers, who then taught them to the young elite. Like Lastarria, Letelier was also active in politics. He served as a Radical Party member of Congress, where he opposed any individual seeking strong executive power and sought reforms to help the working class. He did not seem to believe as much in liberal individualism as Lastarria did.

Nevertheless, both men sponsored important educational reforms from the 1870s to the 1910s. Shaped by the law of three stages, their aim was to eliminate Catholic and metaphysical influences in schools and to infuse secondary school and university curricula with scientific subjects and logic. To control the educational system was to shape the character and morality of society. In addition to secularizing society, especially through schools, as Comte had recommended, they sought to rationalize the economy and thereby ensure Chilean prosperity.[164]

None of these men was much interested in the Religion of Humanity, which was supported chiefly by three brothers, Juan Enrique Lagarrigue (1852–1927), Jorge Lagarrigue (1854–1894), and Luis Lagarrigue (1864–1949). Unlike the more heterodox positivists, these three orthodox positivists approved of the authoritarian government of José Manuel Balmaceda (1840–1891), who sought to secularize the country but did not believe in parliament. They viewed his takeover in 1886 as a step toward sociocracy. Few Chileans adopted Comte's religion and the kind of government it seemed to entail.[165]

Argentina

In Argentina, positivism was a tool to attack reactionary and authoritarian forms of thinking and behavior and to promote civilization. Several groups of positivists spread their philosophy through books and journals, including *La Escuela Positiva* (from 1894 to 1899) and *El Positivismo*, which was published by the Comité Positivista Argentino (Argentine Positivist Committee) from 1925 to 1935.[166] One early group consisted of positivists, such as Domingo Sarmiento (1811–1888), Juan Bautista Alberdi (1810–1884), and Esteban Echeverría (1805–1851), who represented the new "generation of 1837"—the generation opposed to the dictator Juan Manuel de

Rosas (1793–1877). They considered Argentina a barbaric country still suffering from the nefarious effects of Spanish colonialism, which favored strong-man (*caudillo*) rule. They sought through their writing to bring in liberal values to end Rosas's tyranny and modernize their country both economically and intellectually.

Sarmiento became president in 1868 and pushed hard for educational reforms during the six years he was in power and afterward as superintendent of schools; he wanted to make schools secular, scientific in nature, and open to all, including women, in order to eliminate ignorance and to help Argentina become a civilized, wealthy nation. Intellectual advancement and material progress went hand in hand. He was instrumental in setting up the public school system in Argentina. Yet, like Alberdi, he was a racist who wished to "civilize" blacks, Indians, and mestizos, whom he considered primitive. He took Comte's Eurocentrism to a new level, identifying civilization with white Europeans, who were encouraged to immigrate to Argentina in order to revitalize it.

A second group of positivists was devoted solely to educational reforms. Pedro Scalabrini (1848–1916) and his student J. Alfredo Ferreira (1863–1935) made positivism the foundation of the Paraná Teachers College (Escuela Normal de Paraná), established by President Sarmiento in 1870. The first teacher training institution in Argentina, it disseminated positivist ideas to future political leaders, much as Barreda's Escuela Nacional Preparatoria did in Mexico. There were also important positivists at the University of Buenos Aires and the National University of La Plata, who worked in teacher training schools. Connoting action and modernity, positivism permeated Argentine universities at the end of the nineteenth century.

Yet another group of positivists, which included the two doctors Juan Bautista Justo (1865–1928) and José Ingenieros (1877–1925), mixed Comte's ideas with Marx's and Spencer's philosophies to create a theoretical basis for a socialist version of liberalism. In the 1890s they helped found the Argentine Socialist Party, which other positivists soon joined.[167]

Other South and Central American Countries

Positivism had its greatest impact in Brazil, Mexico, Chile, and Argentina, but it had a following in most of the other South American countries. In Uruguay, positivism was seen as a moral doctrine that could be used to stop military coups, eliminate corrupt government, and promote economic development and social welfare. Two of its proponents were José Pedro Varela (1845–1879), a leading educator who pushed for free compulsory secular schools in the 1870s, and José Batlle y Ordóñez (1856–1929), who served several terms as a reform-oriented, populist president in the early twentieth century. In Paraguay, the main positivist was Cecilio Báez (1862–1941), who combined ideas from both Comte and Spencer. A liberal, he served as president from 1905 to 1906.

In Peru and Bolivia positivism was seen as offering a means of rehabilitation especially in the economic sphere after their wars against Chile (1879–1883). In Peru, Manuel González Pardo (1834–1878), who was president of the country in the 1870s, was an admirer of Comte and promoted the sciences and technology (especially railways) to guarantee material progress, which he viewed as crucial to moral progress because financial security led to a more generous spirit. In Bolivia, people blamed the schools for the country's loss of its outlet to the sea in its wars against Chile. The schools were allegedly creating people who were too idealistic. Positivism was embraced for providing a dose of realism. It was promoted by the geographer Agustín Aspiazu (1826–1897), among others. But another Bolivian positivist, Alcides Arguedas (1879–1946), used Spencer's social Darwinism in such books as the *Pueblo enfermo* (1909) to argue that indigenous people and mestizos were unsuited for higher education and for democracy.

In Venezuela, Antonio Guzmán Blanco (1829–1899) was president between 1870 and 1888 and modeled his dictatorship on Comte's ideas of order, modernization, and moral and intellectual improvement. As a self-proclaimed admirer of Comte, he attacked the church and conservatives; modernized Caracas; revitalized various industries; sponsored public works; and encouraged educational reforms. He created a chair in world history at the University of Caracas and gave it to a positivist, Rafael Villavicencio (1838–1920), who also directed his Instituto de Ciencias Sociales.[168]

Finally, positivism had a considerable impact on Central America. In Guatemala, the Sociedad Positivista de Centro-América worked in tandem with the Sociedad Positivist of Mexico to create a community devoted to Humanity. Liberal politicians, including the authoritarian president Justo Rufino Barrios (1835–1885), who ruled in the 1870s and 1880s, used vague positivist rhetoric to support their demands for modernization. Barrios was particularly important in undermining the Catholic church and inaugurating a system of public schools.[169] Tomás Guardia (1831–1882) of

Costa Rica, Rafael Zaldívar (1834–1903) of El Salvador, and Marco Aurelio Soto (1846–1908) of Honduras ruled as liberal presidents in the 1870s and 1880s and embraced positivist notions of order and progress as well. They wanted to free their countries from church influence and encourage economic development.[170] Their liberal followers often personally profited. In Nicaragua, José Santos Zelaya (1853–1919) was president from 1893 to 1909. Zelaya had studied Comte's work in Paris and mixed his ideas with Spencer's ideas of freedom and individualism, which seemed more liberating to many Latin Americans. He was devoted to public education and built 140 public schools. But to promote modernization, he seized not only property of the Catholic church but communal land of indigenous peoples, which was awarded to wealthy individuals.[171]

Positivism's Influence in Latin America

In sum, positivism's impact on Latin America was multifaceted and touched on questions of identity. It was initially attractive as a cutting-edge European philosophy but was paradoxically often employed by Latin Americans to bolster a nationalist agenda, which was supposed to give their countries control over their own resources and a chance to begin a new era in their history. Positivism appealed to those interested in state-generated reforms, economic development, and the preservation of order. Most of all, it presented a way to enter the modern world by cultivating a scientific mindset.[172] In their view, this mindset would free them from colonial and Catholic values, habits, and behaviors that hampered progress. To develop a rational, practical mentality, children were no longer to be given a religious education but were to be taught an encyclopedic array of scientific subjects, usually including logic and history, in an ordered hierarchy. Considering the impact of the Escuela Nacional Preparatoria in Mexico, the Escuela Normal de Paraná in Argentina, the Escola Militar in Brazil, and the Instituto Pedagógico of the University of Chile, it is clear that education was the area in which positivism had the most impact.[173]

The scientific mindset encouraged by positivism was also welcomed as a way to ensure technological, industrial progress based on order. Indeed, one of positivism's greatest legacies in Latin America was a faith in the benefits of scientific and technological change in bringing about economic prosperity, especially if guided by a pragmatic elite. Liberal rulers in particular—men from modest backgrounds—were pleased to have a doctrine that justified their takeover of church property and education and their plans to enrich the country. Positivism served their technocratic, pragmatic purposes, helping them generate change without threatening their domination or social cohesion.[174] As historians Lawrence Clayton and Michael Conniff note, "To achieve material progress, one needed social and political order. To establish a true and enduring order, one needed to make material progress."[175]

The scientific mindset endorsed by positivism was attractive because it seemed to suggest a rational way to reform society and politics. With its emphasis on republicanism and social justice enacted from above, positivism appeared to be a liberating force, a way to change reality, but not in a revolutionary manner that would give power to the masses or threaten order and hierarchy. Positivism indeed strengthened the new elites, that is, the middle-class bankers, lawyers, businessmen, technocrats, engineers, scientists, and philosophers, who could claim that only they had the capacity to rule—a claim that led to the exploitation of the lower classes in the name of order. In many countries, local ethnic groups, especially indigenous peoples, were still oppressed. Paradoxically, in their embrace of social hierarchy (and racial hierarchy in countries influenced by social Darwinism), the positivists often returned to the oppressive practices and social inequities of the past, which some had initially sought to combat. Both conservatives and liberals who embraced Comte followed a similar path. Indeed, one reason positivism had such a large impact in Latin America is that its mix of ideas appealed sufficiently to both liberals and conservatives that it could bring them together. Thanks in part to positivism, liberalism turned against its own values and often became conservative. Unlike liberals in Great Britain and the United States, who emphasized individual rights, liberals in Latin America tended to stress the need for a powerful executive authority, corporate group rights, and emergency laws.[176] They often supported a nondemocratic, dictatorial government just as Comte did. Thus positivism contributed to a situation in Latin America in which, according to Howard Wiarda, "organicism, corporatism, and strong central authority may be viewed historically as the norm in political affairs—which may, incidentally, include democracy, but commonly a top-down, organic . . . form of democracy."[177]

With its ethic of helping society achieve harmony, positivism arguably achieved some humanitarian reforms, as in Brazil, where exponents of the movement helped abolish slavery and attempted to treat Brazilian Indians with respect. But the Religion of Humanity itself had limited appeal.

There were few followers of Comte's religion, which was chiefly promulgated by Augustín Aragón in Mexico, Miguel Lemos and Raimundo Teixeira Mendes in Brazil, and the Lagarrigue brothers in Chile.[178]

Positivism in Asia

Lastly, positivism had an impact in Asia. In Japan, positivist ideas were brought in by the scholar Nishi Amane (1829–1897). In the early 1860s he studied in Leiden, where he came across Comte's books. After his return to Tokyo, Amane gave lectures and wrote articles in the 1870s, promoting positivism at a time when Japan was going through revolutionary changes during the early Meiji period. As in Poland, Russia, and Latin American countries, positivism was used as a weapon to attack the old order. Positivist values of humanism, practicality, and rationality countered traditional feudal and Confucian values, such as subservience, frugality, and blind loyalty. Comte's ideas helped intellectuals in Japan envision a path toward modernization during the Meiji period.[179]

People in India also found sustenance in positivism as they sought to rejuvenate their country and extricate it from the past in a respectful fashion and without admitting any inferiority to the West. Positivism was appealing because in the *Système de politique positive*, Comte had criticized the British domination of India, and Congreve had reiterated his anticolonialism in his book *India* (1857). Indeed, Congreve encouraged positivist "missionaries" to spread positivism in India. These missionaries included Samuel Lobb (1833–1876) in the Indian Educational Service, and James Geddes (1837–1880) and Henry Cotton (1845–1915), who worked in the Indian Civil Service. They spread Comte's ideas, as did Indians drawn to Martineau's translation of the *Cours*.

In the 1850s and1860s positivism and the Religion of Humanity become popular among influential middle-class intellectuals in India, especially in Bengal, where Western ideas often entered the country. Two Indian brothers in particular embraced it: Sanskritist Ram Kamal Bhattacharyya (1838–1860) and Krishna Kamal Bhattacharyya (1840–1932), a Sanskrit professor and attorney who became principal of Ripon College. Both men thought the Religion of Humanity could replace Hinduism as a moral system and glue that held society together.

Reflecting the growing interest in Comte's philosophy, the Positivist Society was created in the late 1870s in Calcutta in Bengal, by Geddes, a magistrate in the Bengal Civil Service. Cotton, a member of the judiciary, took it over around 1880; then it was headed by Jogendro Chandra Ghosh (1842–1902). Given that Geddes was married to Congreve's sister, the Calcutta Positivist Society had a close relationship with Congreve's London Church of Humanity. By 1899 approximately thirty-six Indians, chiefly clerks and teachers, were active in positivist ceremonies, such as the taking of the sacraments, which had been introduced in 1884. The Positivist Society also helped develop the sciences, contributed to the establishment of the first Indian universities, and maintained a fierce critique of British policies in India. Other Indians influenced by positivism and not included in this group included the novelist Bankim Chandra Chatterjee (1838–1894), the journalist Girish Chunder Ghosh (1829–1869), the teacher Guru Das Chatterjee (1814–1882), the high court judge Dwarkanath Mitter (1833–1874), the school superintendent Bhudev Chandra Mukherjee (1827–1898), and the lawyer and educator Satish Chandra Mukherjee (1865–1948).

In general, Indian positivists yearned for scientific, secular progress but did not wish to lose their traditions, religion, and hierarchal society. These people often sought to enhance Hinduism by using Comte's ideas about the need to advance the common good and work for progress. Like Turkish positivists, they appreciated Comte's emphasis on the community and order and his organic view of society. In addition, many Indian positivists adopted Comte's argument that Brahmins held the key to the eventual emancipation of India. Bengal was perhaps the only area of the world where positivism was used to shore up traditional native elites. The reason is that the British, not a group of Indians, were regarded as the enemy. Positivist Indians hoped that if they were enlightened by science and the positivist philosophy, Brahmins, who made up the traditional priestly class, could replace the British and direct public life. Such advances could be made without upsetting the existing social hierarchy.

In their quest to bring together Hinduism and positivism, the Positivist Society celebrated Comte's cult of the dead, created an Indian Positivist Library of Sanskritic books, and paid homage to a painting of a Hindu Goddess of Humanity, wearing a sari and bangles. Brahmins thought that the worship of Humanity, the cult of the dead, and positivist morality were compatible with the principles of the Hindu Shastras. There was speculation that Comte had been a Hindu Rishi in a previous life.

Indian positivists, struck by Comte's emphasis on humanitarianism, self-restraint, duty to others, and intellectual change, tended to be less political and confrontational than positivists in Turkey. However, many Indian positivists used Comte's ideas about science, progress, and social harmony to promote Indian nationalism. Curiously, two Britons, influenced by positivism, served as presidents of the Indian National Congress, which fought for Indian independence: Henry Cotton (elected in 1904) and Annie Besant (elected in 1917).[180] Comte and Congreve would have approved.

Many Positivisms: The Impact on the Disciplines

This altogether too brief overview of the spread of Comte's ideas validates Annie Petit's argument that there are many positivisms. Positivism was a philosophy, religion, and political movement.[181] Thus it was used for a variety of purposes, often contradictory ones. However, at its core was the insistence that people had to think in a more scientific fashion to tackle the subject they were addressing. Whatever explanation seemed to be the most scientific had the most validity—as well as the most prestige. Because this mindset continues to be pervasive, Comte affected modern consciousness on a deep level.[182]

His influence is profound in the area that concerned him the most, education, which was to him the key to transformations first in consciousness and then in action, as many positivists throughout the world recognized. Besides inspiring the establishment of schools around the globe, Comte had a far-reaching effect on academia. He pioneered the idea that each legitimate field of knowledge had to develop its own criteria for investigation and its own methods of proof.[183] Different disciplines then used his theories in a variety of ways, aspiring to be "positivist" in applying the scientific method to their object of study, refusing metaphysical and speculative thinking, and seeking to create laws based on observable phenomena. Comte's academic influence is most evident in the history and philosophy of science, sociology, philosophy, history of religion, anthropology, the study of law, psychology, economics, political economy, and historiography.[184]

Besides historians of sciences, who applaud Comte as the founder of their field largely because of his impressive overview of the development of the sciences in the *Cours de philosophie positive*, philosophers of science have been much marked by his approach. They continue to argue, especially in the so-called Science Wars, about the questions that Comte brought up: the nature of scientific knowledge, the relationships between science and philosophy and between science and society, the epistemic authority and power of scientific thinking, and the classification of the sciences. The latter most reflects Comte's influence. Thinkers since antiquity had tried to classify fields of knowledge, but Comte's argument in favor of classifying the empirical sciences based on the overall development of human thought proved most compelling. Each main science—mathematics, astronomy, physics, chemistry, biology, and sociology—had its own method, and each had to pass through the theological and metaphysical stages before the next, more complex, and more particular (specialized) science could hope to reach the third "positive" stage.

Herbert Spencer, Wilhelm Ostwald, British mathematician Karl Pearson (1857–1936), and other thinkers ended up adopting Comte's basic schema, though they produced important variations. (Mathematics was a slightly unusual case because it was less important as a body of knowledge than as a method of investigation used by other sciences.) The pragmatist American philosopher Charles Peirce (1839–1914) claimed he used for his own classification Comte's idea that the hierarchy of sciences is based on the "order of abstraction of their objects" and that mathematics is the most basic, abstract science.[185] (In paying homage to Comte's insistence that hypotheses could be used if they are verifiable, Peirce also developed his important theory of abduction.[186]) Comte's classification remains the most popular today.[187] Its influence is apparent in the work of one of the most prominent contemporary philosophers of science, Ian Hacking, who freely acknowledges that "there is a lot of Comtianism in my philosophy." Hacking's theory of "styles of reasoning" owes much to Comte's historicist notion that each "branch of knowledge acquires a 'positivity' by the development of a new, positive, style of reasoning associated with it."[188]

Comte's impact on sociology has arguably been even more significant. In 1839 he gave the science of society the name "sociology" and a specific unit of study, social phenomena.[189] He argued that collective groups (families, communities, societies, nations, and so forth) had to be studied holistically, for they had their own behavior, structures, and characters that needed to be explained with scientific methods. Reflecting Comte's view, most sociologists still believe that their task is to discover the laws of society. Comte's division of sociology into social dynamics and social statics has also endured. Evolutionary theorists focus on social dynamics, that is, social change, while structuralists and functionalists embrace social statics, that is, social structure. Many functionalists in particular have adopted

Comte's view of society as an organism with interdependent parts with diverse purposes (functions) that need to be coordinated. In addition, Comte gave sociology a program: to identify and cure social ills, especially the sense of atomization resulting from the breakdown of traditional religious, political, and economic systems. Sociologists, who pride themselves on being scientists not driven by ideology or party politics, still devise ways to rebuild social structures in ways that foster integration or consensus, key concerns of Comte. In short, modern sociologists find themselves marked to varying degrees by his conservatism, reformism, evolutionary worldview, and scientism.[190] Sociology has also influenced other disciplines in the social sciences, such as history and psychology, which maintain the need to study the social context of their respective phenomena.

Yet sometimes positivism as a "value-free," "fact-based" philosophy has affected social scientists and scholars in a manner not truly in keeping with Comte's precepts. At various times, from the mid-nineteenth to mid-twentieth century, they stressed the scientific nature of their studies in order to gain legitimacy and prestige because of the high regard that the sciences traditionally enjoyed. In the late nineteenth century and early twentieth century, history obtained cachet as a discipline and a profession by claiming to be based on the natural sciences. In historiography, the term *positivist history* is still used today to refer to fact-based, event-oriented history; it eschews critical interpretation based on the historian's values and claims to be entirely objective. Comte demanded that historians seek laws of history, but many later positivist historians found that task impossible and endeavored at most to come up with generalizations based on factual evidence. Some positivist historians confined to themselves to simply setting down the facts themselves, a task Comte would have rejected as empiricist and worthless. Historians who are often labeled "positivists" include the Englishman Henry Thomas Buckle; the Americans John Fiske, John Draper (1811–1882), Henry Adams (1838–1918), and Herbert Baxter Adams (1850–1901); and the Frenchmen Hippolyte Taine, Ernest Renan, Fustel de Coulanges (1830–1889), Gabriel Monod (1844–1912), Ernest Lavisse (1842–1922), Charles-Victor Langlois (1863–1929), and Charles Seignobos (1854–1942).[191]

In the mid-twentieth century, quantitative historians revived this empirical approach to the past. Likewise, American sociologists emphasized the importance of statistics in sociology. This use of statistics in this manner helped create the lasting misperception that positivist standards for reliable knowledge demanded quantifiable data and that positivism was synonymous with raw empiricism.[192] This empirical approach became widespread in the social sciences. Legal positivists, who owed more to Jeremy Bentham (1748–1832) and John Austin (1790–1859) than to Comte, treated law "scientifically," put aside questions of morality, and focused simply on the facts relating to legal systems.[193] Positivist economists claimed to focus on laws of the economy, dismissing ethical considerations or any type of social planning to justify their free market liberalism.[194] Positivist psychologists, such as behaviorists, endeavored to be "scientific" in presenting quantifiable data, doing experiments with controls, testing hypotheses, and searching for laws to explain how the mind works.[195] Although they use the term *positivist* or are accused of being positivist, these scholars would not meet with Comte's approval.

Final Evaluation

Positivism has been criticized by contemporary postpositivists and others for espousing a narrow empirical approach and for not recognizing that observations are theory-laden and that what constitutes a fact is problematic.[196] Positivism has also been attacked for buttressing capitalism, discouraging individual action, and supporting resignation to the status quo. Whether for good or for bad, it is clear that Comte's legacy has been significant. If one considers that he first made his appearance on the intellectual scene in 1824 with his *Plan des travaux scientifiques nécessaires pour réorganiser la société* and that people were still acting under his influence in some parts of the globe (France and Brazil, for example) until 1945–1950, he can be seen to have succeeded in challenging traditional ways of thinking and behaving for over a hundred years.

Although such challenges often led to fears of disorder and unease, many people believed that positivism offered certainty.[197] There was the appearance of certainty that came with Comte's unhesitating embrace of dogma, especially his religious dogma. The secular Religion of Humanity, though not as influential as other parts of his doctrine, captured the imagination of serious intellectuals in France, England, Russia, India, Mexico, the United States, and most notably Brazil. These men and women wanted to believe in a creed, especially one that stressed altruism and service to others, concepts at the heart of all religions. But most appealing of all was the supposed certainty that accompanied scientific explanations. In terms of academics and educational reform, Comte promulgated the use of the scientific method in all forms of knowledge, a position that diminished the hold of the church and the appeal of metaphysical speculations and deeply influenced the way we think today.

Positivism not only promoted the use of the intellect but advanced changes in society and the natural world. As a result, Comte's philosophy had a major impact on politics. His ideas inspired a plethora of reformers from the left and the right throughout the world. The many heads of state influenced by Comte and positivists include Ferry, Gambetta, and Clemenceau of France, Branting of Sweden, Masaryk of Czechoslovakia, Atatürk of Turkey, Juárez and Díaz of Mexico, Sarmiento of Argentina, Batlle y Ordóñez of Uruguay, Báez of Paraguay, Pardo of Peru, Guzmán Blanco of Venezuela, Barrios of Guatemala, Guardia of Costa Rica, Zaldívar of El Salvador, Soto of Honduras, Zelaya of Nicaragua, Vargas of Brazil, and Castilhos of Rio Grande do Sul. In general, people in developing countries, such as Poland, Russia, Turkey, Brazil, Mexico, Japan, and India looked at positivism with enthusiasm, for it offered them a vision of inevitable progress and a plan for modernization that was not only economic and political but cultural and intellectual. Unlike Marxism, which also provided a blueprint for modernization, positivism could be interpreted to respect traditional beliefs and practices because it condoned dogmatism, hierarchy, and order. It suited, as Geraldine Forbes asserts, "conservative modernizers," especially those who wished to create a new ruling elite.[198]

Yet positivism did not always lend its weight to buttressing the status quo and authoritarianism, as some critics maintain. In both Britain and the United States, it was a scientific and ethical ideology that a middle-class elite used to give themselves intellectual and moral legitimacy and the tools to enhance their critique of diverse social, political, and economic practices.[199] In Poland and Russia, positivism supported radicals' attempts to overthrow the dominant power. Republicans in France and reformers in Italy, Belgium, Sweden, and the Netherlands wielded its rational, scientific arguments to combat the Catholic and Protestant churches. In different areas of the globe, politicians of different stripes used it to push for educational reforms, industrial productivity, and moral betterment.[200] Positivism in one of its many guises offered people throughout the world on both the right and left some kind of salvation, some hope for a better world.[201]

Notes

Preface

1. Hacking, *Historical Ontology*, 190.
2. Schmaus, "Hypotheses."
3. Schmaus, "Lévy-Bruhl."
4. Hayek, *Road*, 172.
5. For instance, Tocqueville favored the colonization of Algeria. Tocqueville, *L'Algérie*.
6. Schmaus, *Durkheim*, 12–16.
7. Hayek, *Studies*, 285–304.
8. Furthermore, Michel Bourdeau has already published an exhaustive annotated bibliography of works by and about Comte as part of the Oxford Bibliographies series. See "August Comte," *Oxford Bibliographies*, last modified September 29, 2014, doi:10.1093/obo/9780195396577-0246.

Introduction

1. See Kitcher, *Science, Truth, and Democracy* and *Science in a Democratic Society*; and Longino, *Science* and *Fate*.
2. See, for example, Howard, "Two Left Turns"; Reisch, *Cold War*; Uebel, "Political Philosophy."
3. Richardson, "Left Vienna Circle," parts 1 and 2.
4. For more on political activity at the École Polytechnique, see Soltau, *French Political Thought*; Hayek, *Counter-Revolution*; Andrews, *Socialism's Muse*; and Tresch, "Order of the Prophets."
5. *L'esprit*, 42. This work was originally the *Discours préliminaire sur l'esprit positif*, placed at the beginning of his *Traité philosophique de l'astronomie populaire* (1844), which grew out of his free public lectures on the history of astronomy that he began presenting in 1831.
6. *L'esprit*, 41–43; *Système*, I, 58, trans. 45; *Système*, IV, 547, trans. 473.

7. *L'esprit*, 42.

8. *L'esprit*, 41; *Système*, I, 36, trans. 28.

9. *Système*, I, 36–38, trans. 28–29.

10. Hacking, *Historical Ontology*, 190.

11. Kragh, *Introduction*, ch. 1.

12. Littré, *Auguste Comte*; Robinet, *Notice*.

13. Mill, *Auguste Comte*; Morley, "Auguste Comte"; Huxley, "Scientific Aspects"; Lévy-Bruhl, *Philosophy*.

14. Bréhier, *History*, vol. 6, ch. 15; Aron, *Main Currents*, vol. 1; Hayek, *Counter-Revolution*; Manuel, *Prophets*.

15. Schmaus, "Concept of Analysis."

16. *Cours*, I, L 45, 856.

17. *Correspondance*, I, 46.

18. *Cours*, I, L 45, 856.

19. *Cours*, I, L 45, 856.

20. *Correspondance*, III, 61.

21. *Cours*, II, L 54, 362.

22. *Cours*, II, L 57, 696.

23. *Cours*, II, L 46, 15.

24. Comte, 1845 letter to Mill in *Correspondance*, III, 61.

25. *Cours*, II, L 60, 778.

26. Saint-Simon, *Œuvres*, 6:297.

27. *Écrits de jeunesse*, 40.

28. Barrault and Charton, *Religion Saint-Simonienne*, 2:64.

29. *Correspondance*, IV, 22.

Chapter 1. Comte's General Philosophy of Science

1. The *Oxford English Dictionary* gives 1856 as the earliest use of the term *epistemology*.

2. *Cours*, II, L 49, 173n.

3. Chimisso, *Writing*.

4. Kuhn, *Structure*, xl.

5. See Schickore, "More Thoughts."

6. For background on the eclectic spiritualists, see Brooks, *Eclectic Legacy*.

7. *Cours*, II, L 58, 730.

8. Cousin, *Du Vrai*, vi–vii. Although Cousin's lectures on "the True, the Beautiful, and the Good" were first published in 1837, they are based on student notes taken in his classes during the years 1815–1820 (Cousin, *Du Vrai*,

x–xi). According to Pickering (*Comte*, 1:126), Comte was familiar with these very popular lectures.

9. Cousin, *Cours de l'Histoire*, 67, 313. Cousin expressed similar views in his lectures on the True, the Beautiful, and the Good. See previous note.

10. Cousin, *Cours de l'Histoire*, 4, 313.

11. Cousin, *Cours de l'Histoire*, 327–28.

12. Cousin *Cours de l'Histoire*, 49; *Du Vrai*, 19–23, 27–28. For a more detailed account of Cousin and Maine de Biran on the derivation of the categories from internal perception, see Schmaus, "Kant's Reception"; and Schmaus, *Rethinking Durkheim*, ch. 3.

13. *Système*, IV, appendix, 217, trans. 646; *EPW*, 229.

14. See note 12 above.

15. *Cours*, I, L 1, 33; *Introduction*, 20.

16. For a study of Comte's physiological approach to the mind, see Clauzade, *L'Organe de la pensée*.

17. *Cours*, I, L 45, 854.

18. *Système*, IV, appendix, 219–20, trans. 647; *EPW*, 231–32.

19. *Cours*, I, L 45, 854.

20. Comte owned a French translation of the *Enquiry*. Pickering, *Auguste Comte*, 1:306.

21. Locke, *Essay*, bk. 2, ch. 21, sec. 4–5.

22. Hume, *Enquiries*, 68.

23. Maine de Biran, *Essai*, 229ff; Maine de Biran, *Examen*, 369–70. Indeed, to the extent that the eclectic spiritualists were also followers of Thomas Reid's commonsense philosophy, they would have also denied that even external sensation is mediated by mental representations.

24. *Cours*, I, L 1, 33; *Introduction*, 21.

25. Braunstein, "Antipsychologisme," 11.

26. Maine de Biran, *Essai*, 231–32.

27. Mill, *Auguste Comte*, 296–97.

28. According to Lewisohn ("Mill and Comte," 319), Mill found Comte to be ignorant of the principles of associationist psychology.

29. Cousin, *Cours de l'Histoire*, 195.

30. Mill, *Examination*, 453. I would like to thank Laura Snyder for directing my attention to this work.

31. Scharff, *Comte*, 11. Cf. also Scharff, "Mill's Misreading"; Scharff, "Positivism," 260–62.

32. *Cours*, I, L 1, 34; *Introduction*, 21.

33. Scharff, "Mill's Misreading," 565–66; Scharff, *Comte after Positivism*, 33–34.
34. *Cours*, I, L 1, 34; *Introduction*, 21–22.
35. Scharff, "Mill's Misreading," 571–72.
36. *Cours*, I, L 45, 854.
37. In chapter 5 of this volume, Vincent Guillin argues that Comte had additional methodological reasons for dismissing psychology as a science, including its assumption of methodological individualism and its assumption that the human mind could be understood in isolation from both the human body and the larger natural and social world to which it belonged.
38. *Cours*, I, L 1, 22; *Introduction*, 4.
39. *Système*, III, 3, 46, trans. 38–39; *Cours*, I, L 1, 32; *Introduction*, 19.
40. *Cours*, I, L 1, 21; *Introduction*, 1–2. The three-stage law is not entirely original to Comte. It was proposed earlier by Anne-Robert-Jacques Turgot in 1750 in his *Plan de deux discours sur l'histoire* (*Turgot on Progress*, 102). However, Émile Littré argued that, unlike Comte, Turgot never integrated this law into his philosophical system (Littré, *Auguste Comte*, 2nd ed., 48). For a discussion of misunderstandings of the three-stage law by Richard Vernon, Karl Popper, and others, see Schmaus, "Reappraisal." For a more recent and more detailed defense of the three-stage law, see Bourdeau, *Trois Etats*; and Schmaus, "Rescuing."
41. *Cours*, I, L 1, 24–25; *Introduction*, 7–8.
42. *Cours*, I, L 1, 21–22; *Introduction*, 2.
43. *Cours*, I, L 2, 57; *Introduction*, 55.
44. *Cours*, I, L 1, 22, 40; *Introduction*, 2–3, 31.
45. *Cours*, I, L 3, 71.
46. Laudan, "Reassessment," 142–43.
47. *Cours*, II, L 58, 731.
48. *Cours*, I, L 2, 50; *Introduction*, 46–47.
49. *Cours*, I, L 2, 52; *Introduction*, 48.
50. *Cours*, I, L 3, 71.
51. Schmaus, "Concept of Analysis."
52. *L'esprit*, 16.
53. Laudan, "Reassessment," 147.
54. *Cours*, I, L 1, 23–24; *Cours*, I, L 2, 45; *Introduction*, 5–7, 38–39; *Système*, IV, appendix, 142–43, trans. 593–94; *EPW*, 150–51.
55. Hayek, *Counter-Revolution*, 139.

56. *Cours*, II, L 52, 252, 264; *Système*, IV, appendix, 138–39, trans. 591; *EPW*, 147–48.
57. *Cours*, I, 45; *Introduction*, 38.
58. *Cours*, I, L 28, 454; *Système*, I, 10, trans. 7.
59. *Cours*, II, L 58, 730.
60. *Système*, IV, appendix, 145, trans. 595–96; *EPW*, 154.
61. *Système*, IV, appendix, 145, trans. 595; *EPW*, 153.
62. *L'esprit*, 9n.
63. *Cours*, II, L 56, 569.
64. *Système*, III, 541, trans. 459.
65. Scharff, *Comte*, 8.
66. Scharff, *Comte*, 8.
67. Popper, *Poverty*, 117.
68. Popper, *Poverty*, 129n1, 152–53.
69. *Système*, IV, appendix, 164–65, trans. 609; *EPW*, 173–74.
70. *Système*, II, 386, trans. 314.
71. *Système*, II, 386, trans. 314.
72. *Système*, II, 386, trans. 314.
73. *Système*, II, 249, trans. 209.
74. *Système*, II, 249, trans. 209.
75. *Système*, II, 250, trans. 210.
76. *Système*, II, 250, trans. 210.
77. Popper, *Poverty*, 128–29.
78. *Cours*, II, L 51, 227; *Système*, III, 41, trans. 34.
79. Whewell, *On the Philosophy of Discovery*, 227–28.
80. Spencer, "Genesis," 177ff.
81. Renouvier, "Philosophie du XIXe siècle," 29; "Le sens de la méthode phénomeniste," 161–62.
82. Cournot, *Considérations*, 2:223–26.
83. Flint, *Philosophy*, 272ff; *Système*, IV, appendix, 139, trans. 591; *EPW*, 147.
84. *Cours*, I, L 20, 322ff.
85. *Cours*, II, L 58, 728.
86. Hacking, *Historical Ontology*, 164–67, 190.
87. *Cours*, II, L 48, 128–29; *Cours*, II, L 51, 204, 210; *Système*, I, 682, 715–22, trans. 550–51, 578–84. For more on Comte's views on evolution and the development of the human mind, see chapters 4 and 5 in this volume.
88. *Cours*, I, L 2, 48; *Introduction*, 42–43; *Système*, I, 38–39, trans. 30.

89. *Cours*, I, L 2, 48; *Introduction*, 43.
90. *Cours*, I, L 2, 54; *Introduction*, 52.
91. *Cours*, I, L 2, 54, 58; *Introduction*, 51–52, 57.
92. *Cours*, I, L 28, 464.
93. *Cours*, II, L 48, 137–38.
94. *Cours*, I, L 1, 41; *Introduction*, 32.
95. *Cours*, II, L 60, 772.
96. *Cours*, II, L 60, 772.
97. *Cours*, II, L 60, 772–73.
98. *Cours*, II, L 60, 773.
99. However, Clauzade (ch. 4, this volume) argues that there is a sense in which Comte is nevertheless a vitalist.
100. *Cours*, I, L 1, 41; *Introduction*, 32.
101. *Cours*, I, L 1, 26; *Introduction*, 8–9.
102. *Cours*, I, L 28, 460.
103. *Cours*, I, L 30, 487.
104. Benrubi, *Sources*, 16–17.
105. *Système*, IV, 187, trans. 165–66.
106. *Système*, IV, 175, trans. 155.
107. Hacking, *Social Construction*, 83–84.
108. Darwin, *Origin*, 485.
109. *Cours*, I, L 14, 223; *Cours*, I, L 42, 771.
110. *Cours*, II, L 58, 723. Comte made similar claims in the *Discours sur l'esprit positif* (18) and the *Système de politique positive* (IV, 174, trans. 154–55). As Scharff ("Positivism," 259) argues, Comte did not take the idea of an inductive basis for the principle of invariability of laws from Mill, since Comte's earliest statement of it appears in the last volume of the *Cours*, which predates Mill's *Logic*.
111. *Cours*, II, L 48, 123.
112. *Cours*, I, L 2, 62; *Introduction*, 63.
113. *Cours* I, L 3, 72–75; *Cours* I, L 4, 88ff; *Cours* I, L 11, 174; *Cours* I, L 15, 227–28.
114. *Évolution*, 232. For a more complete discussion of Comte's concept of mathematical analysis, see Schmaus, "Concept of Analysis."
115. *Cours*, I, L 10, 154.
116. *Synthèse*, 427.
117. *Cours*, I, L 10, 168.
118. *Cours*, I, L 1, 23; *Introduction*, 5.

119. *Cours*, I, L 19, 305.
120. *Système*, I, 500, trans. 405.
121. *Cours*, II, L 48, 139.
122. *Cours*, II, L 48, 139.
123. *Cours*, II, L 48, 141.
124. *Système*, IV, appendix, 141, trans. 592–93, *EPW*, 149.
125. *Système*, III, 19, trans. 16.
126. *Système*, I, 711–12, trans. 575; *Cours*, II, L 60, 728.
127. Laudan, "Reassessment," 147.
128. *Cours*, I, L 19, 308, 458; *Système*, IV, 210, trans. 185.
129. *Cours*, I, L 28, 461–62. For more on Comte's views on hypotheses in optics and astronomy, see chapters 2 and 3, this volume.
130. *Cours*, I, L 28, 457.
131. *Système*, I, 520, trans. 421.
132. *Système*, I, 520, trans. 421.
133. *Système*, I, 520, trans. 421.
134. *Cours*, II, 736.
135. *Cours*, I, L 40, 720.
136. Laudan, "Reassessment," 155–56.
137. *Cours*, I, L 28, 447; *Cours*, I, L 35, 574; *Système*, I, 519, trans. 420.
138. *Cours*, I, L 28, 447; see also *Cours*, I, L 35, 574; *Cours*, I, L 40, 695; *Cours*, II, L 48, 142.
139. *Cours*, I, L 40, 690.
140. *L'esprit*, 17n1. Other precedents for Mill's method of difference can be found in Herschel, *Preliminary Discourse*, 154–55, and in the first of Bacon's twenty-eight types of prerogative instances, the solitary instance, in his *Novum Organum* (ch. 2, aphorism 22).
141. *Cours*, I, L 35, 574; *Système*, I, 519, trans. 420; *Cours*, I, L 28, 447; *Cours*, I, L 40, 690, 693; *Cours*, II, L 48, 142.
142. *Cours*, I, L 35, 574; *Cours*, I, L 40, 705–6. For more on Comte's understanding of the comparative method in biology, see chapter 4 in this volume.
143. *Cours*, II, L 48, 148.
144. *Cours*, I, L 40, 707.
145. *Cours*, II, L 48, 145.
146. *Cours*, II, L 48, 148; *Cours*, II, L 49, 159. For more on Comte's historical method, see chapter 5 in this volume.
147. *Système*, I, 446, 583, trans. 362, 472; *Système*, III, 567, trans. 483.
148. *Système*, I, 581, trans. 471.

149. *Système*, I, 446–47, trans. 362; *Cours*, II, L 59, 768.
150. *Cours*, II, L 49, 170, 172; *Cours*, II, L 58, 703, 707.
151. *Système*, III, 21ff, 18ff.
152. *Système*, I, 452–53, trans. 367; *Système*, II, 82, trans. 72–73.
153. Mill, *Auguste Comte*, 291–92.
154. Scharff, "Positivism," 256–58, 260; Scharff, *Comte*, 58ff.
155. Renouvier, "Philosophie du XIXe siècle," 35; Renouvier, "Prétentions," 234–35; Renouvier, "Méthode," 402; Renouvier, "*Cours de philosophie positive* est-il au courant," 293–94; Renouvier, *Philosophie analytique*, 227, 663.
156. *Cours*, I, L 19, 304; *Cours*, II, L 58, 708.
157. Savary, "Sur la détermination."
158. *Cours*, I, L 27, 435n; *Cours*, II, L 49, 168ff.
159. Whewell, *On the Philosophy of Discovery*, 231–32.
160. *Système*, IV, 173–74, trans. 154.

Chapter 2. The Analytical Construction of a Positive Science in Auguste Comte

1. *Cours*, I, L 3, 65.
2. *Cours*, I, L 3, 65.
3. On these questions, one may consult Blay, *Naissance*; and Blay, *Reasoning*.
4. *Cours*, I, L 3, 66. Here, Comte refers to Lagrange's *Théorie des fonctions analytiques* and *Mécanique analytique*, subsequently republished in Lagrange's *Œuvres*, vols. 9, 11, and 12.
5. *Cours*, I, L 3, 68.
6. *Cours*, I, L 3, 68–69.
7. On that issue, see Blay and Festa, "Mouvement."
8. On this point, see Blay, "Principe"; and Blay, "Force."
9. Bernier, *Abrégé*, 1:296–99. In this book, Bernier drew particular attention to the fact that the speeds of two moving objects resulted from some "combination of a greater or lesser quantity of small rest." Mariotte, *Traité*. Mariotte argued that, contrary to the Galilean thesis of the continuity of time, "one cannot know whether the falling body goes through a short space without accelerating its first motion, since time is needed to produce most of the natural effects." (This warning appears at the end of Proposition X. See pages 247–49.)
10. Leibniz, *Philosophischen Schriften*, 2:104–5; Leibniz, *Discourse*, 202 [slightly modified translation].
11. Leibniz, *Sämtliche Schriften*, 56; Leibniz, *New Essays*, 56.
12. Comte specifies that "*science* consists in the coordination of facts; if the

various observations were entirely independent, there would be no science. It might even be said that science primarily aims at exempting us, as far as the various phenomena allow it, from any direct observation, by enabling us to deduce from the smallest amount of immediate data the largest possible amount of results" (*Cours*, I, L 3, 71).

13. *Cours*, I, L 3, 70.
14. *Cours*, I, L 3, 70.
15. *Cours*, I, L 15, 227.
16. *Cours*, I, L 15, 227.
17. *Cours*, I, L 15, 228.
18. *Cours*, I, L 17, 266.
19. *Cours*, I, L 15, 240.
20. Huygens, *Discours*, 175–76.
21. Huygens, *Horologium*, 128–30.
22. In his argument, Huygens associates time intervals by pair and considers the respective ratios of the spaces covered. This result can be generalized.
23. Newton, *Principia*, 83 [slightly modified translation].
24. Newton, *Principia*, 89.
25. Newton, *Principia*, 253, 264.
26. See Blay, "Principe"; Blay, "Force"; and Blay, *Principia*.
27. Newton, *Principia*, 99.
28. See Blay, *Naissance*, and Blay, *Reasoning*.
29. *Cours*, I, L 15, 228.
30. D'Alembert, *Traité*, préface.
31. *Cours*, I, L 15, 233.
32. Here, the reference to Kepler rests on some misunderstanding as to his use of the term *inertia*.
33. *Cours*, I, L 15, 235.
34. *Cours*, I, L 15, 236.
35. *Cours*, I, L 17, 266.
36. On that issue, one may consult Jouguet, *Lectures*, 2n58.
37. *Cours*, I, L 17, 268.
38. In the case of continuous action, it is important to consider, so as to avoid the difficulties relative to the appraisal of the space covered, not each of the two time intervals given, but the time interval dt that is equal to the sum of the halves of each of these two time intervals (i.e., a time starting from the middle of the first time interval and ending in the middle of the second). In that very case, Comte, by doing so, avoids using an approach depending on the

idea of limit. From a certain point of view, *mutatis mutandis*, this is a similar approach that can found in Lemma II of book 2 of Newton's *Principia*.

39. *Cours*, I, L 33, 530.
40. *Cours*, I, L 28, 463.
41. *Cours*, I, L 28, 458.
42. *Cours*, I, L 33, 531.
43. *Cours*, I, L 33, 534.
44. *Cours*, I, L 33, 535.
45. Duhem, "Fragments," first frag., 95.
46. Duhem, "Fragments," first frag., 95–96.
47. Duhem, "Fragments," second frag., 28–29.
48. Duhem, "Fragments," first frag., 96.
49. Duhem, "Fragments," second frag., 28–29.
50. In the thirty-third lesson, Comte deals with some "general considerations on mathematical thermology." There, he of course mentions "the admirable doctrine that we owe to the beautiful genius of the great Fourier" (*Cours*, I, L 33, 500). That doctrine is "admirable" because it represents for heat what Lagrange's represented for motion, and because "the analytic object of such an inquiry . . . always consists in discovering the function which expresses, at any time, the temperature of any point of the solid mass" (*Cours*, I, L 33, 501). The general remarks that we have made about the Comtean approach also apply here—hence, no need to repeat them.
51. Duhem, "Fragments," third frag., 105.
52. Rupp's experimental work was subsequently called into question. See, for example, French, "Strange."

Chapter 3. Astronomical Science and Its Significance for Humankind

I am grateful to Annie Petit for sharing with me her intimate knowledge of Auguste Comte's works.

1. Henceforth *Traité*. Translations are mine unless otherwise indicated.
2. The labels "logical positivism" and "logical empiricism" were used by the members of the Vienna Circle themselves. Both raise questions that I shall not go into here. Suffice it to say that I shall use throughout "logical positivism," in order to stress the relation to Comte.
3. This corresponds to the second installment of the *Cours*, L 19–34.
4. *Traité*, part I, corresponds to *Cours*, L 19; part II to L 20; part III to L 21–23; part IV to L 24–26. The *Traité* numbers some 484 pages, whereas the

astronomy lessons of the *Cours* number 378 pages in the first edition and 264 pages in Littré's edition.

5. *Système*, I, 498, trans. 404.

6. *Traité*, 11, 331.

7. Of course, Comte was not able to take into account François Arago's contemporaneous lectures on astronomy, as they were published only posthumously, under the title *Astronomie populaire*, in four volumes between 1854 and 1857 (for a recent comparative study see Christen, "Leçons et traités").

8. See *Traité*, 440. Let us recall that Comte submitted a "Premier mémoire sur la cosmogénie positive" to the academy in 1835. For details see Pickering, *Auguste Comte*, 1:450–53.

9. *Traité*, 250, 262, 287, 291, 431.

10. *Traité*, 146.

11. *Traité*, 127, 189, 207.

12. *Traité*, 177.

13. *Traité*, 217, 300.

14. *Traité*, 157, 260.

15. *Cours*, II, L 57, 681; *Système*, I, 498, trans. 403.

16. *Traité*, 328.

17. *Traité*, 9.

18. *Correspondance*, I, 218.

19. *Cours*, I, L 19, 311. Comte is quoting here Psalm 19: "The heavens declare the glory of God."

20. *Correspondance*, II, 427.

21. See Pickering, *Auguste Comte*, 1:489.

22. Magnin, "Vingt-et-unième anniversaire," 657.

23. *Traité*, 226.

24. *Correspondance*, IV, 66.

25. *Traité*, 455.

26. *Traité*, 455.

27. *Traité*, 483.

28. *L'esprit*, sec. 45; *Traité*, 65.

29. *Traité*, 398.

30. *Cours*, I, L 28, 451. Comte makes a similar remark on page 458. At the end of the last lesson on astronomy, he provides a transition to the effect that physics "derives from the doctrine and method of astronomy a general model and indispensable basis" (*Cours*, I, L 27, 440). For more of Comte's views on hypotheses, see chapter 1 of this volume.

31. Poincaré, *Valeur*, 121, trans. 294. Comte indeed criticized inquiry into the composition of the sun: *Cours*, I, L 19, 301.

32. Milhaud, *Positivisme*, 173.

33. *Cours*, I, L 23, 372, emphasis mine. Compare with *Traité*, 348–49.

34. Weinmann, "Galileo," 9.

35. Brenner, *Raison*.

36. Bachelard, *Essai*, 52.

37. *L'esprit*, sec. 31; *Traité*, 50.

38. *Cours*, I, L 2, 60; *Introduction*, 61.

39. See Daston and Galison, *Objectivity*, 206.

40. *Traité*, 328, emphasis mine.

41. *Traité*, 233.

42. *Traité*, 244.

43. *Traité*, 263.

44. Bachelard, Études, 81.

45. *Cours*, II, L 58, 727. Compare with an earlier reference to Kant, *Cours*, I, L 3, 77.

46. See Pickering, *Auguste Comte*, I, 289–96.

47. *Système*, I, 499, trans. 404. For other uses of the terms *objective* and *subjective*, see *Système*, I, 460, 505, 508, trans. 373, 409, 411.

Chapter 4. Auguste Comte's Positive Biology

I would like to thank Vincent Guillin for translating this chapter into English.

1. See Robin, "Sur la direction."

2. *Cours*, I, L 36, 602n.

3. Blainville, *Cours*, 1:18.

4. Blainville, *Cours*, 1:4.

5. *Cours*, I, L 40, 741; *Positive Philosophy*, 332.

6. Blainville, *Organisation*, viii–ix.

7. *Cours*, I, L 1, 32–33; *Introduction*, 19; see also *Positive Philosophy*, 32.

8. *Cours*, II, L 48, 109; *Positive Philosophy*, 457.

9. *Cours*, I, L 42, 767–68; *Positive Philosophy*, 343.

10. *Cours*, I, L 40, 744; *Positive Philosophy*, 333.

11. *Cours*, I, L 40, 680; *Positive Philosophy*, 309.

12. Blainville, *Organisation*, xxii, 16.

13. *Système*, I, 596, trans. 482.

14. *Cours*, I, L 40, 682n.

15. Blainville, *Cours*, 3:381–432.
16. Lamarck, *Recherches*, 1802, 52.
17. *Cours*, I, L 40, 676; *Positive Philosophy*, 304. For a recent study of the next topic in English, see Tresch, *Romantic Machine*, 270–71.
18. Bichat, *Recherches*, 1; Bichat, *Physiological Researches*, 1.
19. Bichat, *Recherches*, 1; Bichat, *Physiological Researches*, 1.
20. *Cours*, I, L 40, 677.
21. See letter to Valat, May 15, 1819, *Correspondance*, I, 37.
22. See Canguilhem, "L'école," 79.
23. *Cours*, I, L 40, 683; *Positive Philosophy*, 307.
24. *Cours*, I, L 40, 683.
25. *Cours*, I, L 40, 683–84; *Positive Philosophy*, 307, modified.
26. Bichat, *Anatomie*, lxxix; Bichat, *General Anatomy*, 44.
27. Bichat, *Anatomie*, lxxx; Bichat, *General Anatomy*, 45.
28. *Cours*, II, L 50, 183; *Positive Philosophy*, 502.
29. See *Cours*, I, L 40, 764–66; *Positive Philosophy*, 342.
30. See *Cours*, I, L 41, 762; *Positive Philosophy*, 342.
31. See Blainville, *Organisation*, 15–16.
32. See *Cours*, I, L 44, 823; *Positive Philosophy*, 369–70.
33. Bichat, *Anatomie*, lxxii; Bichat, *General Anatomy*, 38.
34. *Cours*, I, L 43, 809; *Positive Philosophy*, 363.
35. *Cours*, I, L 40, 745; *Positive Philosophy*, 333–34.
36. *Cours*, I, L 41, 766.
37. See, e.g., *Cours*, I, L 43, 795.
38. *Système*, I, 642, trans. 519.
39. *Cours*, I, L 40, 685; *Positive Philosophy*, 308; Blainville, *Cours*, 3:381–432; Blainville, *Plan*.
40. See *Système*, I, 665–66, trans. 537–38; Segond, "Examen," 15.
41. *Système*, I, 666–67, trans. 538.
42. See Lévi-Strauss, "La leçon"; Braunstein, *La philosophie*, 174–77.
43. Cuvier, *Règne animale*, 6; *Animal Kingdom*, 3–4.
44. *Cours*, I, L 40, 738; *Positive Philosophy*, 332.
45. *Système*, I, 661, trans. 534.
46. *Système*, I, 661, trans. 534.
47. *Cours*, I, L 40, 721; see *Système*, IV, appendix, 120, trans. 578; *EPW*, 127.
48. *Cours*, I, L 40, 695; *Positive Philosophy*, 312.
49. *Système*, I, 650, trans. 527.
50. See, e.g., *Système*, II, 440–44, 455–58, trans. 359–63, 372–74.

51. See Albury, "Experiment"; Huneman, *Bichat*, 91–96.

52. *Cours*, I, L 40, 698; *Positive Philosophy*, 313.

53. *Cours*, I, L 40, 699; *Positive Philosophy*, 314.

54. See *Cours*, I, L 40, 698–707; *Positive Philosophy*, 313–19; *Système*, I, 654, trans. 528.

55. See *Cours*, I, L 40, 704; *Positive Philosophy*, 316; *Système*, I, 654, trans. 528.

56. *Cours*, I, L 40, 705–6; *Positive Philosophy*, 317.

57. See *Cours*, I, L 42, 770–74; *Positive Philosophy*, 344–45; *Système*, I, 654–45, trans. 529.

58. *Cours*, I, L 42, 773–74; *Positive Philosophy*, 345.

59. See *Cours*, II, L 50, 190–91.

60. See above and *Cours*, I, L 42, 774–80.

61. *Cours*, I, L 42, 775; *Positive Philosophy*, 346.

62. *Cours*, I, L 42, 775; *Positive Philosophy*, 346.

63. *Système*, I, 656, trans. 530. The term *instrument* does not appear in the original French text.

64. See Lamarck, *Philosophie*, 102–3.

65. See *Cours*, I, L 40, 728; *Positive Philosophy*, 327–28; *Système*, I, 657, trans. 531.

66. See *Cours*, I, L 40, 700–701; *Positive Philosophy*, 314; see also Clarke and Jacyna, *Nineteenth-Century Origins*, 38–41.

67. *Cours*, I, L 40, 701; *Positive Philosophy*, 315.

68. *Cours*, I, L 43, 797; *Positive Philosophy*, 356.

69. See *Cours*, I, L 45, 851; *Positive Philosophy*, 382.

70. Bichat, *Recherches*, 3; Bichat, *Physiological Researches*, 3.

71. *Système*, I, 642, trans. 518.

72. *Cours*, I, L 43, 809; *Positive Philosophy*, 362.

73. For properties, see *Cours*, I, L 43, 810–12; *Positive Philosophy*, 363–64; for functions, *Cours*, I, L 43, 812–15; *Positive Philosophy*, 364–65; and for results, *Cours*, I, L 43, 815–19; *Positive Philosophy*, 365–68.

74. *Cours*, I, L 40, 680; *Positive Philosophy*, 309.

75. *Cours*, I, L 43, 813.

76. *Cours*, I, L 43, 818; *Positive Philosophy*, 367.

77. See in particular Canguilhem et al., *Du développement*; Guillo, *Les figures*, 323–27.

78. *Cours*, I, L 43, 810.

79. *Cours*, I, L 2, 55; *Introduction*, 53. See also *Positive Philosophy*, 44.

80. See *Cours*, I, L 44, 836–41; *Positive Philosophy*, 376–80.
81. See Cabanis, *Rapports*, Mémoires II et III, esp. 208–9.
82. *Système*, IV, 237, trans. 209; see also *Système*, II, 437–38, trans. 356.
83. See Blainville, *Plan*.
84. See *Système*, I, 608–9, trans. 492.
85. *Cours*, I, L 44, 840–41; *Positive Philosophy*, 380.
86. *Cours*, I, L 45, 857–58; *Positive Philosophy*, 385.
87. *Cours*, I, L 45, 857–58; *Positive Philosophy*, 385.
88. *Cours*, I, L 44, 841; *Positive Philosophy*, 380.
89. See *Système*, I, 565–670, trans. 456–540.
90. *Système*, I, 402, trans. 323.
91. Comte, *L'esprit*, sec. 20, 24–25; *Discourse*, 39.
92. *Système* I, 585, trans. 474.
93. For vegetality, see *Système*, I, 586–96, trans. 474–82; for animality, *Système*, I, 596–620, trans. 482–501; and for humanity, *Système*, I, 620–39, trans. 501–16.
94. *Système*, I, 589, trans. 477.
95. *Système*, I, 622–23, trans. 503–4.
96. *Système*, I, 616, trans. 498.
97. *Système*, I, 616, trans. 498.
98. *Cours*, I, L 45, 842; *Positive Philosophy*, 380.
99. See letter to Valat, Sept. 8, 1824, *Correspondance* I, 125.
100. *Système*, I, 728, trans. 588.
101. See Clauzade, *L'Organe*.
102. *Cours*, I, L 45, 863; *Positive Philosophy*, 387.
103. *Cours*, I, L 45, 865; *Positive Philosophy*, 388.
104. See Young, *Mind*, 9–53, Lantéri-Laura, *Histoire*.
105. *Cours*, I, L 45, 866; *Positive Philosophy*, 388.
106. See, e.g., Gall, *Sur les fonctions*, 1:320–27.
107. See, e.g., Bain, *Study of Character*.
108. See Spurzheim, *Observation*, 120–313; *Outlines*, 27–76.
109. See Gall and Spurzheim, *Anatomie*, 3:xxv–xxviii.
110. See *Cours*, II, L 56, 487–88; *Positive Philosophy*, 685–86.
111. *Cours*, II, L 51, 204; *Positive Philosophy*, 517.
112. See Guillin, "Le penchant biologique."
113. *Cours*, II, L 48, 153; *Positive Philosophy*, 484.
114. *Cours*, II, L 49, 157–58; *Positive Philosophy*, 487.
115. *Cours*, II, L 49, 159; *Positive Philosophy*, 488.

116. See *Cours*, II, L 48, 153; *Positive Philosophy*, 484; and Canguilhem et al., *Du développement*, 23–24.

117. *Cours*, II, L 58, 714; *Positive Philosophy*, 797.

118. See Guillin, *Auguste Comte*.

Chapter 5. Comte and Social Science

1. In this chapter, Comte's early opuscules are quoted from the appendix to the fourth volume of the *Système* and abbreviated as follows: *Plan de travaux scientifiques nécessaires pour réorganiser la société* (1824) as *PTS*; *Considérations philosophiques sur les sciences et les savants* (1825) as *CPSS*; *Considérations sur le pouvoir spirituel* (1825–1826) as *CPS*; translations for these pieces are taken from *Early Political Writings* (abbreviated as *EPW*).

2. For the classic statement of this position, see Nisbet, *Sociological Tradition*; Karsenti, *Politique*. Karsenti's book, which is undoubtedly the most perceptive study of Comte's sociology to have appeared lately, draws on the same idea.

3. Scharff, *Comte*, 6.

4. Scharff, *Comte*, 6.

5. Scharff, *Comte*, 5.

6. *Cours*, I, L 1, 21; *Introduction*, 1.

7. That is, "au niveau de leur siècle." *Système*, IV, appendix, 217, trans. 646; *EPW*, 229.

8. *PTS*, 54; *EPW*, 57.

9. *PTS*, 48; *EPW*, 50.

10. *PTS*, 63; *EPW*, 65.

11. *PTS*, 63; *EPW*, 65.

12. *PTS*, 72; *EPW*, 75.

13. *PTS*, 72n; *EPW*, 75.

14. *PTS*, 73; *EPW*, 77.

15. *PTS*, 82; *EPW*, 86.

16. *PTS*, 85; *EPW*, 89.

17. *PTS*, 86; *EPW*, 90.

18. *PTS*, 89; *EPW*, 93.

19. *PTS*, 86; *EPW*, 90.

20. *PTS*, 89; *EPW*, 93.

21. As Comte famously put it, "The world is governed and overturned by ideas, or, in other words . . . the whole social mechanism rests finally on opinions" (*Cours*, I, L 1, 38; *Introduction*, 28).

22. Scharff, *Comte*, 6.

23. *PTS*, 79–81; *EPW*, 84–85; *Cours*, I, L 1, 20–41; *Introduction*, 1–33; L 2, 42–64; *Introduction*, 35–67; *Cours*, II, L 46, 13–79.

24. *PTS*, 47; *EPW*, 49.

25. *PTS*, 80; *EPW*, 84; see also *CPSS*, 152; *EPW*, 162, and *Cours*, II, L 47, 80–81.

26. For a book-length analysis of the "law of the three states," see Bourdeau, *Trois États*.

27. *Cours*, I, L 1, 21, *Introduction*, 1–2; see chapter 1 in this volume.

28. About which Comte noted, in the first lesson of the *Cours*, that it was where he "placed on record for the first time [his] discovery of this law" (*Cours*, I, L 1, 22; *Introduction*, 3n1).

29. *PTS*, 78–79, 143–48; *EPW*, 82, 93–97.

30. Rousseau's *Contrat social* (1762) was singled out by Comte as a perfect instance of metaphysical politics (*PTS*, 79; *EPW*, 83), whereas Bossuet's *Discours sur l'Histoire universelle* (1681) can be considered a prime example of theological politics.

31. *PTS*, 81; *EPW*, 85.

32. *Cours*, I, L 2, 59; *Introduction*, 59.

33. An outline of the encyclopedic scale had previously appeared in the *Plan des travaux* (*PTS*, 79–80; *EPW*, 84); see also chapter 1 in this volume.

34. *Cours*, I, L 2, 58; *Introduction*, 57.

35. *Cours*, I, L 1, 28; *Introduction*, 12.

36. *Cours*, I, L 1, 28; *Introduction*, 12.

37. *Cours*, I, L 1, 29; *Introduction*, 13.

38. *Cours*, I, L 1, 22; *Introduction*, 3.

39. While introducing the law of the three states in the *Cours*, Comte made clear that he did not intend to provide a "special demonstration" of this law and "its most important consequences" at that point, but that a "direct study of it" would be found "in the part of this work relating to social phenomena," i.e., the sociological lessons contained in volumes IV, V, and VI of the *Cours* (*Cours*, I, L 1, 22; *Introduction*, 3).

40. *Cours*, II, L 46, 13.

41. See chapter 1 in this volume.

42. *Cours*, II, L 49, 172.

43. *Cours*, II, L 47, 84.

44. *Cours*, II, L 47, 84.

45. See *Cours*, I, L 1, 25–26; *Introduction*, 8.

46. *Cours*, II, L 47, 80; and Braunstein, *Philosophie*, 78–95.

47. Besides the forty-seventh lesson of the *Cours*, historical accounts of the development of sociology can be found in *PTS*, 106–19; *EPW*, 112–26; and *CPSS*, 156–57; *EPW*, 165–66.

48. *Cours*, II, L 47, 85.

49. *Cours*, II, L 47, 84; see also *Système*, III, 309, trans. 259.

50. *Cours*, II, L 47, 85.

51. *Cours*, II, L 47, 85; see also *PTS*, 106–9; *EPW*, 112–15; and *CPSS*, 156; *EPW*, 165.

52. *Cours*, II, L 47, 86.

53. *PTS*, 107; *EPW*, 113.

54. *Cours*, II, L 47, 83.

55. *PTS*, 108; *EPW*, 114; and *Système*, II, 450, trans. 368.

56. *PTS*, 109; *EPW*, 115.

57. *Cours*, II, L 47, 88. Although Comte expressed his "profound reluctance for any habit of systematic neologism," he nonetheless decided to introduce the new term *sociology* as a synonym for *social physics*, so as to mark out the distinctness of "the additional part of natural philosophy that refers to the positive study of all the fundamental laws proper to social phenomena" (*Cours*, II, L 47, 88). But his choice was also motivated by an eagerness to uphold the originality of his contribution against potential rivals: "social physics," just like "positive philosophy," had already been "spoiled by vicious attempts at appropriation from various writers who completely misunderstood its true destination" (*Cours*, II, L 46, 15), most notably the Belgian scientist Adolphe Quetelet, who hijacked the former expression as a title for a "work that only deals with mere statistics" (*Cours*, II, L 46, 15; on Comte's view of "moral arithmetic," see below).

58. Macherey, "Comte"; *Cours*, II, L 47, 89.

59. *PTS*, 109; *EPW*, 116.

60. *PTS*, 118; *EPW*, 125.

61. See chapter 1 in this volume.

62. *Cours*, II, L 47, 89.

63. *Cours*, II, L 47, 90.

64. *PTS*, 119; *EPW*, 126.

65. *Cours*, II, L 47, 90.

66. *PTS*, 114; *EPW*, 121.

67. *Cours*, II, L 47, 99.

68. Brooks, *Eclectic Legacy*.

69. See Leterrier, *L'Institution*, and Le Van-Lemesle, *Le Juste*.

70. Comte unsuccessfully petitioned for the creation of a chair of general history of the sciences at the Collège de France; see Pickering, *Auguste Comte*, 1:445–47; 2:420–22.

71. See for instance Comte to Valat, Sept. 24, 1819, in *Correspondance*, I, 58–59; "Examination of Broussais's Treatise on Irritation" (1828), in *Système*, IV, appendix, 216–28; *EPW*, 228–240; *Cours*, I, L 1, 32–35; *Cours*, I, L 45, 851–63; on Comte and psychology, see Scharff, *Comte*, and Braunstein, "Antipsychologisme."

72. By "ideology," Comte referred to the French Idéologues, heirs to Condillac's sensationalism, such as Destutt de Tracy, Roederer, and Cabanis.

73. Comte to Valat, Sept. 24, 1819, in *Correspondance*, I, 58–59; *Correspondance*, I, 5.

74. See chapter 1 in this volume.

75. *Système*, IV, 217; *EPW*, 229.

76. *Cours*, I, L 45, 852–53.

77. On Comte's naturalistic theory of the mind, see Clauzade, *L'Organe*.

78. *Système*, IV, 221; *EPW*, 233.

79. That position was already advocated in the 1819 "Considérations sur les tentatives qui ont été faites pour fonder la science sociale sur la physiologie et quelques autres sciences" (Comte, *Écrits*, 473–82; see also *Cours*, I, L 2, 58; *Introduction*, 57).

80. *Cours*, I, L 45, 856.

81. *Cours*, I, L 45, 856.

82. *Cours*, I, L 45, 862–63.

83. *Cours*, I, L 45, 862–63.

84. *Cours*, I, L 45, 863.

85. For instance *CPS*, 197–215; *EPW*, 209–27.

86. *Cours*, II, L 47, 93–97; *Système*, I, 155, trans. 124; *Système*, III, 585, trans. 500.

87. Gouhier, *Jeunesse*.

88. On Comte's criticism of political economy, see Maréchal, *Conceptions*; Mauduit, *Auguste Comte*; Arnaud, *Nouveau Dieu*, part I, ch. 1; Weinberg, *Influence*.

89. *CPS*, 209; *EPW*, 22.

90. Nationally or internationally, with respect to free trade and protectionism: see *CPS*, 214–15; *EPW*, 224–26.

91. Such as a responsible control of procreation: see *CPS*, 211; *EPW*, 223.

92. *Cours*, II, L 47, 94.

93. *Cours*, II, L 47, 94.

94. *Cours*, II, L 48, 139.

95. *Correspondance*, I, 58–59.

96. *Cours*, II, L 48, 139.

97. Lessons 48 to 51 (together with lessons 46 and 47) appeared in 1839 as volume 4 of the *Cours*; lessons 52 to 55 were published as volume 5 in 1841; lessons 56 to 60 were gathered in its sixth and final installment in 1842.

98. *Cours*, II, L 48, 108.

99. *Cours*, II, L 48, 109.

100. On Comte's sociology, see Alengry, *Essai*; Wernick, *Auguste Comte*; Gane, *Auguste Comte*; Karsenti, *Politique*.

101. Which he first introduced in *Cours*, I, L 1, 32–33; *Introduction*, 19, and then further developed in *Cours*, I, L 40.

102. *Cours*, II, L 48, 109.

103. See also *Système*, I, 105, trans. 83: "Order is the condition of all Progress; Progress is always the object of Order."

104. *Cours*, II, L 48, 111.

105. *Cours*, II, L 48, 118.

106. *Cours*, II, L 48, 114–15.

107. *Cours*, II, L 48, 119.

108. *Cours*, II, L 50, 183.

109. *Cours*, II, L 50, 191.

110. *Cours*, II, L 48, 120.

111. *Cours*, II, L 48, 123.

112. *Cours*, I, L 2, 57; *Introduction*, 56.

113. *Cours*, II, L 48, 123.

114. *Cours*, II, L 48, 123.

115. *Cours*, II, L 48, 125.

116. *Cours*, II, L 48, 126.

117. See Guillo, *Les figures*.

118. *Cours*, II, L 48, 124.

119. *Cours*, II, L 48, 127–28.

120. *Cours*, II, L 48, 129.

121. *Cours*, II, L 48, 141–42. For Comte's views on current historical scholarship, see *PTS*, 94–95, 207–8; *EPW*, 98–99, 142–43; and *Cours*, II, L 47, 97–98.

122. As illustrated by the very structure of the sociological lessons of the *Cours*. Take for instance the fifty-third lesson, which dealt with the second phase of the theological stage, polytheism: after having considered its influence on the evolution of human intelligence (scientifically, artistically, and practically), Comte then detailed the various moral, social, and political consequences of this theoretical worldview (*Cours*, II, L 53).

123. See Canguilhem, *Normal*.

124. *Cours*, II, L 48, 143.

125. *Cours*, II, L 48, 144.

126. Comte would later refer to the "occidental disease" that plagued the modern world: see Comte to Audiffrent, July 12, 1855, in *Correspondance*, VIII, 5.

127. As is well known, although highly critical of Comte, Durkheim was nonetheless influenced by his ideas about the application of the pathological method in sociology (see Durkheim, *Rules*).

128. *Cours*, II, L 48, 146.

129. *Cours*, II, L 48, 147.

130. *Cours*, II, L 48, 148.

131. *Cours*, II, L 48, 150.

132. *Cours*, II, L 48, 150.

133. *Cours*, II, L 48, 149.

134. *Cours*, II, L 48, 151.

135. *Cours*, II, L 48, 153.

136. *Cours*, II, L 49.

137. *Cours*, II, L 49, 156.

138. See chapter 4 in this book.

139. *Cours*, II, L 49, 157.

140. *Cours*, II, L 50, 177; see also *Cours*, II, L 51, 212.

141. *Cours*, II, L 50, 161.

142. However, it could be argued that although Comte theoretically advocated a nonreductionist view of sociology, on certain occasions he himself could fall prey to the charge of reducing sociology to biology, a case in point being his controversy with John Stuart Mill about gender equality; on this topic, see Guillin, *Auguste Comte*.

143. *Cours*, II, L 51, 209.

144. *Cours*, II, L 51, 205–9.

145. *Cours*, II, L 51, 204.

146. *Cours*, II, L 51, 204.

147. *Cours*, II, L 51, 223.

148. *Cours*, II, L 51, 204.

149. *Cours*, II, L 58, 701, 707.

150. *Cours*, II, L 58, 699–700.

151. *Cours*, II, L 58, 699.

152. *Cours*, II, L 58, 702–3. Comte's violent attacks on "geometers" also had to with his repeated failure to secure a permanent position as a mathematics professor at the École Polytechnique: see Pickering, *Auguste Comte*.

153. *Cours*, II, L 49, 168.

154. *Cours*, II, L 58, 703.

155. *Cours*, II, L 49, 167–68. On Comte's proscription of mathematics from biology, see chapter 4.

156. *Cours*, II, L 49, 168.

157. On Comte's opposition to probability theory, see Hacking, *Taming*, 143–45.

158. Bernoulli, *Ars conjectandi*.

159. Condorcet, *Mémoire*; Condorcet, *Essai*; Condorcet, *Tableau général*.

160. *Cours*, II, L 49, 168.

161. Laplace, *Essai philosophique*.

162. *Cours*, II, L 58, 704.

163. Comte later developed (*Système*, I, 49–54, trans. 39–42) an idiosyncratic conception of "materialism" that captured, among others, the encyclopedic infringement characteristic of the "alleged calculus of chances" (*Cours*, II, L 58, 703).

164. *Cours*, II, L 49, 169.

165. *Cours*, II, L 58, 703.

166. *Cours*, II, L 49, 170.

167. *Cours*, II, L 58, 707.

168. Bourdeau, "L'Idée."

169. *Cours*, II, L 49, 172.

170. *Cours*, II, L 58, 707.

171. *Cours*, II, L 49, 173; *Cours*, II, L 58, 708.

172. *Cours*, II, L 58, 709.

173. *Cours*, II, L 57, 634.

174. *Cours*, II, L 49, 172.

175. *Cours*, II, L 49, 173.

176. Mill, *Auguste Comte*, 291–92.

Chapter 6. Comte's Political Philosophy

Epigraph: Correspondance, I, 148. For Comte's quotations, we cite the French original first, followed by the English translation where it is available, and we use the following abbreviations: *PTS* refers to the *Plan des travaux scientifiques nécessaires pour réorganiser la société* (1822–1824); *CPSS* refers to the *Considérations philosophiques sur la science et les savants* (1825); *CPS* refers to the *Considérations sur le pouvoir spirituel* (1826); *CP* refers to the *Catéchisme positiviste*. The essays appear in the *Écrits de jeunesse* for the French original and the *EPW* for the English translation.

1. *Système*, I, 73, trans. 58; this formulation does not appear in the 1848 edition of the *Discours sur l'ensemble du positivisme*.
2. *Synthèse*, 196.
3. *PTS*, 247, 57.
4. Mill, *Autobiography*, 211.
5. *CP*, 205–6, 278–79.
6. *CP*, 264, 259; and *Système*, III, 10, trans. 9.
7. Mill, *Auguste Comte*, 332.
8. *Système*, II, 314, trans. 257, mod. trans.
9. *Système*, II, 295, trans. 244, mod. trans.
10. *Système*, II, 301, trans. 248, mod. trans.
11. *Système*, II, 310, 319–20, trans. 254, 261–62.
12. *Cours*, II, L 52, 263.
13. From the second volume of the *Système* onward, politics and ethics become the two branches of the Religion of Humanity, the former being designed to bring individuals together ("rallier les diverses individualités") while the latter aims at regulating their individual lives (*Système*, II, 9, trans. 8–9). Therefore, to consider politics independently from its relations with ethics and religion amounts, to a certain extent, to betraying the spirit of positive politics, at least in its final formulation. We will nevertheless adopt such a perspective, for, given the intellectual habits of the contemporary reader, the use of a religious vocabulary in this context is an obstacle one might wish to avoid.
14. Tocqueville, *De la démocratie*, 210; Tocqueville, *Democracy*, 2:9–10.
15. *Système*, II, 281, trans. 234.
16. *L'esprit*, 183.
17. *Système*, I, 158, trans. 126–27; see also *Cours*, II, L 48, 145.
18. See, respectively, *CPS*, 380–84, 209–14; *Cours*, II, L 50, 266–72; *Système*, II, 293–99, trans. 242–47; for Aristotle, see *Système*, II, 293, trans. 281.

19. *Cours*, II, L 50, 190–91; Hayek quotes this passage and is surprised to find the idea of spontaneous social order in Comte, without apparently realizing that it is the very basis of his political thought. Hayek, *Counter-Revolution*, 244n381. But the title of lesson 50 could not be more explicit: "Social statics, or general theory of the spontaneous order of human society."

20. *Cours*, II, L 50, 196.

21. *Cours*, II, L 50, 196; cf *Système*, II, 294–95, trans. 243–44.

22. *CP*, 205, 193.

23. *Système*, II, 295–97, trans. 244–45; see also *Cours*, II, L 50, 198.

24. Pickering, *Auguste Comte*, 1:301. On the importance of Comte's debt toward the "Scottish school," see his letter to Mill from Oct. 21, 1844, in *Correspondance*, 3:291.

25. See, for instance, what Comte wrote in 1826 with respect to the economic crisis that had just struck England:

> The commercial and manufacturing crisis which is currently devastating the country in which industrial activity is most developed—a crisis which could at any moment assume a more or less serious political power—is well suited to prove to impartial observers the need for government to exert some action on industrial relations as it did in the past on military relations. No doubt these irritants are in their nature transient. But social order and individual happiness by common accord demand more direct, more explicit, in a word more regular guarantees against the ever-imminent renewal of these grievous oscillations—guarantees which do not make each individual judge in his own case, and which do not require minds that are customarily situated at very particular vantage-points spontaneously and consistently to adopt a general point of view. (*CPS*, 291, 221n)

26. See for instance *Cours* II, L 48, 130–36.

27. *Système*, II, 302, trans. 249.

28. Montesquieu, *Esprit des lois*, vol. XI, ch. 4.

29. *Cours*, II, L 48, 136; cf. *Système*, I, 244, trans. 196: The rule applies to all human effort [*ce principe universel de l'art humain*]: namely that the order of things instituted by man [*l'ordre artificiel*] ought to be simply a consolidation and improvement of the natural order.

30. *Système*, IV, 38–39, trans. 35.

31. *Cours*, II, L 48, 135; *Positive Philosophy*, 2:94–95.

32. *Cours*, II, L 57, 655, and *Système*, I, 113, trans. 89–90.

33. See, for instance, *CPS*, 365–66, 192–93. Later on, Comte took pains to answer the objection raised by Constant, who accused the modern advocates

of spiritual power of restoring theocracy (To John Stuart Mill, Dec. 25, 1844, *Correspondance*, II, 308); see also *Cours*, II, L 57, 652; and Bourdeau and Fink, "De l'industrie."

34. *Système*, I, 324, trans. 259, mod. trans.

35. *Système*, I, 214, trans. 173, mod. trans.

36. *Cours*, II, L 54, 325–26; *Cours*, II, L 57, 656.

37. *Cours*, II, L 57, 653.

38. CPS, 376, 205. Comte also speaks of a consultative influence in one case, and of an imperative influence in the other (*Système*, I, 89, trans. 71); later (*Système*, II, 355, 288), he claims that the spiritual power "is to change the will without compelling actions."

39. PTS, 246, 56; on that issue, see Bourdeau, "Pouvoir spirituel." It is worth noting that the idea is old; it appears in the *Logique* of Port Royal, 4th part, ch. 12: "Of what we know through faith, whether human or divine."

40. *Cours*, II, L 57, 652; cf. 657.

41. *Cours*, II, L 57, 667–68; see also *Cours*, II, L 48, 115: "Authority really derives from cooperation and not cooperation from authority."

42. *Cours*, II, L 57, 680.

43. See for instance *Discours sur l'esprit positif* (1844), 212; and particularly the third chapter of *General View of Positivism*, where workers are described as "spontaneous philosophers" and philosophers as "systematic workers." Some years later, when he saw that workers were more attracted to socialism, Comte became disappointed and decided to launch his *Appel aux conservateurs* (1855).

44. *Cours*, II, L 57, 680.

45. See *Système*, I, 139–50, trans. 111–19; and Reynié, "L'opinion publique."

46. *Système*, I, 140, trans. 112.

47. *Système*, I, 146, trans. 116, mod. trans.

48. *Système*, II, 303, 312, trans. 256, 249.

49. See *CPS*, 377, 385, 206 (due to the existence of several editions, the second passage is not translated in *EPW*). About positivism and education, see Arbousse-Bastide, *Doctrine*.

50. *Système*, I, 169, trans. 135–36.

51. *Système*, IV, 199–248, trans. 176–218.

52. *Système*, II, 314–15, trans. 257–58.

53. "Auguste Comte avait le sens de l'égalité des êtres, mais d'une égalité fondée sur la différenciation radicale des fonctions et des dispositions." Aron, *Les Étapes*, 112.

54. *Système*, I, 379, trans. 305.

55. *Cours*, II, L 50, 198.

56. *Système*, III, 57–60, trans. 47–50.

57. *Cours*, II, L 57, 670–80; *Système*, I, 358–60, trans. 287–89; *Système*, II, 324–35, trans. 264–72; and *CPS*, 393, 223. Should we relate this continuous interest to the fact that Comte, as an entrance examiner for the École Polytechnique, had spent several months each year ranking students in order to establish a list of admission?

58. *Système*, II, 328, trans. 267, mod. trans.

59. *CP*, 246, 240.

60. *Cours*, II, L 57, 679; cf. *CPS*, 393, 223.

61. *Cours*, II, L 57, 680.

62. *Système*, I, 359, trans. 288; cf. *Système*, II, 324–25, trans. 264–66.

63. "We have directly acknowledged that, in the modern sociability, consideration and power are necessarily distributed according to laws that are so different that their superior degrees are mutually exclusive. Now, we deal here with an order of dignity, not an order of power, with the universal consideration enjoyed by an individual, not with the direct influence he has on real actions" (*Cours*, II, L 57, 674).

64. *Système*, I, 359, trans. 288.

65. *Système*, II, 318, trans. 260. As was already noticed in 1826, "the spiritual power increases its domain as society gets more complicated, instead of the domain of temporal power seeing its own diminish. In fact only those who cannot be ruled spiritually are ruled temporally; that is, we govern by force only those who cannot be adequately ruled by opinion" (*CPS*, 384, 213).

66. *Système*, I, 195, trans. 157, mod. trans.

67. *Système*, II, 334–35, trans. 288.

68. See note 11. One of the few contemporary studies of the idea of spiritual power in Comte (Acton, *Idea*) ends with a long discussion of the function of the BBC and its independence from temporal power.

69. *Système*, I, 596, trans. 482.

70. *Système*, II, 285, trans. 237.

71. *PTS*, 288, 105; cf *Système*, II, 364, trans. 295–96. Here, Comte merely applies a rule that he would regard as one of the fifteen laws of his first philosophy and "which represents the intermediate as in all cases subordinate to the extremes which it brings into connection" (*Système*, IV, 180, trans. 160).

72. See *Système*, IV, 273–75, trans. 239–42. On utopias in Comte, see De Boni, *Descrivere*; Gane, *Auguste Comte*, and Braunstein, *La philosophie*.

73. *Système*, I, 110–18, trans. 87–93.

74. *Système*, IV, 421, trans. 366.

75. It is worth remembering that Mill and Tocqueville almost shared this view that the English case was somewhat exceptional (Mill, *Auguste Comte*, 122; and Tocqueville, *L'Ancien Régime*, 147). It is of course a completely different question to know whether the exception can serve as a model.

76. On the correlation spiritual-central/global, temporal-local, see for instance *Système*, II, 319–20, trans. 261–62, and *Système*, III, 482, trans. 405–6.

77. *Système*, IV, 305, trans. 267.

78. See Vernon, "Auguste Comte."

79. *Cours*, II, L57, 696.

80. *Cours*, II, L 48, 123; *Système*, II, 369, trans. 452.

81. *Système*, IV, 10, trans. 8.

82. *Cours*, II, L 52, 263.

83. Lévi-Strauss, "Leçon," 158; see also Braunstein, *La philosophie*, 167–82.

84. *Système*, I, 615, trans. 497.

85. *Cours*, L 52, 263.

86. *Système*, IV, 359, trans. 311–12.

87. "When we have learnt to avail ourselves of the heart and intelligence of our allies to such a degree as to entrust them with the main superintendence of inorganic forces, we shall find a better use for the human agents thus set free, and be able to develop on a larger scale the sense of fraternity upon earth" (*Système*, IV, 359–60, trans. 313). Such a collaboration being "one of the essential sources of human greatness . . . positivism will conveniently extend the elementary feeling of human brotherhood to all beings worthy of being associated with Man. And we shall be benefited by the association no less than they" (*Système*, I, 614, trans. 497).

88. *Système*, I, 619, trans. 500.

89. *Système*, I, 615, trans. 498.

90. The social question is primarily moral; it is political only in a derivative manner; hence the downplaying of politics and the criticism of "the attempts made to satisfy popular demands by measures of a purely political kind," "a course of action" Comte judged "as useless as it is destructive" (*Système*, I, 169, trans. 136, mod. trans.). Ten years before the promotion of ethics as the seventh science, the conclusions of the *Cours* already considered that "the legitimate social supremacy belongs, properly speaking, neither to force nor reason, but to morals, which dominates both the acts of the former and the advice of the latter; such is at least the ideal limit that reality must approximate, although it will never be able to reach it" (*Cours*, II, L 57, 657).

91. Tocqueville, *De la démocratie*, bk. 2, 17; Tocqueville, *Democracy*, 2:9–10.

92. *Cours*, I, L1, 38; *Introduction*, 28.

93. Houellebecq, "Préliminaires," 7. Almost fifty years ago, in a chapter that reveals a remarkable understanding of Comte's project, E. Gilson underlined that "Comte's work defies all attempts at classification. No simple formula can summarize the multiplicity of its aspects and the history of its influence easily demonstrates that it contained the germs of various possible systems." *Les métamorphoses*, 251.

94. *Système*, I, 361, trans. 289–90. The temporal version of the law of the three states associates the metaphysical state with jurists and lawyers.

95. *Cours*, II, L 48, 114.

96. *Système*, II, 306, trans. 251; cf. *CP*, 248, 243–44.

97. *Système*, I, 119, trans. 94.

98. Casanova, Introduction, 1095. On the meaning of the term *dictature* in Comte, see Cassina, "Comte face à la dictature," and Nicolet, *La fabrique*, 149–53.

99. Hayek, *Counter-Revolution*, 194.

Chapter 7. Art, Affective Life, and the Role of Gender in Auguste Comte's Philosophy and Politics

1. On the complex publication and distribution history of this work, which Comte presented in three different formats over the course of his intellectual career, see Pickering, *Auguste Comte*, 1:224–39, 245–49; Jones, Notes; Grange et. al., Notice.

2. *Plan*, 104. The English translations of specific passages from the *Plan des travaux scientifiques nécessaires pour réorganiser la société* are my own. For a complete English translation of the *Plan*, see *EPW*.

3. *Cours*, II, L 50, 176–201. The English translations of specific passages from the *Cours de philosophie positive* are my own. In addition to the two-volume Hermann edition of 1975, I have used the six-volume Littré edition of 1869, which includes some additional prefatory material from the original six-volume Comte edition of 1830–1842. There is no complete English translation of the *Cours*, but for the abridged English translation by Harriet Martineau that Comte himself approved, see *Positive Philosophy*. For an assessment of Martineau's work compared to Comte's original, see Pickering, *Auguste Comte*, 3:132–56.

4. *Système*, II, 178. The English translations of specific passages from the *Système de politique positive* are my own. For a complete English translation of the *Système*, see Comte, *System of Positive Polity*.

5. *Discours sur l'ensemble*, 46–47; *Système*, I, 4–5. The English translations

of specific passages from the *Discours sur l'ensemble du positivisme* are my own. For an analysis of the differences between Comte's initial version of the *Discours* and the revised version that he reprinted in the *Système*, see Petit, Présentation to *Discours* and Notes to *Discours*.

6. *Système*, IV, 550.
7. Gane, *Auguste Comte*, 52.
8. Jones, Notes, xxix.
9. Saint-Simon, "Avant-propos"; Jones, Notes, xxix–xxx.
10. Pickering, *Auguste Comte*, 1:231–39.
11. [Rodrigues], "Digression," 233. On Comte and the Saint-Simonians more generally, see Gane, "Dans le gouffre"; Gane, *Auguste Comte*, 52–54; Le Bras-Chopard, "L'effervescence"; Le Bras-Chopard, "Une statue"; Pickering, "Comte and the Saint-Simonians"; Picon, *Les Saint-Simoniens*.
12. Pedersen, "Sexual Politics"; Guillin, *Auguste Comte*. For Comte's views on women and gender more generally, see also note 14 and later sections of this essay.
13. Lepenies, *Between Literature and Science*; Pozzi, "Comte." On the wide variety among Comte's initial and eventual followers, see Lazinier and de Acevedo, "Quelques disciples"; and chapter 10 of this volume.
14. For historical approaches see Pedersen, "Sexual Politics"; Pickering, "Angels and Demons"; Pickering, "New Look." For philosophical approaches, see Guillin, *Auguste Comte*; Kofman, *Aberrations*; Le Bras-Chopard, "Une statue"; Petit and Bensaude, "Le féminisme." For approaches from social theory, see Gane, *Harmless Lovers*; Lepenies, *Between Literature and Science*. For a literary critical approach with close ties to political philosophy, see Moscovici, *Gender*.
15. *Cours*, II, L 50, 186.
16. *Discours sur l'ensemble*, 38–39.
17. *Système*, I, xv.
18. I thank Theodore M. Brown, Brian Ogilvie, Rachel Remmel, and Laura Smoller for many fascinating conversations about the history of the distinction between art and science. On the history and philosophy of art and its changing definitions in the French intellectual tradition, see Becq, *Genèse*; Mustoxidi, *Histoire*; Saisselin, *Rule of Reason*; Shroder, *Icarus*; and, in the European tradition more generally, Shiner, *Invention of Art*. On the history of science and its relationship to the history of artists and artisans in the European context, see, for example, Long, *Artisan/Practioners*; Shapin, *Scientific Revolution*; Smith, *Body of the Artisan*.

19. *Cours*, I, L 2, 44, 45, 47.

20. *Cours*, II, L 60, 772, 774, 775. For a similar usage in Comte's later comments on the progress of "the medical art" and "the social art, whether moral or political," see *Discours sur l'ensemble*, 71; *Système*, I, 31. For further analysis of Comte's views on medicine as an art, see Grange, *Auguste Comte*, 244–53.

21. *Discours sur l'ensemble*, 45; *Système*, I, 3. For Comte's definition of "the social art" as an art with both "moral" and "political" dimensions, see *Discours sur l'ensemble*, 71; *Système*, I, 31.

22. *Discours sur l'ensemble*, 141; *Système*, I, 106.

23. *Plan*, 106.

24. *Cours*, II, L 53, 277, 279.

25. *Discours sur l'ensemble*, 47; 309–11; *Système*, I, 5–6; 282–84. The complete hierarchy of the fine arts included, in descending order from highest to lowest, poetry, music, painting, sculpture, and architecture. See *Cours*, II, L 53, 281; *Discours sur l'ensemble*, 317–21; *Système*, I, 291–95.

26. *Plan*, 103–4.

27. *Plan*, 104–6.

28. *Cours*, II, L 51, 210.

29. *Cours*, II, L 53, 275–76.

30. *Cours*, II, L 53, 276. "Any real inversion of this elementary relationship," Comte insisted, "would tend directly to the fundamental disorganization of the human economy, whether individual or social, by abandoning the general conduct of our life to that which can only embellish and sweeten it: from whence would come a sort of chronic alienation." *Cours*, II, L53, 276.

31. *Cours*, II, L 55, 452–53.

32. *Cours*, II, L 57, 675.

33. *Cours*, II, L 60, 787. On the distinction between mathematical and sociological positivism, see also *Cours*, II, L 58, 711–12.

34. *Cours*, II, L 60, 786.

35. *Cours*, II, L 60, 786.

36. *Cours*, II, L 60, 786–87.

37. *Discours sur l'ensemble*, 39. In the introduction to the general preamble that he wrote for the *Discours* and reprinted in the first volume of the *Système*, he promised similarly that his new work would work "to convince any well-prepared reader that the new general doctrine, which still seems to be able to satisfy only reason, is not, at base, less favorable to feeling, and even to imagination." *Discours sur l'ensemble*, 43; *Système*, I, 1.

38. Petit, Présentation to *Discours*.

39. *Discours sur l'ensemble*, 304; *Système*, I, 277.

40. *Discours sur l'ensemble*, 306–8; *Système*, I, 279–81. The use of the term "aesthetic pedantocracy" is especially insulting here as it parallels and recalls the term that Comte regularly used to criticize his chief intellectual and political opponents, the "scientific pedantocracy." Petit suggests that Comte was probably thinking of "romantics who willingly play the role of the marginal and the misunderstood." Mike Gane notes that romantic composers such as Berlioz, Liszt, and Chopin also had ties to the Saint-Simonians. See Petit, Notes to *Discours*, 451; Gane, "Dans le gouffre," 150; and, on Comte's own ambiguous relationship to romanticism, Petit, "Le romantisme social."

41. *Discours sur l'ensemble*, 335–37; *Système*, I, 309–11.

42. Petit, Présentation to *Discours*, 27.

43. *Discours sur l'ensemble*, 340–43; *Système*, I, 315–18.

44. See, for example, *Cours*, II, L 53, 283; *Cours*, II, L 58, 738; *Cours*, II, L 60, 785. On the larger contemporary debates over the social role of art, see McWilliam, *Dreams*; Needham, *Le développement*; Thibert, *Le rôle*.

45. *Discours sur l'ensemble*, 326; *Système*, I, 301.

46. *Discours sur l'ensemble*, 420; *Système*, I, 398.

47. *Plan*, 104, 105.

48. *Discours sur l'ensemble*, 35.

49. *Système*, I, title page.

50. *Système*, II, 65.

51. *Système*, I, title page.

52. *Cours*, II, L 50, 181.

53. *Cours*, I, L 45, 866–67.

54. *Cours*, II, L 50, 186.

55. *Cours*, II, L 51, 210.

56. *Cours*, II, L 57, 675.

57. *Cours*, II, L 60, 770.

58. *Discours sur l'ensemble*, 53; *Système*, I, 12.

59. *Discours sur l'ensemble*, 55; *Système*, I, 14.

60. *Discours sur l'ensemble*, 58; *Système*, I, 17.

61. *Discours sur l'ensemble*, 418; *Système*, I, 397.

62. *Système*, II, 19.

63. *Système*, II, 432–33.

64. *Système*, IV, 7.

65. *Discours sur l'ensemble*, 60; *Système*, I, 19, 20.

66. Bourdeau, "L'esprit ministre," 184.

67. *Discours sur l'ensemble*, 235; *Système*, I, 204.
68. *Discours sur l'ensemble*, 240, 241; *Système*, I, 210.
69. Moses, *French Feminism*, 42.
70. *Cours*, II, L 50, 184–85.
71. *Discours sur l'ensemble*, 276–77; *Système*, I, 248.
72. See Pilbeam, *French Socialists*, 77–78. On the difficulties of reconstructing Saint-Simon's ideas about women from these fragmentary elements, see also Grogan, *French Socialism*, 67–69; on the nature and import of Saint-Simon's ideas more generally, Picon, *Les Saint-Simoniens*, 35–57.
73. On women's engagement with Saint-Simonian socialism, see, for example, Gordon and Cross, *Early French Feminisms*, 3–6, 19–141; Grogan, *French Socialism*, 67–154; Moses, *French Feminism*, 41–87; Moses and Rabine, *Feminism, Socialism*; Pilbeam, *French Socialists*, 78–88; Riot-Sarcey, *La démocratie*.
74. Although the Saint-Simonians recognized women's public roles in unprecedented new ways, women never became entirely equal in the movement either in theory or in practice. See further Andrews, *Socialism's Muse*; Gordon and Cross, *Early French Feminisms*, 263–64; Grogan, *French Socialism*, 67–154; Moses, *French Feminism*, 41–59; Moses and Rabine, *Feminism, Socialism*; Riot-Sarcey, *La démocratie*.
75. See Moses, *French Feminism*, 55–56; Pilbeam, *French Socialists*, 78–80; and on women's roles in the Saint-Simonian movement more generally, notes 73 and 74 above.
76. On the complexities and ambiguities of Enfantin's position, see Andrews, *Socialism's Muse*, 33–37; Grogan, *French Socialism*, 107–10, 131–54, 124–27; Moses, *French Feminism*, 45–49, 56; Moses and Rabine, *Feminism, Socialism*, 32–42; Picon, *Les Saint-Simoniens*, 131–38; Pilbeam, *French Socialists*, 81–82; Riot-Sarcey, *La démocratie*, 57, 71–74.
77. Enfantin's theories attracted the attention of the French authorities, who put him on trial and sent him to prison. While he caused a public sensation when he tried to rely for his defense on two Saint-Simonian women, Aglaé Saint-Hilaire and Cécile Fournel, he excluded women from the movement even more definitively upon his release from prison, when he joined his remaining male followers in an all-male retreat to "wait for the woman" at Ménilmontant. See Moses, *French Feminism*, 49–50; Moses and Rabine, *Feminism, Socialism*, 42–44; Picon, *Les Saint-Simoniens*, 138–64; Pilbeam, *French Socialists*, 86–87.
78. On Saint-Simonian and other forms of early feminist socialism and so-

cialist feminism, see Andrews, *Socialism's Muse*; Gordon and Cross, *Early French Feminisms*; Grogan, *French Socialism*; Moses, *French Feminism*, 41–149; Moses and Rabine, *Feminism, Socialism*; Pilbeam, *French Socialists*, 75–106; Riot-Sarcey, *La démocratie*.

79. On the work of this pioneering journal in its successive incarnations, see Moses, *French Feminism*, 63–87; Grogan, *French Socialism*, 100–106; and on Véret and her work more extensively, Riot-Sarcey, *La démocratie*.

80. See Moses, *French Feminism*, 127–49, esp. 128–30; Pilbeam, *French Socialists*, 93–94; and on Niboyet and her work more extensively, Riot-Sarcey, *La démocratie*.

81. See Moses, *French Feminism*, 127–49, esp. 142–49; Pilbeam, *French Socialists*, 75, 93–106; and on Deroin and her work more extensively, Gordon and Cross, *Early French Feminisms*, 1–6, 59–141, 261–66; and Riot-Sarcey, *La démocratie*.

82. Quoted in Pickering, "Angels and Demons," 12–13.

83. Pickering, "Angels and Demons," 13.

84. Préface personnelle, *Cours*, II, 467.

85. *Cours*, II, L 60, 777.

86. *Cours*, II, L 50, 183, 184, 186.

87. *Cours*, II, L 50, 186–87.

88. See, for example, his praise of Roman Catholicism for instituting monogamy, his criticisms of Protestantism for instituting divorce, and his praise of monotheism for replacing primitive fetishistic societies in which women served as warriors and ancient polytheistic societies in which women served as priestesses with more advanced medieval Christian societies in which women gave up both royal and sacred authority to stay secluded in the home. See, for example, *Cours*, II, L 53, 299–301, 304; *Cours*, II, L 54, 364–66; *Cours*, II, L 55, 436–39.

89. *Cours*, ed. Littré, I, 1.

90. *Système*, I, 25.

91. *Cours*, ed. Littré, I, 3.

92. *Système*, I, 11–13.

93. *Discours sur l'ensemble*, 239; *Système*, I, 209.

94. *Discours sur l'ensemble*, 263, 265, 269; *Système*, I, 234, 235, 240. For Comte's further views of marriage as "the most powerful of all the domestic affections," see *Système*, II, 186–89.

95. *Discours sur l'ensemble*, 263, 270, 271; *Système*, I, 234, 241, 242. For

Comte's further views on the proper nature and social influence of relationships between parents and children, see, for example, *Système*, II, 189–90, 196–99; *Système*, IV, 298–301.

96. *Discours sur l'ensemble*, 261–62; *Système*, I, 232.

97. *Discours sur l'ensemble*, 408; *Système*, I, 384–85. This marks a shift from the final volume of the *Cours*, where Comte had imagined that the council would consist only of thirty men. See *Cours*, II, L 57, 696.

98. *Discours sur l'ensemble*, 288, 289; *Système*, I, 259, 261. See further *Système*, II, 63–64, 121–22, 203–5; *Système*, IV, 298–99.

99. *Discours sur l'ensemble*, 409.

100. *Système*, I, 387–88.

101. *Discours sur l'ensemble*, 267–69; *Système*, I, 238–24. See further *Système*, IV, 300–301.

102. *Discours sur l'ensemble*, 278; *Système*, I, 249–50. See further *Système*, II, 194; *Système*, IV, 301. For Comte's special mistrust of independently wealthy women, see *Discours sur l'ensemble*, 275–76; *Système*, I, 246–47; *Système*, II, 193; Petit and Bensaude, "Le féminisme."

103. *Discours sur l'ensemble*, 278–79; *Système*, I, 249–51. See further *Système*, IV, 301.

104. *Discours sur l'ensemble*, 270–71, 279–80; *Système*, I, 241–42, 250–52. See further *Système*, II, 206; *Système*, IV, 301.

105. *Discours sur l'ensemble*, 279; *Système*, I, 250.

106. *Discours sur l'ensemble*, 408, 410–14; *Système*, I, 384–85, 388–92.

107. *Discours sur l'ensemble*, 337–38, 383, 385; *Système*, I, 311–12, 358, 361. See further *Système*, II, 208–9, *Système*, IV, 69–74. As Andrew Wernick has recently put it, "Within the spiritual power, nonetheless, even the saintliest women were subordinate to (exclusively male) priests." Wernick, *Auguste Comte*, 4.

108. Petit and Bensaude, "Le féminisme," 305–6. For the calendars themselves, see *Cours*, II, 799–803; *Système*, IV, 159; table inserted at *Système*, IV, 402–3.

109. A partial list of those who have stressed discontinuity, an approach that dates all the way back to some of Comte's earliest disaffected disciples, would include Littré, *Auguste Comte*; Mill, *Auguste Comte*; Durkheim, *Le socialisme*; Simon, *European Positivism*; Aron, *Main Currents*. Those who now focus on continuity instead include Wright, *Religion*; Pickering, *Auguste Comte*; Scharff, *Comte*; Jones, Introduction; Wernick, *Auguste Comte*.

110. See, for example, Jones, Introduction, xl; Pickering, *Auguste Comte*, 1:3, 691–93, 2:4; Wright, *Religion*, 15.

111. *Système*, IV, 74.

112. *Cours*, I, L 2, 45. For further discussion of the "more direct and more elevated" function of science, its ability to "satisfy the fundamental need of our intelligence to know the laws of phenomena," see *Cours*, I, L 2, 45.

113. *Discours sur l'ensemble*, 57–58; *Système*, I, 16–17. See further Bourdeau, "L'esprit ministre," 181–82.

114. *Système*, IV, inserted between 402 and 403.

115. *Système*, IV, 159.

116. *Cours*, II, 799; *Système*, IV, 159.

117. *Discours sur l'ensemble*, 338; *Système*, I, 312. Comte acknowledges women's innate poetic and musical ability, but he excludes them from "the special arts," especially those of painting, sculpture, and architecture, as "requiring a technical facility that is little suited to them, and in which the slow apprenticeship would stifle their admirable spontaneity." *Discours sur l'ensemble*, 337; *Système*, I, 312.

118. *Discours sur l'ensemble*, 337–38; *Système*, I, 312–13.

119. *Discours sur l'ensemble*, 336–38; *Système*, I, 311–13.

120. *Discours sur l'ensemble*, 344; *Système*, I, 320.

Chapter 8. The Religion of Humanity and Positive Morality

1. *Système*, I, 6–7ff, trans. xiii–xivff.

2. *CPR*, 66. For references to Comte's *The Catechism of Positive Religion* [1852], I shall use the abbreviation *CPR* and refer to page numbers in the 1858 first edition of Richard Congreve's translation. The entire second volume of the *Système de politique positive* was devoted to the theory of individual and collective unity, taken to be the foundation of social statics.

3. Initially published in 1848 as *Discours préliminaire sur l'ensemble du Positivisme*.

4. *CPR*, 70.

5. The *Catechism* is written as if in "the incomparable year" of 1845. *CPR*, 23–24.

6. A full account is given in volume 3 of *Système de politique positive* and a summary version is in *CPR*, 368 *et seq*. According to Manuel, the term *fetishism* was invented by Charles de Brosses in his *Culte des dieux fétiches* and came to Comte through Constant's *De la religion*. The term was superseded in

nineteenth-century anthropology by *totemism* and later by *animism*. Manuel, *Prophets*, 186–87.

7. *Système*, II, 8, 18, trans. 8, 17.

8. The Western Republic comprised the populations of France, England, Germany, Spain, and Italy—not coincidentally major territories of medieval (Catholic) Christendom. *Système*, I, 82–83, trans. 66.

9. *Système*, II, 18, trans. 17.

10. To which corresponds the distinction between the sciences constituting cosmology and those making up biology. *Système*, I, 573ff, trans. 464ff.

11. *Système*, I, 573ff, trans. 464ff.

12. *Système*, I, 586, trans. 474–75.

13. *Système*, II, 239–40, trans. 288–30.

14. "The three [previous] transitions . . . had more and more threatened the destruction of the fundamental order of society; but their organic character had always saved them from such a catastrophe. . . . It was only at the close of the medieval period, when Theologism was worn out, that the West began to lose all general direction." *Système*, III, 503, trans. 424.

15. "In fact the anarchy of the West consists chiefly in the interruption of human continuity; first Catholicism cursing Antiquity; then Protestantism reprobating the Middle Ages; and lastly Deism denying filiation altogether." *Système*, III, 2, trans. 2.

16. *Système*, III, 585–87, trans. 500–502.

17. "The growing perfection of the animal organism consists above all in the increasingly pronounced specialty of diverse functions accomplished by increasingly distinct organs." *Cours*, II, L 50, 190. However, there was an essential rider: growing differentiation would "contradict existence if it were not always accompanied by a perfecting of general unity." *Système*, III, 9, trans. 8.

18. *Système*, III, 567–68, trans. 484.

19. Both advances, however, are attributed more to feudalism than to Christianity, which merely ratified them. *CPR*, 397; *Système*, III, 451–52, 491–92, trans. 381–82, 413–14.

20. *Système*, I, 701, trans. 566.

21. *Système*, I, 608, trans. 492.

22. For Comte's critique of the inherent egoism of salvation religions, see *Cours*, II, L 55, 504. He did not, however, acknowledge the egoism that remained when the Christian schema was transposed to that of the "subjective" immortality promised to Humanity's faithful servants.

23. *CPR*, 243.

24. *Système*, I, 608–9, trans. 492–93.

25. "In a complicated organism the harmony of the whole must always be dependent upon adequate subordination of all spontaneous impulses to one preponderating principle." *Système*, I, 700, trans. 565.

26. *Système*, I, 610, trans. 494.

27. *Système*, IV, 530, trans. 460–61.

28. "Between these two conflicting elements there can never be a perfect balance. Equilibrium of the whole is possible only by one of the two gaining the preponderance." *Système*, I, 691, trans. 558.

29. Rule 4 of Comte's "first philosophy" that summarized the methodological principles of positivism. See *Système*, IV, 176, trans. 156.

30. "The individual must subordinate himself to an existence outside itself in order to find in it the source of his own stability. And this condition cannot be effectually realized except under the impulse of propensities prompting him to live for others." *Système*, I, 700, trans. 565–66.

31. *CPR*, 369–70.

32. *CPR*, 79.

33. *Système*, I, 82–83, trans. 66.

34. "The Great Being is the whole constituted by the beings, past present and future, which cooperate willingly in perfecting the order of the world. Every gregarious animal race has a natural tendency to such cooperation. But it is only the paramount race on each planet that can attain unity as a race, for its ascent to power necessarily checks that of the lower animals." *Système*, IV, 30, trans. 27.

35. Though Comte often calls the idea of it "eternal," as for example in the *Catechism*, 63 ("an immense and eternal being").

36. *Système*, II, 37–40, trans. 34–36.

37. Comte cites the *Imitatio Christi* on this point: "I am necessary to thee; thou art useless to me." *CPR*, 80.

38. "The development . . . and preservation of the Great Being must then depend on the free activity of its different children." *CPR*, 81.

39. *CPR*, 496.

40. "Habitant une tombe anticipée, je puis désormais tenir aux vivants un langage posthume qui sera mieux affranchi des divers préjugés, surtout théoriques, dont nos descendants se trouveront préservés." (Dwelling in an anticipated tomb, I can henceforth hold out to the living a posthumous language that is more fully liberated from the various prejudices, above all theoretical, from which our descendants will find themselves saved.) Comte, *Testament*, 24.

41. *Système*, I, 364, trans. 292.

42. *CPR*, 90.

43. *CPR*, 77; *Synthèse*, 68.

44. *CPR*, 141.

45. For the wider shift in "collective mentalities" associated with the nineteenth-century cemeteries movement, see Ariès, *Hour*.

46. *Système*, IV, 103, trans. 91.

47. *CPR*, 135–36.

48. The impact of Leibniz's *Theodicy* on Comte's philosophy of history is evident but, in contrast with Leibniz's importance (together with Descartes) as a modern forerunner of the positivist synthesis of knowledge, is hardly acknowledged. To be sure, in the 1822 *Plan général des travaux nécessaires pour réorganiser la société* Comte had noted the "resemblance" and "deep analogy" between "the spirit of positive politics" and "the famous theological and metaphysical doctrine of optimism" (*Système*, IV, appendix, 116, trans. 575)—a clear allusion to Leibniz. However, the latter's doctrine is immediately criticized as static and is not even mentioned in book 3 of the *Système* where each epoch in the sequence of progress is "appreciated" as the best possible world given its determinate historical limits.

49. *Système*, I, 703, trans. 568.

50. *Système*, IV, 304, trans. 266.

51. *Synthèse*, 10.

52. *Système*, I, 204ff, trans. 164ff.

53. *CPR*, 137.

54. *Système*, IV, 67, trans. 59

55. Olympe de Gouges led a movement for women's suffrage, which was suppressed during the Terror. For Comte's estimate of the "rights of women" and other "modern sophisms" see *Système*, I, 244–47, trans. 196–98.

56. *Système*, I, 696, trans. 562.

57. *Système*, IV, 67, trans. 59.

58. *CPR*, 327.

59. *CPR*, 285.

60. *Système*, I, 227–28, cf. 250, 326, trans. 183, 201, 260–61.

61. *CPR*, 142. This image is beautifully captured in the 1900 painting by Eduardo de Sa which still hangs in the Chapelle de l'Humanité at 5, Rue Payenne in Paris.

62. *CPR*, 108.

63. "The Positivist shuts his eyes during private prayer, the better to see

the internal image; the believer in theology opened his, to enable him to perceive outside an object that was an illusion." *CPR*, 93.

64. *CPR*, 139 *et seq.*

65. *CPR*, 128 *et seq.*

66. *CPR*, 137.

67. At the personal level, this includes a prescribed regimen of instinct control aiming to moderate the strength of the nutritive and sexual impulses, with the former taken as key to the latter. *Système*, IV, 283–86, trans. 248–51.

68. "The Positivist regime will always require full liberty of exposition and even of discussion. . . . The only admissible restriction of this liberty is public opinion." *CPR*, 340.

69. *Système*, I, 137ff, trans. 108ff.

70. *CPR*, 300 *et seq.*

71. *CPR*, 81–82.

72. *CPR*, 287 *et seq.*

73. *Système*, I, 431–34, trans. 349–52.

74. *Système*, I, 321, trans. 257.

75. Comte's letters of appeal to Czar Nicholas II, and Reschid Pacha, Grand Vizier of the Ottoman Empire, were reprinted in the prefatory material to volume 3 of the *System of Positive Polity*.

76. This is marked by the alignment of "the long axis of the temple and sacred wood toward the metropolis of the race, which, as the result of the whole past, is, for a long time, fixed at Paris." *Système*, IV, 156, trans. 139.

77. *CPR*, 359–66.

78. Huxley, *Physical Basis*, 28.

79. *Système*, III, 614–15, trans. 526–27.

80. The *Synthèse subjective* was dedicated to Daniel Encontre, Comte's mathematics teacher at his *lycée* (high school), who had taken Comte under his wing and helped prepare him for the entrance exams for the École Polytechnique.

81. Pickering, *Auguste Comte*, 2:566–71.

82. His scorn for the "philosophical and metaphysical agitators" at the Sorbonne was unbounded. "The whole of modern history teaches us to see in the abolition of the University [of Paris] the consequence and complement of the abolition of the Parliamentary regime." But all "special schools" were to be abolished too, except those teaching "veterinary science." *Système*, IV, 337.

83. *CPR*, 304–5.

84. *Synthèse*, 12–13.

85. *CPR*, 154. For a good discussion of this shift in Comte's theory of historical stages see Gane, *Comte*, 110 *et seq.*

86. Nietzsche, *Daybreak*, 215.

87. Popper, *Open Society.*

88. Hayek, *Counter-Revolution.*

89. Marcuse, *Reason and Revolution.*

90. Bourdeau notes that the *Philosophie positive* did "not belong to Comte's initial program and . . . originally was meant as a parenthesis, or prelude, that was supposed to take a few years at most." Bourdeau, "Auguste Comte."

91. Pickering, *Auguste Comte*, vol. 1, chs 3–5.

92. Taylor, *Henri Saint-Simon*, 124–28.

93. *Système*, III, 511–12, trans. 431–32.

94. *Synthèse*, 25.

95. *Système*, I, 711, trans. 574.

96. Heidegger's most succinct critique of Nietzsche in these terms is in his essay "On the Saying of Nietzsche, 'God is Dead.'" Heidegger, *Question*, 53–114.

97. Wartelle, *L'Héritage.*

Conclusion

1. *Revue philosophique*, cited by Gérard, "1852–1902," 143.

2. Nordmann, "Taine," 23–24.

3. Bensaude-Vincent and Simon, *Chemistry*, 176.

4. Renouvier, "Positivismes, " 33–34. Renouvier pointed out that it was sometimes difficult to tease out what was clearly factual in interpretations, inductions, and generalizations. For more on Renouvier's commentary on Comte's law of three states, see his "Y a-t-il une loi du progrès?" 65–68; "La loi des trois états," 81–86; "Contradictions positivistes," 129–32.

5. Stengers, "Intervention," 295. On the "revolt against positivism," see Hughes, *Consciousness*, 29–30, 33–66.

6. Bensaude-Vincent and Simon, *Chemistry*, 176, 182–83.

7. "Ten Most Harmful Books of the 19th and 20th Centuries," *Human Events*, May 31, 2005, http://humanevents.com/2005/05/31/ten-most-harmful-books-of-the-19th-and-20th-centuries/.

8. Chabert, *Un nouveau pouvoir*, 14.

9. Singer, *Legacy*, 87.

10. For a list of positivist sympathizers throughout Europe and the Americas, see Plé, *Die "Welt,"* 501–61. See also Simon, *European Positivism*, 10–11;

Lenzer, introduction to *Auguste Comte*, xxiv, xxxii. It is worth noting that Simon tends to understate the influence of Comte, whom he dislikes.

11. Delfau, "Le positivisme," 235.

12. Renouvier asserted that there were six forms of positivisms based on the "simple *pretention* of relying on facts and experience and of emanating from the sciences." These were Comte's version in the *Cours*; that version plus his Religion of Humanity; Mill and Bain's psychological positivism; Spencer's evolutionary positivism; the attitude of those who deny the existence of God, a soul, natural morality, and innate reason; and the attitude of those who disdain philosophy and trust only the sciences to explain facts and laws. See Renouvier, "Les Positivismes," 33–37.

13. Littré, *Principes*, 14, 29, 30, 75.

14. Pickering, *Auguste Comte*, 2:252–57, 3:33–52; Singer, *Legacy*, 77–83; Thompson, *Durkheim*, 30; Fox, *Savant*, 143–44. For Littré's effort to make history an important part of sociology, see Petit, "Comte revu."

15. Petit and Bourdeau, "Pierre Laffitte," 7–20.

16. Dr. Georges Audiffrent (1823–1909), another close disciple of Comte's, carried on the religious side of positivism in France. After he had a rupture with Laffitte over the latter's lack of engagement on this issue, the Positivist Society expelled him and another doctor, Eugène Sémérie, in the late 1870s. Audiffrent set up his headquarters at 30 rue Jacob and made an alliance with Richard Congreve, an orthodox English positivist. Fox, *Savant*, 178–81.

17. Harvey, *Almost a Man*, 90, 150.

18. Pickering, *Auguste Comte*, 3:568; Fox, *Savant*, 181–82; Wartelle, *L'Héritage*, 209.

19. France, Notes on "Auguste Comte."

20. Fox, *Savant*, 141.

21. Charlton, *Positivist Thought*, 224.

22. LeGouis, *Positivism*, 17, 22–23, 40, 61, 171, 183–87.

23. Cointet, *Hippolyte Taine*, 161–65. See also Nordmann, "Taine," 22. On how Comte and Taine differed, see also Grondeux, "Taine," 180.

24. Richard, *Hippolyte Taine*, 9, 139, 201. See also Seys, *Hippolyte Taine*, 157.

25. Watson, *Ideas*, 708, 802n45.

26. Charlton, *Positivist Thought*, 88–93, Simon, *European Positivism*, 94–99; Reardon, *Religion*, 251; Stanguennec, *Ernest Renan*, 13, 91–94; Petit, "Le prétendu Positivisme," 73–101. Petit argues that the positivist label is "inadequate" in describing Renan, who fought positivism more than he followed it (100).

27. Brenner, "Reflections," 58.

28. These include the naturalist Charles Darwin (1809–1882), the historian of science William Whewell (1794–1866), the philosopher Antonio Rosmini (1797–1855), the physicist John Tyndall (1820–1893), and the chemists Michel Chevreul (1786–1889) and Justus von Liebig (1803–1873). Schmaus, *Rethinking Durkheim*, 78–79; Schmaus, "Renouvier," 132–35.

29. Lindenfeld, *Transformation*, 2.

30. Schmaus, "Renouvier," 132, 136, 138–39.

31. Schmaus, *Rethinking Durkheim*, 79. See also Copleston, *History of Philosophy*, vol. 9, part 1, 162; Fox, *Savant*, 141–42.

32. Nye, "Boutroux Circle," 110–13.

33. Sánchez, *Pity*, 49.

34. Schmaus, "Lévy-Bruhl," 424, 430–31, 436; Schmaus, *Rethinking Durkheim*, 98; Chimisso, *Writing*, 23, 64, 69.

35. Charlton, *Positivist Thought*, 74, 75, 77.

36. Kremer-Marietti, "Comte," 503.

37. Chimisso, *Writing*, 161.

38. Petit, "Claude Bernard," 208, 215. See also Petit, "D'Auguste Comte à Claude Bernard," 45–62.

39. Crosland, *Science*, 7. See also Bensaude-Vincent and Simon, *Chemistry*, 193. For a list of scholars who consider him a positivist, see Petit, "Marcelin Berthelot," 81n1.

40. Laudan, "Positivism," 670; Crosland, *Science*, 194.

41. Bensaude-Vincent, "Berthelot," 74. Also see Bensaude-Vincent and Simon, *Chemistry*, 181.

42. Petit, "Marcelin Berthelot,"103.

43. Zammito, *Derangement*, 8; Brenner, "Reflections," 61.

44. Anastasios Brenner argues that the philosophy of conventionalism, which he says these two physicists upheld, "deliberately" distanced itself from positivism yet was indebted to it nonetheless. See Brenner, *Les origines*, 12, 31.

45. Édouard Le Roy, quoted in Brenner, "Reflections," 61.

46. Bensaude-Vincent and Simon, *Chemistry*, 194.

47. Anastopoulos, *Particle*, 382n21.

48. Nye, "Boutroux Circle," 108, 119.

49. De Paz, "Third Way," 49.

50. Brenner, *Les origines*, 28–34.

51. Brenner and Gayon, Introduction, 2, 8. Brenner and Gayon also point out Canguilhem's influence on Michel Foucault (1926–1984), who pushed the

philosophy of science in new directions. Foucault was indeed a close reader of Comte's work. See Braunstein, "Auguste Comte," 176. For more on Comte's later influence, see Wernick, "Auguste Comte," 86; Wernick, "Comte, Auguste," 133.

52. Brenner, "Reflections," 58.

53. Brenner and Gayon, Introduction, 1–2, 7; Gutting, Introduction, 4.

54. Daston, "History of Science," 6842–43; Pyenson, *Passion*, 187, 276, 445–47.

55. Harvey, "Evolutionism," 289–90. On Robin and Comte's influence on medicine and the spread of "biology," see Braunstein, *La philosophie*, 121–24; Fox, *Savant*, 164–65.

56. Blanckaert, "Un Artefact historiographique?" 257, 268–69.

57. Brooks, *Eclectic Legacy*, 21, 70, 97; Charlton, *Positivist Thought*, 228. Charlton also points out that Georges Dumas and Pierre Janet injected positivist ideas into psychology and that Charles Richet did the same with regard to physiology.

58. Charlton, *Positivist Thought*, 84.

59. Goldstein, *Console*, 326–28.

60. Brooks, *Eclectic Legacy*, 23, 97, 98, 194.

61. Schmaus, *Durkheim's Philosophy*; Schmaus, "Hypotheses," 1–30; Petit, "De Comte à Durkheim," 42–70; Gane, "Durkheim," 32, 38; Heilbron, "Ce que Durkheim doit à Comte," 62, 65; Levine, *Visions*, 166; Ferrarotti, *Invitation*, 43; Thompson, *Emile Durkheim*, viii–ix, 30.

62. Schmaus, *Rethinking Durkheim*, 98.

63. Simon, *European Positivism*, 167; Claridge, "Naturalism," 916; Ripoll, "Zola," 125–35.

64. Barrès, *Mes cahiers*, vol. 2, 69. The other "gods" were Pascal and Voltaire.

65. Barrès, *Mes cahiers*, vol. 1, 110.

66. Barres, *Mes cahiers*, vol. 2, 95.

67. Houellebecq, "Préliminaires," 7–12; Sartori, "Michel Houellebecq," 143–51.

68. Betty, "Classical Secularisation Theory," 111.

69. Pickering, *Auguste Comte*, 3:302, 416.

70. Wartelle, *L'Héritage*, 181, 191, 194, 325–37; Perrot, "Note," 201–4; Schkolnyk, *Victoire Tinayre*.

71. Grondeux, *La France*, 91.

72. Fox, *Savant*, 228–39.

73. Robin and Littré asked Ferry to act as a lawyer for Comte's wife, Caroline Massin, a request he refused.

74. Foley and Sowerine, *Political Romance*, 146–47; Barral, "Ferry et Gambetta," 149–60; Mayeur, "Le positivisme," 137–47; Legrand, *L'influence*, 192–92, 252; Nicolet, "Jules Ferry," 23–48; Nicolet, *L'idée républicaine*, 187–248; Deroisin, *Notes*, 78n1; Varley, *Under the Shadow*, 67, 73–74; Kselman, *Death*, 140–41.

75. Ariès, *Hour*, 543.

76. Curiously, one subscriber to the fund for the statue was Marcelin Berthelot. Petit, "Marcelin Berthelot," 93.

77. LeGouis, *Positivism*, 41.

78. Corra, "Pierre Laffitte," 21.

79. Coubertin, *La Chronique*, 142–45; Wernick, *Auguste Comte*, 14n40.

80. Howell, "Philosopher Alain," 603.

81. Amster, *Medicine*.

82. Claeys, *Imperial Sceptics*.

83. Sutton, *Nationalism*, 11–45; Lepenies, *Between Literature and Science*, 42–43; Gérard, "Les Disciples," 301; Leymarie, *De la Belle Epoque*, 137.

84. Wright, *Religion*, 276.

85. Cashdollar, *Transformation*, 37; Wright, *Religion*, 129.

86. Scharff, *Comte*, x, 36–72, 120–22. See also Pickering, *Auguste Comte*, 2:70–113; Bourdeau, introduction to *Mill*, 3–4; Guillin, *Auguste Comte*.

87. Cashdollar, *Transformation*, 74–89.

88. Hawkins, *Positivism*, 15; Forbes, *Positivism*, 1–23; Wright, *Religion*, 73–124, 240.

89. Singer, *Legacy*, 87.

90. Besant, *Auguste Comte*, 27–39.

91. Collyer, *Mapping*, 55–56; Pickering, "Curious Case."

92. Wright, *Religion*, 135, 143, 157, 173–29, 269–70.

93. Taylor, *Philosophy*, 29, 43–46, 93; Wright, *Religion*, 165; Offer, *Herbert Spencer*, 6; Duncan, *Life*, 1:206–8; Pickering, *Auguste Comte*, 3:137, 152n346, 154, 484–86; Becquemont, "Positivisme," 61–68; Becquemont, "Auguste Comte," 322–28.

94. Hurd, *Public Spheres*, 194. See also 103–4, 109–10, 114, 117, 125, 192–97.

95. Gérard, "1852–1902," 159.

96. Wils, "Les Sympathisants," 333–49 ; Wils, *De omweg van de wetenschap*; Fedi, "Avant-propos," 13–15.

97. Sarti, *Italy*, 498; Donzelli, "Comte, L'Italie," 351–62; Donzelli, *Origini*, 76, 81; Plé, *Die "Welt,"* 387–440; Garin, *History*, xxxv, 981–88, 998–99; Cohen, "Italian Contributions, " 582–83.

98. Estes, "Ferri," 335–37; Patuelli, "Garofalo," 352–54.

99. Femia, *Pareto*, 16–19; Parsons, *Structure*, 181.

100. Bourdeau, "La réception," 7; Fedi, "Avant-propos," 16–18, 21, 25, 26. See the first issue of *Les Cahiers philosophiques de Strasbourg* of 2014 for more information on Comte and Germany.

101. Marx to Frederick Engels, July 7, 1866, in *Letters of Marx*, 213. See also Sperber, *Marx*, 402.

102. Lafargue was also close to Beesly. Derfler, *Paul Lafargue*, 19, 21, 22, 51, 54.

103. Marx to Engels, July 7, 1866, in *Letters of Marx*, 213.

104. Bratton, Denham, and Deutschmann, *Capitalism*, 63.

105. Beiser, *After Hegel*, 174–78.

106. Barkan, "Usable Past," 178–79; Pyenson, *Passion*, 204.

107. LeGouis, *Positivism*, 17, 22, 51, 105, 142, 147. Scherer was much influenced by Mill and Thomas Buckle.

108. Lindenfeld, *Transformation*, 47.

109. Smith, *Austrian Philosophy*, 43.

110. The sociologists Karl Mannheim (1893–1947) and Norbert Elias (1897–1990) are also said to have been influenced by Comte's approach to society. Much of the information about Germany in this section comes from Fuchs, "Positivism and History," 154–55, 158; Simon, *European Positivism*, 245–63; LeGouis, *Positivism*, 100–105; Wernick, "Auguste Comte," 86; Fedi, *La Réception germanique*.

111. Lindenfeld, *Transformation*, 51, 90.

112. Simon, *European Positivism*, 247.

113. Coen, *Vienna*, 12.

114. Boll also contributed articles to *L'Echo de Paris*, *Mercure de France*, *La Science & et la Vie*, and *Cahiers rationalistes*. In addition, he was instrumental in getting seven texts of the Vienna Circle translated into French and published. He thus had a major role in the reception of their ideas in France. Schöttler, "From Comte to Carnap,"14, 17–18, 20, 23, 25. I thank Bernadette Bensaude-Vincent for this reference.

115. In recent years, whether logical positivism had a moral and/or political agenda has become controversial. See Uebel, "Political Philosophy," 754–73. Uebel argues that some members of the Vienna Circle, such as Otto

Neurath and Rudolf Carnap, hoped that a philosophy that facilitated scientific progress would advance an emancipatory political agenda. Sarah Richardson disagrees, asserting that members of the Vienna Circle could not have embraced a political philosophy of science because of their ethical noncognitivism. See Richardson, "Left Vienna Circle, Part 1," 14–24; "Left Vienna Circle, Part 2," 167–74. I thank Warren Schmaus for these references.

116. On logical positivism and Comte in general, see Fedi, *Auguste Comte*, 15; Turner, "Positivism," in *Encyclopedia of Sociology*, 1510–11; Peterson, *Revoking*, 71; Singer, *Legacy*, xv, 88; Bryant, *Positivism*, 109–32; Kincaid, "Positivism," 558–59; Dunning, "Figurational Sociology." On Neurath's admiration for Comte, see Kremer-Marietti, "De l'unité," 189–203.

117. Szporluk, *Political Thought*, 1–9, 16–17; Sebestik, "Thomas Garrigue Masaryk," 102–23.

118. Milosz, *History*, 283. See his chapter "Positivism," 281–321.

119. Milosz, *History*, 284–85, 291–308; Walicki, *History*, 349; Tatarkiewicz, *Outline*, 41–44. I thank Lukasz Zatorski for information about Polish positivism.

120. Quinn, *Marie Curie*, 64. I thank Bernadette Bensaude-Vincent for this reference.

121. LeGouis, *Positivism*, 52.

122. Frank, *Dostoevsky*, 253; Nemeth, *Early Solov'ëv*, 229–31.

123. Walicki, *History*, 273, 349–51, 362–70; Copleston, *Philosophy in Russia*, 101, 127, 127–31, 139, 400; Berlin, *Russian Thinkers*, 5, 282; LeGouis, *Positivism*, 14, 52–53, 87. 107; Nemeth, *Early Solov'ëv*, 231–32; Scott, *Social Theory*, 82–83.

124. Walicki, *History*, 351–60, 376, 371, 387; Copleston, *Philosophy in Russia*, 227–29; Clauzade, "Grégoire Wyrouboff," 298–302. This issue of the *Archives de Philosophie* (Summer 2016) is devoted to positivism in Russia.

125. Forbes, *Positivism*, 150, 153; Billington, "Intelligentsia," 807–21; LeGouis, *Positivism*, 87–100.

126. Information on Turkey is taken from Gawrych, *Crescent*, 141–42; Hanioğlu, *Young Turks*, 203–5; Ramsaur, *Young Turks*, 22–26; Okyar, "Atatürk's Quest," 46; Akural, "Kemalist Views"; Bourdeau, "La réception," 3.

127. Cashdollar, *Transformation*, 262–67; Hawkins, *Positivism*, 3–4, 102–3; Wunderlich, *Low Living*, 171.

128. Pickering, *Auguste Comte*, 2:531–32; 3:443–45.

129. The books were *Sociology for the South*, by Fitzhugh, and *A Treatise on Sociology*, by Hughes.

130. Ritzer, *Classical Sociological Theory*, 40; Laudan, "Positivism," 670.

131. Hawkins, *Auguste Comte*, 14; Cashdollar, *Transformation*, 179, 441–48.

132. Cashdollar, *Transformation*, 93–101; 121–23; 140, 215–25, 262–67, 281–28; Hawkins, *Auguste Comte*, 14–25; Hawkins, *Positivism*, 100, 215–25; Pickering, *Auguste Comte*, 2:530–37.

133. Information about the Positive Society and its members comes from: Hawkins, *Auguste Comte*, 18–26, 38–48, 61; Hawkins, *Positivism*, 90–99, 101–4, 212–25; Harp, *Positivist Republic*, 30–36, 46, 48, 52–55, 72; Croly, *Memories*, 74–75; Pickering, *Auguste Comte*, 3:105–111, 433–49, 472–73.

134. Information about progressives comes from Ritzer, *Classical Sociological Theory*, 45–46; Harp, *Positivist Republic*, xiv, 109–76, 180, 213.

135. Harp, *Positivist Republic*, xvi, 3, 70, 88, 159–62, 195, 213, 215.

136. Forbes, *Positivism*, 147–49; Gilson and Levinson, Introduction, viii.

137. Ardao, "Positivism in Latin America," 150; Nuccetelli and Seay, *Latin American Philosophy*, 143, 145; Chevalier, *L'Amérique Latine*, 435; Zea, *Latin-American Mind*, 20–26, 130–31; Wiarda, *Soul of Latin America*, 146–47; Clark, "Emergence," 54; Nuccetelli, *Latin American Thought*, 180–83.

138. Dussel, "Philosophy in Latin America," 11; Wiarda, *Soul of Latin America*, 14.

139. Zea, *Latin-American Mind*, 28; Nuccetelli and Seay, *Latin American Philosophy*, 145; Hoeg, "Rebellion," 85; Nuccetelli, *Latin American Thought*, 185; Costa, *History*, 85–100, 326n46; Arbousse-Bastide, *Le positivisme*, 34–35; Diacon, *Stringing Together*, 82; Burns, *History*, 255, 255, De Boni, *Storia*, 404–5.

140. Costa, *History*, 99, 329n83, 334n111; Burns, *History*, 253; Rose, "Brazil's Military Positivists," 134; Pickering, *Auguste Comte*, 3:453–55, 575–78. Other founding members in 1876 were Antônio Carlos de Oliveira Guimarães, Joaquim Ribeiro de Mendonça, Oscar de Araújo, Alvaro de Oliveira, Miguel Lemos, and Teixeira Mendes.

141. The website is *Templo da Humanidade*, at http://templodahumanidade.org.br/. See also Simon Romero, "Nearly in Ruins: The Church Where Sages Dreamed of a Modern Brazil," *New York Times*, Dec. 25, 2016, https://www.nytimes.com/2016/12/25/world/americas/nearly-in-ruins-the-church-where-sages-dreamed-of-a-modern-brazil.html?mcubz=0; Costa, *History*, 87, 94–108, 125–26, 135; Forbes, *Positivism*, 153; Gollo, "Birth," 160–64.

142. Bourne, *Getulio Vargas*, 3–4; Nuccetelli and Seay, *Latin American Philosophy*, 145; Grange, *Auguste Comte*, xv–xvi.

143. The influence of the positivists was not immense, given that in 1889

there were only fifty-three members of the Positivist Society, most of whom were concerned with problems relating to Laffitte and religious worship. Costa, *History*, 143–45.

144. Costa, *History*, 147.

145. Nuccetelli, *Latin American Thought*, 184; Burns, *History*, 287–93; Costa, *History*, 147, 151, 158, 169, 173.

146. Nuccetelli, *Latin American Thought*, 185; Costa, *History*, 113–14, 139–40, 152, 170; Burns, *History*, 254, 295, 341; Diacon, *Stringing Together*, 83–84; Gollo, "Birth,"165.

147. Bourne, *Getulio Vargas*, 4. Positivist elements in the Constitution included freedom of education and free schools.

148. Burns, *History*, 399; Bourne, *Getulio Vargas*, 13, 211–22; Chevalier, *L'Amérique Latine*, 440.

149. Kury, "Nation," 125, 134–36.

150. Diacon, *Stringing Together*, 1–7, 79–85, 93, 103–8, 114–18, 129, 160–61; Gomes, *Indians*, 78–82.

151. For more on positivism and Brazil, see Hilton, "Positivism," 540–45; Forbes, *Positivism*, 147, 152; Gérard, "1852–1902," 159.

152. Arbousse-Bastide, *Le positivisme*, 453.

153. Zea, *Positivism in Mexico*, 39n1, 41, 43; Zea, "Positivism," 221–26; Sherman, *Mexican Right*, 2, 6; Quirk, *Mexican Revolution*, 14–17; Mendieta, "Death of Positivism," 4.

154. Zea, *Positivism in Mexico*, 39–40; Zea, "Positivism," 230–31.

155. Romanell, *Making of the Mexican Mind*, 45.

156. Cuspinera, "Positivism," 1179; Hale, "Political and Social Ideas," 84; Meyer and Sherman, *Course*, 407–8.

157. Cuspinera, "Positivism," 1179.

158. Raat, "Augustín Aragón," 241–59.

159. Hale, *Transformation*, 20, 246; Hale, "Political and Social Ideas," 387–88.

160. Clayton and Conniff, *History*, 123–24, 169; Nuccetelli, *Latin American Thought*, 189–90; Chevalier, *L'Amérique Latine*, 440; Raat, "Antipositivist Movement," 83; Mendieta, "Death of Positivism," 5.

161. Zea, "Positivism," 223–28, 238. Charles Hale also states that Spencer was "more influential in Mexican social thought than Auguste Comte." Hale, *Transformation*, 251.

162. Hale, *Transformation*, 247; Clayton and Conniff, *History*, 123–24, 169; Nuccetelli, *Latin American Thought*, 189–90; Butler, *Popular Piety*, 38, 81;

Sherman, *Mexican Right*, 5; Raat, "Augustín Aragón," 254; Lipp, *Leopoldo Zea*, 61; Cockcroft, *Intellectual Precursors*, 56; Meyer and Sherman, *Course*, 440–42, 457.

163. Fehrenbach, *Fire*, 456; Shorris, *Life*, 196; Raat, "Antipositivist Movement," 83, 196; Forbes, *Positivism*, 151; Zea, "Positivism and Porfirism," 216.

164. Jaksić, *Academic Rebels*, 8–9, 43–66; Clayton and Conniff, *History*, 123; Hale, "Political and Social Ideas," 384–85, 389–91; Martí, "Positivist Thought," Clark, "Emergence," 61; Lipp, *Three Chilean Thinkers*, 54, 144.

165. Zea, *Latin-American Mind*, 29–31. See also Jaksić, *Academic Rebels*, 49; Martí, "Positivist Thought," 568; Costa, *History of Ideas*, 103.

166. Rabossi, "Latin American Philosophy," 509.

167. Lipp, *Leopoldo Zea*, 61, 62; Zea, *Latin-American Mind*, 28, 30–31; Wiarda, *Soul*, 153, 201; Clayton and Conniff, *History*, 123, 145–6; Sáenz, *Identity*, 159; Spektorowski, *Origins*, 38–39, 47; Hale, "Political and Social Ideas," 384, 406–7; Clark, "Emergence," 58–59.

168. Wiarda, *Soul*, 168; López Alves, "Authoritarian Roots," 129; Lipp, *Leopoldo Zea*, 62, Zea, *Latin-American Mind*, 28, 31–33; Clayton and Conniff, *History*, 126, 178; Sáenz, *Identity*, 225–26; Chevalier, *L'Amérique Latine*, 436, 441–42.

169. Raat, "Aragón," 251; McCreery, "Coffee and Class," 438, 441.

170. Dym, "Central America," 310.

171. Lipp, *Leopoldo Zea*, 62; Zea, *Latin-American Mind*, 28, 32, 257; Jaksić, *Academic Rebels*, 41; Staten, *History*, 32–33.

172. Ardao, "Positivism," 151, Clark, "Emergence," 62.

173. Zea, *Positivism in Mexico*, xx.

174. Chevalier, *L'Amérique Latine*, 323, 668; Zea, *Positivism in Mexico*, xiii–xv, 39n1; Hilton, "Positivism," 544–45; Collier, "Positivism," 457–58; Wartelle, *"L'Héritage,* 233–36; Eastwood, "Positivism," 331–57; Forbes, *Positivism*, 147–50, 153; Raat, "Augustín Aragón," 441–57; Capurro, *Le positivisme*, 141–54; Martí, "Positivist Thought," 568.

175. Clayton and Conniff, *History*, 123.

176. Zea, *Latin-American Mind*, 33; Wiarda, *Soul*, 6, 14, 172–74, 347–49, 352. Lipp, *Leopoldo Zea*, 62, Ardao, "Positivism," 151; Hale, *Transformation*, 62–63; Martí, "Positivist Thought," 568; Nuccetelli, *Latin American Thought*, 189.

177. Wiarda, *Soul*, 7.

178. Hale, "Political and Social Ideas," 385; Wiarda, *Soul*, 171.

179. Havens, "Comte," 223; Bourdeau, "La réception," 4; Forbes, *Positivism*, 153.

180. Information on India is from Forbes, *Positivism*, 1, 15, 30, 51, 57, 71, 56, 99–101, 105–9, 123–30, 147–48, 150, 154–58; Flora, *Evolution*, 4; Dasgupta, *Social Thought*, 32; Halbfass, *India*, 242–43, 256, 428; Bourdeau, "La réception," 4; Claeys, *Imperial Sceptics*, 67–70; De Boni, *Storia*, 299–315.

181. Petit, "Des sciences positives," 87.

182. Singer, *Legacy*, 90–91.

183. Singer, *Legacy*, xvi, 94; Scharff, *Comte*, x, 36–72, 120–22.

184. Dale, *In Pursuit*, 9, 10; Fedi, *Auguste Comte*, 14; Singer, *Legacy*, 84–142.

185. Peirce, *Reasoning*, 114. See also Atkins, "Restructuring," 483–500.

186. Peirce believed that a hypothesis should be able to be verified by induction or experimental procedures. Peirce, *Essential Peirce*, 225, 236. See also McCarthy, "Pragmatism," 176.

187. Laudan, "Classification," 155; Laudan, "Philosophy," 632–34; and Laudan, "Positivism," 670; Bourdeau, "Auguste Comte."

188. Hacking, *Historical Ontology*, 164, 190.

189. Recently it has come to light that Abbé Sieyès first used the term *sociologie* in an unpublished manuscript from the 1780s. See Guilhaumou, "Sieyès," 117–34.

190. Turner, Beeghley, and Powers, *Emergence*, 53–54, 434–36, 472–73, 472–77; Parsons, "General Interpretation," 92; Ritzer, *Classical Sociological Theory*, 10, 16, 107, 111; Boudon, "Sociology," 21:14581; Manicas, *History*, 281.

191. Fuchs, "Positivism," 148–153, 159; DiVanna, *Writing History*; Singer, *Legacy*, 86; Steinmetz, "Introduction," 9–11; Carbonell, "L'Histoire," 173–85; Tosh, *Pursuit*, 166, 171.

192. Fuller, "Positivism," 11823. See also Turner, "Positivism: Sociological," 11829.

193. Jori, "Legal Positivism," 514.

194. Fuller, "Positivism," 11825; Alvey, "Ethics," 5–34.

195. Sullivan, "Method," 19–20; Kincaid, "Positivism," 558.

196. Laudan, "Positivism and Scientism," 671.

197. LeGouis, *Positivism*, 1997.

198. Forbes, *Positivism*, 147.

199. Kent, *Brains*, xiii, 59; Harp, *Positivist Republic*, xv, 21.

200. LeGouis, *Positivism*, 15.

201. Fuller, "Positivism," 11821.

Bibliography

Auguste Comte's Works

Comte, Auguste. *Appel aux conservateurs*. Paris: L'auteur, Victor Dalmont, 1855.

Comte, Auguste. *Auguste Comte. Évolution originale*, edited by Raimundo Teixeira Mendes, vol. 1. Rio de Janeiro: Au siège central de l'église positiviste du Brésil, 1913.

Comte, Auguste. *Catéchisme positiviste, ou Sommaire exposition de la religion universelle en treize entretiens systématiques entre une femme et un prêtre de l'humanité*. Paris, 1852. Translated as *The Catechism of Positive Religion*.

Comte, Auguste. "Considérations philosophiques sur la science et les savants" (1825). In *Système de politique positive*, vol. 4, appendix, 137–175. Paris, 1854. Translated in *Early Political Writings*.

Comte, Auguste. "Considérations sur le pouvoir spirituel" (1826). In *Système de politique positive*, vol. 4, appendix, 176–215. Paris, 1854. Translated in *Early Political Writings*.

Comte, Auguste. *Correspondance générale et confessions*. 8 vols. Paris: Éditions de L'École des Hautes Études en Sciences Sociales–Vrin, 1973–90.

Comte, Auguste. *Cours de philosophie positive*. Paris, 1830–42. Freely translated and condensed in *The Positive Philosophy of Auguste Comte*; the first two lessons are translated in *Introduction to Positive Philosophy*.

Comte, Auguste. *Cours de philosophie positive*. Edited by Emile Littré. Paris, 1869.

Comte, Auguste. *Cours de philosophie positive*. Edited by Michel Serres, François Dagognet, and Allal Sinaceur. 2 vols. Paris: Hermann, 1975.

Comte, Auguste. *Discours sur l'ensemble du positivisme*. Paris, 1848. Reprinted with preface and notes by Annie Petit. Paris: Flammarion, 1998. Reproduced, with minor changes, as part one (*Discours préliminaire*) of vol. 1 of

Système de politique positive by Auguste Comte. Paris, 1851–54. Translated as *A General View of Positivism*, subsequently incorporated into vol. 1 of *System of Positive Polity*.

Comte, Auguste. *Discours sur l'esprit positif*. Paris, 1844. Reprinted with preface and notes by Annie Petit. Paris: Vrin, 1995. Translated as *A Discourse on the Positive Spirit*.

Comte, Auguste. *Du pouvoir spirituel*. Edited by Pierre Arnaud. Paris: Librairie Générale de France, 1978.

Comte, Auguste. *Écrits de jeunesse, 1816–1828: suivis du mémoire sur la cosmogonie de Laplace, 1835*. Edited by Paulo E. Carneiro and Pierre Arnaud. Paris: École Pratique des Hautes Études, 1970.

Comte, Auguste. "Examen du Traité de Broussais sur l'irritation" (1828). In *Système de politique positive*, vol. 4, appendix, 216–228. Paris, 1854. Translated in *Early Political Writings*.

Comte, Auguste. *Plan des travaux scientifiques nécessaires pour réorganiser la société* (1824). In *Système de politique positive*, vol. 4, appendix, 47–136. Paris, 1854. Translated in *Early Political Writings*.

Comte, Auguste. *Science et politique. Les conclusions générales du "Cours de philosophie positive."* Edited by Michel Bourdeau. Paris: Pocket, 2003.

Comte, Auguste. "Sommaire appréciation de l'ensemble du passé moderne" (1820). In *Système de politique positive*, vol. 4, appendix: 4–46. Paris, 1854. Translated in *Early Political Writings*.

Comte, Auguste. *Synthèse subjective ou Système universel des conceptions propres à l'état normal de l'humanité*. Paris, 1856. Translated as *Subjective Synthesis*.

Comte, Auguste. *Système de politique positive*. 4 vols. Paris, 1851–54. Reprint, Paris: Au siège de la Société positiviste, 1929. Translated as *System of Positive Polity*.

Comte, Auguste. *Testament d'Auguste Comte avec les documents qui s'y rapportent*. Paris, 1884.

Comte, Auguste. *Testament d'Auguste Comte*. 2nd ed. Paris, 1896.

Comte, Auguste. *Traité élémentaire de géométrie analytique à deux et à trois dimensions, contenant toutes les théories générales de géométrie accessibles à l'analyse ordinaire*. Paris, 1843.

Comte, Auguste. *Traité philosophique d'astronomie populaire*. Paris, 1844. Reprint, Paris: Fayard, 1985.

English Translations

Comte, Auguste. *A Discourse on the Positive Spirit*. Translated by Edward S. Beesley. London: William Reeves, 1903.

Comte, Auguste. *Early Political Writings*. Translated by H. Stuart Jones. Cambridge, UK: Cambridge University Press, 1998.

Comte, Auguste. *Introduction to Positive Philosophy*. Translated by Frederick Ferré. Indianapolis: Hackett, 1988.

Comte, Auguste. *Subjective Synthesis*. Translated by Richard Congreve. London, 1891.

Comte, Auguste. *System of Positive Polity*. Translated by Edward S. Beesly, John. H. Bridges, Richard Congreve, Frederic Harrison, and Henry Dix Hutton. 4 vols. London, 1875–77. Reprint, New York: Burt Franklin, 1966.

Comte, Auguste. *The Catechism of Positive Religion*. Translated by Richard Congreve. 2nd ed. London, 1883.

Comte, Auguste. *The Positive Philosophy of Auguste Comte: Freely Translated and Condensed by Harriet Martineau*. London, 1853. Reprint, New York, 1858.

Other Works

Acton, Harry B. *The Idea of a Spiritual Power*. London: Athlone Press, 1974.

Akural, Sabri M. "Kemalist Views on Social Change." In *Atatürk and the Modernization of Turkey*, edited by John M. Landau, 125–52. Leiden: E. J. Brill, 1984.

Albury, William R. "Experiment and Explanation in the Physiology of Bichat and Magendie." *Studies in History of Biology* 1 (1977): 47–131.

Alengry, Franck. *Essai historique et critique sur la sociologie chez Auguste Comte* (1899). Geneva: Slatkine, 1984.

Alvey, James E. "Ethics and Economics, Today and in the Past." *Journal of Philosophical Economics* 5 (2011): 5–34.

Amster, Ellen. *Medicine and the Saints: Science, Islam, and the Colonial Encounter in Morocco, 1877–1956*. Austin: University of Texas Press, 2013.

Anastopoulos, Charis. *Particle or Wave: The Evolution of the Concept of Matter in Modern Physics*. Princeton, NJ: Princeton University Press, 2008.

Andrews, Naomi. *Socialism's Muse: Gender in the Intellectual Landscape of French Romantic Socialism*. Lanham, MD: Lexington Books, 2006.

Arbousse-Bastide, Paul. *La doctrine de l'éducation universelle dans la philosophie d'Auguste Comte*. 2 vols. Paris: Presses Universitaires de France, 1956.

Arbousse-Bastide, Paul. *Le positivisme politique et religieux au Brésil*. Edited by Annie Petit and Francis Utéza. Turnhout: Brepols, 2010.

Ardao, Arturo. "Positivism in Latin America." Translated by Solena V. Bryant. In *Latin American Philosophy: An Introduction with Readings*, edited by Susana Nuccetelli and Gary Seay, 150–56. Upper Saddle River, NJ: Pearson Prentice Hall, 2004.

Ariès, Philippe. *The Hour of Our Death*. Translated by Helen Weaver. New York: Oxford University Press, 1981.

Arnaud, Pierre. *Le Nouveau Dieu. Préliminaires à la politique positive*. Paris: Vrin, 1973.

Aron, Raymond. *Les Étapes de la pensée sociologique*. Paris: Gallimard, 1967.

Aron, Raymond. *Main Currents in Sociological Thought*. Translated by Richard Howard and Helen Weaver. New York: Basic Books, 1975.

Atkins, Richard Kenneth. "Restructuring the Sciences: Peirce's Categories and His Classifications of the Sciences." *Transactions of the Charles S. Peirce Society* 42, no. 4 (2006): 483–500.

Bacon, Francis. *Novum Organum*. 1620. Reprint, London: Colonial Press, 1900.

Bachelard, Gaston. *Essai sur la connaissance approchée*. 1928. Reprint, Paris: Vrin, 1987.

Bachelard, Gaston. *Études*. 1970. Reprint, Paris: Vrin, 2002.

Bain, Alexander. *On the Study of Character including an Estimate of Phrenology*. London, 1861.

Barkan, Diana Kormos. "A Usable Past: Creating Disciplinary Space for Physical Chemistry." In *The Invention of Physical Science: Intersections of Mathematics, Theology, and Natural Philosophy Since the Seventeenth Century: Essays in Honor of Erwin N. Hiebert*, edited by Mary J. Nye, Joan L. Richards, and Roger H. Stuewer, 175–202. Dordrecht: Springer, 2012.

Barral, Pierre. "Ferry et Gambetta face au positivisme." *Romantisme* 8, no. 21–22 (1978): 149–60.

Barrault, Émile, and Édouard Charton. *Religion Saint-Simonienne. Recueil de prédications*. 2 vols. Paris, 1832.

Barrès, Maurice. *Mes cahiers, vol. 1 (Janvier 1896–Novembre 1904)*. Paris: Equateurs, 2010.

Barrès, Maurice. *Mes cahiers, vol. 2 (Novembre 1904–Juin 1908)*. Paris: Equateurs, 2011.

Becq, Annie. *Genèse de l'esthétique française moderne de la raison classique à l'imagination créatrice, 1680–1814*. Pisa: Pacini Editore, 1984.

Becquemont, Daniel. "Auguste Comte et l'Angleterre." In Petit, *Auguste Comte: Trajectoires positivistes*, 317–31.

Becquemont, Daniel. "Positivisme et utilitarisme: Regards croisés, Comte, Spencer, Huxley." *Revue d'histoire des sciences humaines* 8, no. 1 (2003): 57–72.

Beiser, Frederick C. *After Hegel: German Philosophy, 1840–1900*. Princeton, NJ: Princeton University Press, 2014.

Benrubi, Isaak. *Les Sources et les courants de la philosophie contemporaine en France*. Paris: Alcan, 1933.

Bensaude-Vincent, Bernadette. "Berthelot, un chimiste positiviste attardé?" In *Marcelin Berthelot (1827–1907): Sciences et politique*, edited by Jean Balcou, 71–80. Rennes: Presses Universitaires de Rennes, 2010.

Bensaude-Vincent, Bernadette, and Jonathan Simon. *Chemistry: The Impure Science*. 2nd ed. London: Imperial College Press, 2012.

Bernier, François. *Abrégé de la Philosophie de M. Gassendi*. Lyon, 1678.

Berlin, Isaiah. *Russian Thinkers*. Edited by Henry Hardy and Aileen Kelly. London: Hogarth, 1978.

Bernoulli, Jakob. *Ars conjectandi, opus posthumum. Accedit Tractatus de seriebus infinitis, et epistola gallicé scripta de ludo pilae reticularis*. Basel, 1713.

Bertrand, Joseph. "Souvenirs académiques: Auguste Comte et l'École polytechnique." *Revue des deux mondes* 6 (1896): 528–48.

Besant, Annie. *Auguste Comte: His Philosophy, His Religion, and His Sociology*. 1881. Reprint, Whitefish, MT: Kessinger, 2006.

Betty, Louis. "Classical Secularisation Theory in Contemporary Literature—The Curious Case of Michel Houellebecq." *Literature and Theology* 27, no. 1 (March 2013): 98–115.

Bichat, Xavier. *Anatomie générale appliquée à la physiologie et à la médecine*. Vol. 1. Paris, 1801.

Bichat, Xavier. *General Anatomy*. Translated by Georges Hayward. Boston, 1822.

Bichat, Xavier. *Physiological Researches on Life and Death*. Translated by Tobias Watkins. Philadelphia, 1809.

Bichat, Xavier. *Recherches physiologiques sur la vie et la mort*. Paris, 1800.

Billington, James H. "The Intelligentsia and the Religion of Humanity." *American Historical Review* 65, no. 4 (1960): 807–21.

Blainville, Henri Marie Ducrotay de. *Cours de physiologie générale et comparée*. 3 vols. Paris, 1833.

Blainville, Henri Marie Ducrotay de. *De L'Organisation des animaux, ou Principes d'anatomie comparée*. Paris, 1822.

Blainville, Henri Marie Ducrotay de. *Plan du cours de physiologie générale et comparée, fait à la faculté des sciences de Paris pendant les années 1829, 1830, 1831 et 1832*. Reprinted in *L'Organe de la pensée: Biologie et philosophie chez Auguste Comte*, by Laurent Clauzade. Besançon: Presses Universitaires de Franche-Comté, 2009.

Blanckaert, Claude. "Un Artefact historiographique? L'anthropologie 'positiviste' en France dans la seconde moitié du XIXe siècle." In Petit, *Auguste Comte: Trajectoires positivistes*, 253–84.

Blay, Michel. "Force, Continuity and the Mathematization of Motion at the End of the Seventeenth Century." In *Isaac Newton's Natural Philosophy*, edited by Jed Z. Buchwald and I. Bernard Cohen, 225–48. Cambridge, MA: MIT Press, 2000.

Blay, Michel. *La naissance de la mécanique analytique: la science du mouvement au tournant des XVIIe et XVIIIe siècles*. Paris: Presses Universitaires de France, 1992.

Blay, Michel. *Les "Principia" de Newton*. Paris: Presses Universitaires de France, 1995.

Blay, Michel. "Principe de continuité et mathématisation du mouvement dans la deuxième moitié du XVIIe siècle." *Studia Leibniziana* 24 (1992): 191–204.

Blay, Michel. *Reasoning with the Infinite: From the Closed World to the Mathematical Universe*. Chicago: University of Chicago Press, 1998.

Blay, Michel, and Egidio Festa. "Mouvement, continu et composition des vitesses au XVIIe siècle." *Archives Internationales d'Histoire des Sciences* 48, no. 140 (1998): 65–118.

Boudon, Raymond. "Sociology." In *International Encyclopedia of the Social and Behavioral Sciences*, edited by Neil Smelser and Paul Bates, vol. 21, 14581–85. New York: Elsevier, 2001.

Bourdeau, Michel. "Auguste Comte." *The Stanford Encyclopedia of Philosophy*, edited by Edward N. Zalta. Winter 2015. https://plato.stanford.edu/archives/win2015/entries/comte/.

Bourdeau, Michel. "L'esprit ministre du cœur: Auguste Comte et la place de l'affectivité dans la vie morale." *Revue de théologie et de philosophie* 132 (2000): 175–92.

Bourdeau, Michel. "L'idée de point de vue sociologique: La philosophie des sciences comme sociologie des sciences chez Auguste Comte." *Cahiers internationaux de sociologie* 117 (2004): 225–38.

Bourdeau, Michel. Introduction to *Auguste Comte et le positivisme*, by John

Stuart Mill. Translated by Georges Clemenceau, 3–4. Paris: L'Harmattan, 1999.

Bourdeau, Michel. "Pouvoir spirituel et fixation de croyances." *Commentaire* 136, no. 4 (2011): 1095–1104.

Bourdeau, Michel. "La réception du positivisme (1843–1928)." *Revue d'histoire des sciences humaines* 8, no. 1 (2003): 3–8.

Bourdeau, Michel. *Les Trois États: Science, théologie et métaphysique chez Auguste Comte*. Paris: Éditions du Cerf, 2006.

Bourdeau, Michel, Jean-François Braunstein, and Annie Petit, eds. *Auguste Comte aujourd'hui*. Paris: Kimé, 2003.

Bourdeau, Michel, and Béatrice Fink. "De l'industrie à l'industrialisme: Benjamin Constant aux prises avec le Saint-Simonisme." *Œuvres et critique* 33, no. 1 (2008): 61–78.

Bourne, Richard. *Getulio Vargas of Brazil, 1883–1954: Sphinx of the Pampas*. London: Charles Knight, 1974.

Bratton, John, David Denham, and Linda Deutschmann. *Capitalism and Classical Sociological Theory*. Toronto: Toronto University Press, 2009.

Braunstein, Jean-François. "Antipsychologisme et philosophie du cerveau chez Auguste Comte." *Revue internationale de philosophie* 52, no. 1 (1998): 7–28.

Braunstein, Jean-François. "Auguste Comte et la philosophie de la médecine." In Petit, *Auguste Comte: Trajectoires positivistes*, 159–76.

Braunstein, Jean-François. *La philosophie de la médecine d'Auguste Comte: Vaches carnivores, Vierge Mère et morts vivants*. Paris: Presses Universitaires de France, 2009.

Bréhier, Émile. *Histoire de la philosophie*. Paris: Presses Universitaires de France, 1932.

Bréhier, Émile. *The History of Philosophy*. Translated by Wade Baskin. Chicago: University of Chicago Press, 1968.

Brenner, Anastasios. *Les origines françaises de la philosophie des sciences*. Paris: Presses Universitaires de France, 2003.

Brenner, Anastasios. "A Problem in General Philosophy of Science: The Rational Criteria of Choice." In *French Studies in the Philosophy of Science: Contemporary Research in France*, edited by Anastasios Brenner and Jean Gayon, 73–90. Vienna: Springer, 2009.

Brenner, Anastasios. *Raison scientifique et valeurs humaines: Essai sur les critères du choix objectif*. Paris: Presses Universitaires de France, 2011.

Brenner, Anastasios. "Reflections on Chimisso: French Philosophy of Science and the Historical Method." In *The Present Situation in the Philosophy of Science*, edited by Friedrich Stadler, 57–65. Heidelberg: Springer, 2010.

Brenner, Anastasios, and Jean Gayon. Introduction to *French Studies in the Philosophy of Science: Contemporary Research in France*, edited by Anastasios Brenner and Jean Gayon, 13–24. Dordrecht: Springer, 2009.

Brooks, John I., III. *The Eclectic Legacy: Academic Philosophy and the Human Sciences in Nineteenth-Century France*. Newark: University of Delaware Press, 1998.

Bryant, Christopher G. A. *Positivism in Social Theory and Research*. London: Macmillan, 1985.

Burns, E. Bradford. *A History of Brazil*. 2nd. ed. New York: Columbia University Press, 1980.

Butler, Matthew. *Popular Piety and Political Identity in Mexico's Cristero Rebellion: Michoacán, 1927–29*. Oxford: Oxford University Press, 2004.

Cabanis, Pierre-Jean-Georges. *Rapports du physique et du moral de l'homme*. 1802. Edited by Claude Lehec and Jean Cazeneuve. Paris: Presses Universitaires de France, 1956.

Canguilhem, Georges. *Études d'histoire et de philosophie des sciences*. 5th ed. Paris: Vrin, 1983.

Canguilhem, Georges. "L'école de Montpellier jugée par Auguste Comte." In *Études d'histoire et de philosophie des sciences*, by Georges Canguilhem, 75–80. 5th ed. Paris: Vrin, 1983.

Canguilhem, Georges. *The Normal and the Pathological*. Translated by Carolyn R. Fawcett with Robert S. Cohen, with an introduction by Michel Foucault. New York: Zone Books, 1989.

Canguilhem, Georges, Georges Lapassade, Jacques Piquemal, and Jacques Ulmann. *Du développement à l'évolution au XIXe siècle*. Paris: Presses Universitaires de France, 1962.

Capurro, Raquel. *Le Positivisme est un culte des morts: Auguste Comte*. Paris: Epel, 2001.

Carbonell, Charles-Olivier. "L'histoire dite positiviste en France." *Romantisme* 8, no. 21–22 (1978): 173–85.

Casanova, Jean-Claude. Introduction to "Pouvoir spirituel et fixation de croyances," by Michel Bourdeau. *Commentaire* 136 (2011): 1095.

Cashdollar, Charles. *The Transformation of Theology, 1830–1890: Positivism and Protestant Thought in Britain and America*. Princeton, NJ: Princeton University Press, 1989.

Cassina, Christina. "Comte face à la dictature." In Bourdeau, Braunstein, and Petit, *Auguste Comte aujourd'hui*, 184–97.

Chabert, Georges. *Un nouveau pouvoir spirituel: Auguste Comte et la religion scientifique au XIXe siècle*. Caen: Presses Universitaires de Caen, 2004.

Charlton, D. G. *Positivist Thought in France during the Second Empire, 1852–1870*. Oxford: Clarendon, 1959.

Chevalier, François. *L'Amérique Latine: De l'indépendance à nos jours*. Paris: Presses Universitaires de France, 1977.

Chimisso, Cristina. *Writing the History of the Mind: Philosophy and Science in France, 1900 to 1960s*. Aldershot, UK: Ashgate, 2008.

Christen, Carole. "Les leçons et traités d'astronomie populaire dans le premier XIXe siècle." *Romantisme* 166 (2014): 8–20.

Claeys, Gregory. *Imperial Sceptics: British Critics of Empire, 1850–1920*. Cambridge, UK: Cambridge University Press, 2010.

Claridge, Henry. "Naturalism." In *Encyclopedia of the Novel*, vol. 2, edited by Paul Schellinger, 914–18. New York: Routledge, 1998.

Clark, Meri L. "The Emergence and Transformation of Positivism." In *A Companion to Latin American Philosophy*, edited by Susana Nuccetelli, Ofelia Schutte, and Otávio Bueno, 53–67. Oxford: Blackwell, 2010.

Clarke, Edwin, and Leon S. Jacyna. *Nineteenth-Century Origins of Neuroscientific Concepts*. Berkeley: University of California Press, 1987.

Clauzade, Laurent. "Grégoire Wyrouboff: Penser la Russie. Essais de sociologie positive appliquée." *Archives de Philosophie* 79 (2016): 297–315.

Clauzade, Laurent. *L'Organe de la pensée: Biologie et philosophie chez Auguste Comte*. Besançon: Presses Universitaires de Franche-Comté, 2009.

Clayton, Lawrence A., and Michael L. Conniff. *A History of Modern Latin America*. New York: Harcourt Brace, 1999.

Cockcroft, James D. *Intellectual Precursors of the Mexican Revolution, 1900–1913*. Austin: University of Texas Press, 1968.

Coen, Deborah R. *Vienna in the Age of Uncertainty: Science, Liberalism, and Private Life*. Chicago: Chicago University Press, 2007.

Cohen, Morris R. "Italian Contributions to the Philosophy of Law." *Harvard Law Review* 59, no. 4 (1946): 577–89.

Cointet, Jean-Paul. *Hippolyte Taine: Un regard sur la France*. Paris: Perrin, 2012.

Collier, Simon. "Positivism." In *Encyclopedia of Latin American History and Culture*, edited by Barbara A. Tenenbaum, vol. 4, 457–58. New York: Charles Scribner's Sons, 1996.

Collyer, Fran. *Mapping the Sociology of Health and Medicine: America, Britain, and Australia Compared*. New York: Palgrave Macmillan, 2012.

Condorcet, Jean-Antoine-Nicolas de Caritat. *Essai sur l'application de l'analyse à la probabilité des décisions rendues à la pluralité des voix*. Paris, 1785.

Condorcet, Jean-Antoine-Nicolas de Caritat. *Mémoire sur le calcul des probabilités*. *Mémoires de l'Académie royale des sciences*. 1784.

Condorcet, Jean-Antoine-Nicolas de Caritat. *Tableau général de la science, qui a pour objet l'application du calcul aux sciences politiques et morales*. Edited by A. Condorcet O'Connor and F. Arago. Paris, 1847.

Copleston, Frederick Charles. *A History of Philosophy*. Vol. 9, *Modern Philosophy, from the French Revolution to Sartre, Camus, and Lévi-Strauss*. Garden City: Doubleday Image Books, 1974.

Copleston, Frederick Charles. *Philosophy in Russia: From Herzen to Lenin to Berdyaev*. Notre Dame, IN: University of Notre Dame Press, 1986.

Corra, Émile. *Pierre Laffitte: Successeur d'Auguste Comte*. Paris: Revue Positiviste Internationale, 1923.

Costa, João Cruz. *A History of Ideas in Brazil: The Development of Philosophy in Brazil and the Evolution of National History*. Translated by Suzette Macedo. Berkeley: University of California Press, 1964.

Coubertin, Pierre de. *La Chronique de France*. Vol. 3. Paris: A. Lanier, 1902.

Cournot, Antoine Augustin. *Considérations sur la marche des idées et des événements dans les temps modernes*. 2 vols. Paris, 1872.

Cousin, Victor. *Cours de l'Histoire de la Philosophie Moderne. Histoire de la Philosophie au XVIIIe siècle. École de Kant*. New edition, first series, vol. 5. Paris: 1846.

Cousin, Victor. *Du Vrai, de beau et du bien*. Paris, 1860.

Croly, Jane Cunningham. *Memories of Jane Cunningham Croly*. New York: G. P. Putnam's Sons, 1904.

Crosland, Maurice. *Science under Control: The French Academy of Sciences 1795–1914*. Cambridge, UK: Cambridge University Press, 2002.

Cuspinera, Margarita Vera. "Positivism." In *Encyclopedia of Mexico: History, Society and Culture*, edited by M. S. Werner, vol. 2, 1178–80. Chicago: Fitzroy Dearborn, 1997.

Cuvier, Georges. *The Animal Kingdom Arranged in Conformity with Its Organization*. Translated by H. McMurtrie. New York, 1831.

Cuvier, Georges. *Le règne animale distribué d'après son organisation*. Paris, 1817.

Dale, Peter Allan. *In Pursuit of a Scientific Culture: Science, Art, and Society in the Victorian Age*. Madison: University of Wisconsin Press, 1989.

Bibliography

D'Alembert, Jean Le Rond. *Traité de dynamique*. Paris, 1743.

Darwin, Charles. *On the Origin of Species*. London, 1859.

Dasgupta, Tapati. *Social Thought of Rabindranath Tagore: A Historical Analysis*. New Delhi: Abhinav, 1993.

Daston, Lorraine. "History of Science." In *International Encyclopedia of the Social and Behavioral Sciences*, edited by Neil Smelser and Paul Bates, vol. 10, 6842–48. New York: Elsevier, 2001.

Daston, Lorraine, and Peter Galison. *Objectivity*. New York: Zone Books, 2007.

De Boni, Claudio. *Descrivere il futuro, Scienza e utopia in Francia nell'età del positivismo*. Florence: Firenze University Press, 2003.

De Boni, Claudio. *Storia di un'utopia, la religione dell'Umanità di Comte e la sua circulazione nel mondo*. Milan: Mimesis, 2013.

Delfau, Gérard. "Le Positivisme, l'histoire de la critique et nous." *Romantisme* 21–22 (1978): 233–38.

De Paz, María. "The Third Way Epistemology: A Recharacterization of Poincaré's Conventionalism." In *Poincaré, Philosopher of Science: Problems and Perspectives*, edited by María de Paz and Robert DiSalle, 47–66. Dordrecht: Springer, 2014.

Derfler, Leslie. *Paul Lafargue and the Founding of French Marxism, 1842–1882*. Cambridge, MA: Harvard University Press, 1991.

Deroisin, Hippolyte Philémon. *Notes sur Auguste Comte par un de ses disciples*. Paris: G. Crès, 1909.

Diacon, Todd A. *Stringing Together a Nation: Cândido Mariano da Silva Rondon and the Construction of a Modern Brazil, 1906–1930*. Durham: Duke University Press, 2004.

DiVanna, Isabel Noronha. *Writing History in the Third Republic*. Newcastle upon Tyne: Cambridge Scholars, 2010.

Dixon, Thomas. *The Invention of Altruism: Making Moral Meanings in Victorian Britain*. Oxford: Oxford University Press, 2008.

Donzelli, Maria. "Comte, L'Italie et la France entre le XIXe et le XXe siècles." In Petit, *Auguste Comte: Trajectoires positivistes*, 351–62.

Donzelli, Maria. *Origini e declino del positivismo: Saggio su Auguste Comte in Italia*. Napoli: Liguori, 1999.

Duhem, Pierre. "Fragments d'un cours d'optique." *Annales de la Société scientifique de Bruxelles* 18, 2nd part (1894): 95–107 (1st fragment).

Duhem, Pierre. "Fragments d'un cours d'optique." *Annales de la Société scientifique de Bruxelles* 19, 2nd part (1895): 27–94 (2nd fragment).

Duhem, Pierre. "Fragments d'un cours d'optique." *Annales de la Société scientifique de Bruxelles* 20, 2nd part (1896): 27–105 (3rd fragment).

Duncan, David. *Life and Letters of Herbert Spencer*. 2 vols. New York: D. Appleton, 1908.

Dunning, Eric. "Figurational Sociology and the Sociology of Sport." *Blackwell Encyclopedia of Sociology*, edited by George Ritzer. Oxford: Blackwell, 2007. doi:10.1111/b.9781405124331.2007.x.

Durkheim, Émile. *Les Formes élémentaires de la vie religieuse*. Paris: Alcan, 1912.

Durkheim, Émile. *Le socialisme*. Edited by Marcel Mauss. Paris: Alcan, 1928.

Durkheim, Émile. *The Rules of Sociological Method*. Edited by Steven Lukes. Translated by W. D. Halls. London: Macmillan, 1982.

Dussel, Enrique. "Philosophy in Latin America in the Twentieth Century: Problems and Currents." In *Latin American Philosophy: Currents, Issues, Debates*, edited by Eduardo Mendieta, 11–53. Bloomington: Indiana University Press, 2003.

Dym, Jordana. "Central America." In *Nations and Nationalism: A Global Historical Overview*, edited by Guntram Herb and David H. Kaplan, vol. 1, 309–22. Santa Barbara: ABC-CLIO, 2008.

Eastwood, Jonathan. "Positivism and Nationalism in 19th Century France and Mexico." *Journal of Historical Sociology* 17, no. 4 (2004): 331–57.

Estes, Angela N. "Ferri, Enrico: Positivist School." In *Encyclopedia of Criminological Theory*, edited by Francis T. Cullen and Pamela Wilcox, vol. 1, 335–37. London: Sage, 2010.

Fedi, Laurent, ed. *Auguste Comte: Cours sur l'histoire de l'humanité (1849–1851), Manuscrit de César Lefort*. Geneva: Droz, 2017.

Fedi, Laurent. "Avant-propos. De Paris à Vienne. Quelques jalons." In *La réception germanique d'Auguste Comte*, edited by Laurent Fedi, 9–36. *Les Cahiers philosophiques de Strasbourg* 35. Strasbourg: Presses Universitaires de Strasbourg, 2014.

Fedi, Laurent. *Comte*. Paris: Les Belles Lettres, 2000.

Fehrenbach, T. R. *Fire and Blood: A History of Mexico*. New York: Da Capo, 1995.

Femia, Joseph V. *Pareto and Political Theory*. New York: Routledge, 2006.

Ferrarotti, Franco. *An Invitation to Classical Sociology: Meditations on Some Great Social Thinkers*. New York: Lexington, 2003.

Flammarion, Camille. *Astronomie populaire*. Paris: Flammarion, 1880.

Flint, Robert. *The Philosophy of History in France and Germany*. Edinburgh, 1874.

Flora, Giuseppe. *The Evolution of Positivism in Bengal: Jogendra Chandra Ghosh, Bankimchandra Chattopadhyay, Benoy Kumar Sarkar*. Napoli: Istituto Universitario Orientale, 1993.

Foley, Susan K., and Charles Sowerine. *A Political Romance: Léon Gambetta, Léonie Léon and the Making of the French Republic, 1872–82*. New York: Palgrave Macmillan, 2012.

Fontenelle, Bernard Le Bouvier de. *Entretiens sur la pluralité des mondes*. 1686. Reprint, Paris: Flammarion, 1998.

Forbes, Geraldine Hancock. *Positivism in Bengal: A Case Study in the Transmission and Assimilation of an Ideology*. Calcutta: Minerva, 1975.

Foucault, Michel. *The Archeology of Knowledge*. Translated by A. M. Sheridan Smith. New York: Pantheon, 1971.

Foucault, Michel. *L'archéologie du savoir*. Paris: Gallimard, 1969.

Fox, Robert. *The Savant and the State: Science and Cultural Politics in Nineteenth-Century France*. Baltimore, MD: Johns Hopkins University Press, 2012.

France, Anatole. Notes on "Auguste Comte et le Positivisme." N.a.fr. 15395, Bibliothèque Nationale, Paris, France.

Frank, Joseph. *Dostoevsky: The Seeds of Revolt, 1821–1849*. Vol. 10. Princeton, NJ: Princeton University Press, 1979.

French, A. P. "The Strange Case of Emil Rupp." *Physics in Perspective* 1 (1999), 3-21.

Fuchs, Eckhardt. "Positivism and History in the XIXth Century." In *Positivismes: Philosophie, Sociologie, Histoire, Sciences*, edited by Andrée Despy-Meyer and Didier Devriese, 147–62. Turnhout: Brepols, 1999.

Fuller, Steve. "Positivism, History of." In *International Encyclopedia of the Social and Behavioral Sciences*, edited by Neil Smelser and Paul Bates, vol. 17, 11821–27. New York: Elsevier, 2001.

Gall, Franz Joseph. *Sur les fonctions du cerveau et sur celles de chacune de ses parties*. 6 vols. Paris, 1822–1825.

Gall, Franz Joseph, and Johann Gaspar Spurzheim. *Anatomie et physiologie du système nerveux en général et du cerveau en particulier*. 4 vols. Paris, 1810–1819.

Gane, Mike. *Auguste Comte*. London: Routledge, 2006.

Gane, Mike. "Dans le gouffre: Entre science et religion, les premiers sociologues français, de Bazard à Littré." In Bourdeau, Braunstein, and Petit, *Auguste Comte aujourd'hui*, 151–69.

Gane, Mike. "Durkheim contre Comte dans *Les Règles*." In *Durkheim d'un siècle à l'autre: lectures actuelles des "Règles de la méthode sociologique,"* edited by Charles-Henry Cuin, 32–38. Paris: Presses Universitaires de France, 1997.

Gane, Mike. *Harmless Lovers? Gender, Theory, and Personal Relationships.* London: Routledge, 1993.

Garin, Eugenio. *History of Italian Philosophy.* Vol. 1. Translated by Giorgio Pinton. Amsterdam: Rodopi, 2008.

Gawrych, George. *The Crescent and the Eagle: Ottoman Rule, Islam and the Albanians, 1874–1913.* London: I. B. Tauris, 2006.

Gérard, Alice. "1852–1902: Auguste Comte au purgatoire." In *Auguste Comte: Qui êtes-vous?* edited by Gérard de Ficquelmont, 143–79. Lyon: La Manufacture, 1988.

Gérard, Alice. "Les Disciples 'complets' de Comte et la Politique Positive (1870–1914)." In Petit, *Auguste Comte: Trajectoires positivistes*, 285–302.

Gilson, E. *Les métamorphoses de la cité de Dieu.* Paris: Vrin, 1952.

Gilson, Gregory D., and Irving W. Levinson. Introduction to *Latin American Positivism: New Historical and Philosophical Essays*, edited by Gregory Gilson and Irving Levinson, vii–xi. New York: Lexington Books, 2013.

Goldstein, Jan. *Console and Classify: The French Psychiatric Profession in the Nineteenth Century.* Cambridge, UK: Cambridge University Press, 1997.

Gollo, Rodney Rhodes. "The Birth of a New Political Philosophy: Religion and Positivism in Nineteenth-Century Brazil." In Gilson and Levinson, *Latin American Positivism*, 153–69.

Gomes, Mércio Pereira. *The Indians and Brazil.* Translated by John W. Moon. Gainesville: University Press of Florida, 2002.

Gordon, Felicia, and Máire Cross. *Early French Feminisms, 1830–1940: A Passion for Liberty.* Cheltenham, UK: Edward Elgar, 1996.

Gouhier, Henri. *La Jeunesse d'Auguste Comte et la formation du positivisme.* 2nd ed. 3 vols. Paris: Vrin, 1964–1970.

Gouhier, Henri. *La Vie d'Auguste Comte.* Paris: Gallimard, 1931.

Grange, Juliette. *Auguste Comte: la politique et la science.* Paris: Odile Jacob, 2000.

Grange, Juliette, Pierre Musso, Philippe Régnier, and Franck Yonnet. Notice to *Catéchisme des industriels* in *Henri Saint-Simon: Œuvres complètes*, edited by Juliette Grange, Pierre Musso, Philippe Régnier, and Franck Yonnet, 4:2865–69. Paris: Presses Universitaires de France, 2012.

Grogan, Susan. *French Socialism and Sexual Difference: Women and the New Society, 1803–44.* New York: Saint Martin's Press, 1992.

Grondeux, Jérôme. *La France entre en République (1870–1893).* Paris: Libraire Générale Française, 2000.

Grondeux, Jérôme. "Taine et Comte face à l'histoire." In Petit, *Auguste Comte: Trajectoires positivistes*, 177–88.

Guilhaumou, Jacques. "Sieyès et le non-dit de la sociologie: du mot à la chose." *Revue d'histoire des sciences humaines* 15, no. 2 (2006): 117–34.

Guillin, Vincent. *Auguste Comte and John Stuart Mill on Sexual Equality: Historical, Methodological and Philosophical Issues*. Boston: Brill, 2009.

Guillin, Vincent. "Le penchant biologique de la sociologie comtienne: la question de l'égalité des sexes." *Revue d'histoire des sciences* 65, no. 2 (2012): 259–86.

Guillo, Dominique. *Les figures de l'organisation: Sciences de la vie et sciences sociales au XIXe siècle*. Paris: Presses Universitaires de France, 2003.

Gutting, Gary. Introduction, "*What Is Continental Philosophy of Science?*" In *Continental Philosophy of Science*, 1–15. Oxford: Blackwell, 2005.

Hacking, Ian. *Historical Ontology*. Cambridge, MA: Harvard University Press, 2002.

Hacking, Ian. *The Social Construction of What?* Cambridge, MA: Harvard University Press, 1999.

Hacking, Ian. "'Style' for Historians and Philosophers." *Studies in History and Philosophy of Science* 23, no. 1 (1992): 1–20. Republished in Hacking, *Historical Ontology*, 178–99.

Hacking, Ian. *The Taming of Chance*. Cambridge, UK: Cambridge University Press, 1990.

Halbfass, Wilhelm. *India and Europe: An Essay in Understanding*. Albany: State University of New York Press, 1988.

Hale, Charles A. "Political and Social Ideas in Latin America, 1870–1930." In *The Cambridge History of Latin America*, edited by Leslie Bethell, vol. 4, 367–442. Cambridge, UK: Cambridge University Press, 1986.

Hale, Charles A. *The Transformation of Liberalism in Late Nineteenth-Century Mexico*. Princeton, NJ: Princeton University Press, 1989.

Hanioğlu, M. Şükrü. *The Young Turks in Opposition*. Oxford: Oxford University Press, 1995.

Harp, Gillis. *Positivist Republic: Auguste Comte and the Reconstruction of American Liberalism, 1865–1920*. University Park: Pennsylvania State University Press, 1994.

Harvey, Joy. "*Almost a Man of Genius*": Clémence Royer, Feminism, and Nineteenth-Century Science. New Brunswick, NJ: Rutgers University Press, 1997.

Harvey, Joy. "Evolutionism Transformed: Positivists and Materialists in the Société d'Anthropologie de Paris from Second Empire to Third Republic." In *The Wider Domain of Evolutionary Thought*, edited by D. R. Oldroyd and Ian Langham, 289–310. Dordrecht: D. Reidel, 1983.

Havens, Thomas R. H. "Comte, Mill, and the Thought of Nishi Amane in Meiji Japan." *Journal of Asian Studies* 27, no. 2 (1968): 217–28.

Hawkins, Richmond Laurin. *Auguste Comte and the United States, 1816–1853*. Cambridge, MA: Harvard University Press, 1936.

Hawkins, Richmond Laurin. *Positivism in the United States (1853–1861)*. Cambridge, MA: Harvard University Press, 1938.

Hayek, Friedrich August von. *Studies on the Abuse and Decline of Reason*. Vol. 13 of *The Collected Works of F. A. Hayek*. Edited by Bruce Caldwell. Chicago: University of Chicago Press, 2010.

Hayek, Friedrich August von. *The Counter-Revolution of Science: Studies on the Abuse of Reason*. Glencoe, IL: Free Press, 1952.

Hayek, Friedrich August von. *The Road to Serfdom*. London: Routledge, 2001.

Heidegger, Martin. *The Question Concerning Technology, and Other Essays*. Translated by William Lovitt. New York: Harper & Row, 1977.

Heilbron, Johan. "Ce que Durkheim doit à Comte." In *Division du travail et lien social: La thèse de Durkheim un siècle après*, edited by Philippe Besnard, Massimo Borlandi, and Paul Vogt, 59–66. Paris: Presses Universitaires de France, 1993.

Herschel, John. *A Preliminary Discourse on the Study of Natural Philosophy*. London, 1830. Reprint, Chicago: University of Chicago Press, 1987.

Hilton, Ronald. "Positivism in Latin America." In *Dictionary of the History of Ideas*, edited by Philip P. Wiener, vol. 3, 540–45. New York: Charles Scribner's Sons, 1973.

Hoeg, Jerry. "Rebellion in the Badlands (Os Sertões): The Darwinian Landscape." In *Reading and Writing the Latin American Landscape*, edited by Beatriz Rivera-Barnes and Jerry Hoeg, 83–100. New York: Palgrave Macmillan, 2009.

Houellebecq, Michel. "Préliminaires au positivisme." In Bourdeau, Braunstein, and Petit, *Auguste Comte aujourd'hui*, 7–12.

Howard, Don. "Two Left Turns Make a Right: On the Curious Political Career of North American Philosophy of Science at Midcentury." In *Logical Empiricism in North America*, edited by Gary L. Hardcastle and Alan W. Richardson, 25–93. Minneapolis: University of Minnesota Press, 2003.

Howell, Ronald F. "The Philosopher Alain and French Classical Radicalism." *Western Political Quarterly* 18, no. 3 (1965): 594–614.

Hughes, H. Stuart. *Consciousness and Society: The Reorientation of European Social Thought, 1890–1930*. New York: Random House/Vintage, 1958.

Hume, David. *Enquiries Concerning the Human Understanding and Concerning the Principles of Morals*. 1748. Reprint, Oxford: Clarendon Press, 1966.

Huneman, Philippe. *Bichat, la vie et la mort*. Paris: Presses Universitaires de France, 1998.

Hurd, Madeleine. *Public Spheres, Public Mores, and Democracy: Hamburg and Stockholm, 1870–1914*. Ann Arbor: University of Michigan Press, 2000.

Huxley, Thomas Henry. *On the Physical Basis of Life*. New Haven, CT, 1869.

Huxley, Thomas Henry. "The Scientific Aspects of Positivism." In *Lay Sermons*, 162–91. London, 1870.

Huygens, Christiaan. *Discours de la cause de la pesanteur*. Leiden, 1690. Reprint, Paris: Dunod, 1992.

Huygens, Christiaan. *Horologium oscillatorium sive de motu pendulorum ad horologia aptato demonstrationes geometricae*. 1673. Vol. 18 of the *Œuvres complètes de Christiaan Huygens*. The Hague: M. Nijhoff, 1934.

Jaksić, Iván. *Academic Rebels in Chile: The Role of Philosophy in Higher Education and Politics*. Albany: State University of New York Press, 1989.

Jones, H. S. Introduction to *Early Political Writings*, by Auguste Comte, vii–xxviii. Translated and edited by H. S. Jones. Cambridge, UK: Cambridge University Press, 1998.

Jones, H. S. Notes on text and translation to *Early Political Writings*, by Auguste Comte, xxix–xxxii. Translated and edited by H. S. Jones. Cambridge, UK: Cambridge University Press, 1998.

Jori, Mario. "Legal Positivism." In *Routledge Encyclopedia of Philosophy*, edited by E. Craig, vol. 5, 514–21. London: Routledge, 1998.

Jouguet, Jacques Charles Émile. *Lectures de mécanique*. Paris: Gauthier-Villars, 1924.

Karsenti, Bruno. *Politique de l'esprit. Auguste Comte et la naissance de la science sociale*. Paris: Hermann, 2006.

Kent, Christopher. *Brains and Numbers: Elitism, Comtism, and Democracy in Mid-Victorian England*. Toronto: University of Toronto Press, 1978.

Kincaid, Harold. "Positivism in the Social Sciences." In *Routledge Encyclopedia of Philosophy*, edited by E. Craig, vol. 7, 558–61. London: Routledge, 1998.

Kitcher, Philip. *Science in a Democratic Society*. Amherst, NY: Prometheus Books, 2011.

Kitcher, Philip. *Science, Truth, and Democracy*. Oxford, UK: Oxford University Press, 2001.

Kofman, Sarah. *Aberrations: Le devenir-femme d'Auguste Comte*. Paris: Flammarion, 1978.

Kragh, Helge. *An Introduction to the Historiography of Science*. Cambridge, UK: Cambridge University Press, 1987.

Kuhn, Thomas S. *The Structure of Scientific Revolutions*. 4th ed. Chicago: University of Chicago Press, 2012.

Kury, Lorelai. "Nation, races et fétichisme: La religion de l'humanité au Brésil." *Revue d'histoire des sciences humaines* 8, no. 1 (2003): 125–37.

Kremer-Marietti, Angèle. "Comte, Isidore-Auguste-Marie-François-Xavier." In *Routledge Encyclopedia of Philosophy*, edited by E. Craig, vol. 2, 497–504. London: Routledge, 1998.

Kremer-Marietti, Angèle. "De l'unité de la science à la science unifiée." In Petit, *Auguste Comte: Trajectoires positivistes*, 189–204.

Kselman, Thomas. *Death and the Afterlife in Modern France*. Princeton, NJ: Princeton University Press, 1993.

Lagrange, Joseph-Louis. *Œuvres de Lagrange*. Edited by J.-A. Serret. Paris, 1867–92.

Lamarck, Jean-Baptiste. *Philosophie zoologique*. Paris, 1809.

Lamarck, Jean-Baptiste. *Recherches sur l'organisation des corps vivants*. Paris, 1802.

Lantéri-Laura, Georges. *Histoire de la phrénologie. L'homme et son cerveau selon F. J. Gall*. Paris: Presses Universitaires de France, 1970.

Laplace, Pierre Simon, Marquis de. *Essai philosophique sur les probabilités*. Paris, 1814.

Laplace, Pierre Simon, Marquis de. *Exposition du système du monde*. Paris, 1808.

Laudan, Larry. "Towards a Reassessment of Comte's 'Méthode Positive.'" In *Science and Hypothesis: Historical Essays on Scientific Methodology*, 141–62. Dordrecht: D. Reidel, 1981.

Laudan, Rachel. "Classification of the Sciences." In *The Oxford Companion to the History of Modern Science*, edited by J. L. Heilbron, 154–55. Oxford: Oxford University Press, 2003.

Laudan, Rachel. "Philosophy and Science." In *The Oxford Companion to the History of Modern Science*, edited by J. L. Heilbron, 632–34. Oxford: Oxford University Press, 2003.

Laudan, Rachel. "Positivism and Scientism." In *The Oxford Companion to the History of Modern Science*, edited by J. L. Heilbron, 670–71. Oxford: Oxford University Press, 2003.

Lazinier, Emmanuel, and Sybil de Acevedo. "Quelques disciples et sympathisants célèbres." In *Auguste Comte: Qui êtes-vous?*, edited by Gérard de Ficquelmont, 327–50. Lyon: La Manufacture, 1988.
Le Bras-Chopard, Armelle. "L'effervescence des idées socialistes au début du XIXe siècle." In Petit, *Auguste Comte: Trajectoires positivistes*, 53–70.
Le Bras-Chopard, Armelle. "Une statue de marbre, l'idéal féminin d'Auguste Comte: Convergences et dissonances avec ses contemporains socialistes." In Bourdeau, Braunstein, and Petit, *Auguste Comte aujourd'hui*, 170–83.
LeGouis, Catherine. *Positivism and Imagination: Scientism and Its Limits in Emile Hennequin, Wilhelm Scherer, and Dmitrii Pisarev*. Lewisburg, PA: Bucknell University Press, 1997.
Legrand, Louis. *L'influence du positivisme dans l'œuvre scolaire de Jules Ferry: Les origines de la laïcité*. Paris: Marcel Rivière, 1961.
Leibniz, Gottfried Wilhelm. *Die philosophischen Schriften von Leibniz*. Edited by Christian. I. Gerhardt. Berlin, 1875–1890.
Leibniz, Gottfried Wilhelm. *Discourse on Metaphysics, Correspondence with Arnauld, and Monadology*. Translated by George R. Montgomery. Chicago: Open Court, 1908.
Leibniz, Gottfried Wilhelm. *New Essays on Human Understanding*. Translated and edited by Peter Remnant and Jonathan Bennett. Cambridge, UK: Cambridge University Press, 1996.
Leibniz, Gottfried Wilhelm. *Sämtliche Schriften und Briefe, Sechster Band: Nouveaux Essais*. Berlin: Akademie, 1962.
Lenzer, Gertrude. Introduction to *Auguste Comte and Positivism: The Essential Writings*, edited by Gertrude Lenzer, xvii–lxviii. New York: Harper & Row, 1975.
Lepenies, Wolf. *Between Literature and Science: The Rise of Sociology*. Translated by R. J. Hollingdale. Cambridge, UK: Cambridge University Press, 1988.
Le Roy, Édouard. "La science positive et les philosophies de la liberté." In *Congrès international de philosophie*, 313–41. Paris: Armand Colin, 1900.
Leterrier, Sophie-Anne. *L'Institution des sciences morales. L'Académie des sciences morales et politiques, 1795–1850*. Paris: L'Harmattan, 1995.
Le Van-Lemesle, Lucette. *Le Juste ou le Riche. L'enseignement de l'économie politique, 1815–1950*. Paris: Comité pour l'histoire économique et financière de la France, 2004.
Lévi-Strauss, Claude. "La leçon de sagesse des vaches folles." *Études rurales* 157–58 (2001): 9–14. Republished in *Nous sommes tous des cannibales*. Paris: Le Seuil, 2013.

Lévi-Strauss, Claude. *Les Structures élémentaires de la parenté.* Paris: Presses Universitaires de France, 1949.

Levine, Donald. *Visions of the Sociological Tradition.* Chicago: University of Chicago Press, 1995.

Lévy-Bruhl, Lucien. *The Philosophy of Auguste Comte.* Authorized translation. New York: G. P. Putnam's Sons, 1903.

Lewisohn, David. "Mill and Comte on the Methods of Social Science." *Journal of the History of Ideas* 33, no. 2 (1972): 315–24.

Leymarie, Michel. *De la Belle Epoque à la Grande Guerre: Le triomphe de la République. (1893–1918).* Paris: Librairie Générale Française, 1999.

Lindenfeld, David F. *The Transformation of Positivism: Alexius Meinong and European Thought, 1880–1920.* Berkeley: University of California Press, 1980.

Lipp, Solomon. *Leopoldo Zea: From Mexicanidad to a Philosophy of History.* Waterloo, Ontario: Wilfrid Laurier University Press, 1980.

Lipp, Solomon. *Three Chilean Thinkers.* Waterloo, Ontario: Wilfrid Laurier University Press, 1975.

Littré, Émile. *Auguste Comte et la philosophie positive.* Paris, 1863.

Littré, Émile. *Auguste Comte et la philosophie positive.* 2nd. ed. Paris, 1864.

Littré, Émile. *Conservation, révolution et positivisme.* Paris, 1852.

Littré, Émile. *Principes de philosophie positive.* Paris, 1868.

Locke, John. *An Essay Concerning Human Understanding.* 1690. London: Dent, 1961.

Long, Pamela O. *Artisan/Practitioners and the Rise of the New Sciences, 1400–1600.* Corvallis: Oregon State University Press, 2011.

Longino, Helen. *Science as Social Knowledge: Values and Objectivity in Scientific Inquiry.* Princeton, NJ: Princeton University Press, 1990.

Longino, Helen. *The Fate of Knowledge.* Princeton, NJ: Princeton University Press, 2002.

López Alves, Fernando. "The Authoritarian Roots of Liberalism: Uruguay, 1810–1886." In *Liberals, Politics, and Power: State Formation in Nineteenth-Century Latin America*, edited by Vicente C. Peloso and Barbara A. Tenenbaum, 111–34. Athens: University of Georgia Press, 1996.

Macherey, Pierre. "Comte dans la querelle des anciens et des modernes. La critique de la perfectibilité." In *L'Homme perfectible*, edited by Bertrand Binoche, 274–92. Seyssel: Champ Vallon, 2004.

Magnin, Fabien. "Le vingt-et-unième anniversaire de la mort d'Auguste Comte." *La Revue Occidentale philosophique, sociale et politique* 1 (1878): 655–67.

Maine de Biran, François-Pierre-Gonthier. *Essai sur les fondements de la psychologie et sur ses rapports avec l'étude de la nature.* 1810–12. Vol. 8 of the *Œuvres complètes de Maine de Biran.* Paris: Alcan, 1932.

Maine de Biran, François-Pierre-Gonthier. *Examen des leçons de Laromiguière.* 1817. Vol. 11 of the *Œuvres complètes de Maine de Biran.* Paris: Alcan, 1939.

Manicas, Peter T. *A History and Philosophy of the Social Sciences.* Oxford: Basil Blackwell, 1987.

Manuel, Frank. *Prophets of Paris.* Cambridge, MA: Harvard University Press, 1962.

Marcuse, Herbert. *Reason and Revolution: Hegel and the Rise of Social Theory.* Boston: Beacon, 1968.

Maréchal, Henry. *Les Conceptions économiques d'A. Comte.* Bar-sur-Seine: Saillard, 1919.

Mariotte, Edme. *Traité de la percussion ou chocq des corps.* Paris, 1673.

Martí, Oscar R. "Positivist Thought in Latin America." In *Routledge Encyclopedia of Philosophy*, edited by E. Craig, vol. 7, 565–70. London: Routledge, 1998.

Mauduit, Roger. *Auguste Comte et la science économique.* Paris: Alcan, 1929.

Marx, Karl. *The Letters of Karl Marx.* Translated and edited by S. K. Padover. Englewood Cliffs, NJ: Prentice Hall, 1979.

Mayeur, Françoise. "Le positivisme et l'école républicaine." *Romantisme* 8, no. 21 (1978): 137–47.

McCarthy, Jeremiah. "Pragmatism, Abduction, and Weak Verification." In *Charles S. Peirce and the Philosophy of Science*, edited by Edward C. Moore, 175–85. Tuscaloosa: University of Alabama Press, 1993.

McCreery, David J. "Coffee and Class: The Structure of Development in Liberal Guatemala." *Hispanic American Historical Review* 56, no. 3 (1976): 438–60.

McWilliam, Neil. *Dreams of Happiness: Social Art and the French Left, 1830–1850.* Princeton, NJ: Princeton University Press, 1993.

Mendieta, Eduardo. "The Death of Positivism and the Birth of Mexican Phenomenology." In Gilson and Levinson, *Latin American Positivism*, 1–25.

Meyer, Michael C., and William L. Sherman. *The Course of Mexican History.* 4th ed. New York: Oxford University Press, 1991.

Milhaud, Gaston. *Le positivisme et le progrès de l'esprit.* Paris: Alcan, 1902.

Mill, John Stuart. *Auguste Comte and Positivism.* In *Essays on Ethics, Religion, and Society*, vol. 10 of *The Collected Works of John Stuart Mill*, 263–368. Edited by John M. Robson. Toronto: University Press of Toronto, 1985.

Mill, John Stuart. *Autobiography*. Vol. 1 of *The Collected Works of John Stuart Mill*. Edited by John M. Robson and Jack Stillinger. Toronto: University Press of Toronto, 1963.

Mill, John Stuart. *An Examination of Sir William Hamilton's Philosophy*. Vol. 9 of *The Collected Works of John Stuart Mill*. Edited by John M. Robson. Toronto: University Press of Toronto, 1979.

Mill, John Stuart. *On the Subjection of Women*. In *Essays on Equality, Law, and Education*, vol. 21 of *The Collected Works of John Stuart Mill*, edited by John M. Robson, 259–340. Toronto: University Press of Toronto, 1984.

Mill, John Stuart. *A System of Logic*. Vols. 7–8 of *The Collected Works of John Stuart Mill*. Edited by John M. Robson. Toronto: University Press of Toronto, 1974.

Milosz, Czeslaw. *The History of Polish Literature*. London: Collier-Macmillan, 1969.

Morley, John. "Auguste Comte." In *Critical Miscellanies*, vol. 3, 337–84. London, 1886.

Montesquieu, Charles-Louis de Secondat, baron de La Brède et de. *L'Esprit des lois*. Paris: Classiques Garnier, 2011.

Moscovici, Claudia. *Gender and Citizenship: The Dialectics of Subject-Citizenship in Nineteenth-Century French Literature and Culture*. Lanham, MD: Rowman and Littlefield, 2000.

Moses, Claire Goldberg. *French Feminism in the Nineteenth Century*. Albany: State University of New York Press, 1984.

Moses, Claire Goldberg, and Leslie Wahl Rabine. *Feminism, Socialism, and French Romanticism*. Bloomington: Indiana University Press, 1993.

Mustoxidi, T. M. *Histoire de l'esthétique française, 1700–1900*. 1920. Reprint, New York: Burt Franklin, 1968.

Needham, H. A. *Le développement de l'esthétique sociologique en France et en Angleterre au XIXe siècle*. Paris: Librairie ancienne Honoré Champion, 1926.

Nemeth, Thomas. *The Early Solov'ëv and His Quest for Metaphysics*. Dordrecht: Springer, 2013.

Newton, Isaac. *Newton's Principia. The Mathematical Principles of Natural Philosophy*. Translated by A. Motte. New York: Daniel Adee, 1848.

Nicolet, Claude. "Jules Ferry et la tradition positiviste." In *Jules Ferry: Fondateur de la République: Actes du colloque*, edited by François Furet, 23–48. Paris: L'École des Hautes Études en Sciences Sociales, 1985.

Nicolet, Claude. *La fabrique d'une nation: La France entre Rome et les Germains*. Paris: Perrin, 2003.

Nicolet, Claude. *L'idée républicaine en France: Essai d'histoire critique*. Paris: Gallimard, 1982.

Nietzsche, Friedrich. *Daybreak: Thoughts on the Prejudices of Morality*. Cambridge: Cambridge University Press, 1997.

Nisbet, Robert. *The Sociological Tradition*. New York: Basic Books, 1966.

Nordmann, Jean-Thomas. "Taine et le positivisme." *Romantisme* 8, no. 21 (1978): 21–33.

Nuccetelli, Susana. *Latin American Thought: Philosophical Problems and Arguments*. Boulder, CO: Westview Press, 2002.

Nuccetelli, Susana, and Gary Seay, eds. *Latin American Philosophy: An Introduction with Readings*. Upper Saddle River, NJ: Pearson Prentice Hall, 2004.

Nye, Mary Jo. "The Boutroux Circle and Poincaré's Conventionalism." *Journal of the History of Ideas* 40, no. 1 (1979): 107–20.

Offer, John. *Herbert Spencer and Social Theory*. New York: Palgrave Macmillan, 2010.

Okyar, Osman. "Atatürk's Quest for Modernism." In *Atatürk and the Modernization of Turkey*, edited by Jacob M. Landau, 45–53. Leiden: E. J. Brill, 1984.

Patuelli, Francesca. "Garofalo, Raffaele: Positivist School." In *Encyclopedia of Criminological Theory*, edited by Francis T. Cullen and Pamela Wilcox, vol. 1, 352–54. London: Sage, 2010.

Parsons, Talcott. "The General Interpretation of Action." In *Theories of Society: Foundation of Modern Sociological Theory*, edited by Talcott Parsons, Edward Shils, Kaspar Naegele, and Jesse Pitts, vol. 1, 85–97. Glencoe, IL: Free Press, 1961.

Parsons, Talcott. *The Structure of Social Action*. Vol. 1. New York: Free Press, 1968.

Pedersen, Jean Elisabeth. "Sexual Politics in Comte and Durkheim: Feminism, History, and the French Sociological Tradition." *SIGNS: A Journal of Women in Culture and Society* 27, no. 1 (2001): 229–63.

Peirce, Charles Sanders. *Reasoning and the Logic of Things: The Cambridge Conferences Lectures of 1898*. Edited by K. L. Ketner. Cambridge, MA: Harvard University Press, 1992.

Peirce, Charles Sanders. *The Essential Peirce: Selected Philosophical Writings*. Vol. 2, *(1893–1913)*. Edited by the Peirce Edition Project. Bloomington: Indiana University Press, 1998.

Perrot, Michelle. "Note sur le positivisme ouvrier." *Romantisme* 8, no. 21 (1978): 201–4.

Peterson, David J. *Revoking the Moral Order: The Ideology of Positivism and the Vienna Circle*. New York: Lexington Books, 1999.

Petit, Annie, ed. *Auguste Comte: Trajectoires positivistes: 1798–1998*. Paris: L'Harmattan, 2003.

Petit, Annie. "Claude Bernard and the History of Science." *Isis* 78 (1987): 201–19.

Petit, Annie. "Comte revu et corrigé: Le cas Littré." *Revue européenne des sciences sociales* 54, no. 2 (2016): 69–88.

Petit, Annie. "D'Auguste Comte à Claude Bernard, un positivisme déplacé." *Romantisme* 8, no. 21 (1978): 45–62.

Petit, Annie. "De Comte à Durkheim: Un héritage ambivalent." In *La Sociologie et sa méthode. Les "Règles" de Durkheim un siècle après*, edited by Massimo Borlandi and Laurent Mucchielli, 42–70. Paris: L'Harmattan, 1995.

Petit, Annie. "Des sciences positives à la politique positive." In Petit, *Auguste Comte: Trajectoires positivistes*, 87–116.

Petit, Annie. "Le prétendu positivisme d'Ernest Renan." *Revue d'Histoire des Sciences Humaines* 8, no. 1 (2000): 73–101.

Petit, Annie. "Le romantisme social d'Auguste Comte." In *Romantismes et socialismes en Europe (1800–1848)*, edited by A. Billaz, 171–206. Paris: Didier Érudition, 1988.

Petit, Annie. *Le système d'Auguste Comte: De la science à la religion par la philosophie*. Paris: Vrin, 2016.

Petit, Annie. "Marcelin Berthelot et le positivisme." In *Marcelin Berthelot (1827–1907): Sciences et politique*, edited by Jean Balcou, 81–103. Rennes: Presses Universitaires de Rennes, 2010.

Petit, Annie. Notes to *Discours sur l'ensemble du positivisme*, by Auguste Comte, 427–62. Paris: G. F. Flammarion, 1998.

Petit, Annie. Présentation to *Discours sur l'ensemble du positivisme*, by Auguste Comte, 9–34. Paris: G. F. Flammarion, 1998.

Petit, Annie, and Bernadette Bensaude. "Le féminisme militant d'un auguste phallocrate." *Revue philosophique de la France et de l'étranger* 166, no. 3 (1976): 293–31.

Petit, Annie, and Michel Bourdeau. "Pierre Laffitte, un disciple très discipliné." *Revue internationale d'histoire et de philosophie des sciences et des techniques* 2nd ser., 8, no. 2 (2004): 7–20.

Pickering, Mary. "Angels and Demons in the Moral Vision of Auguste Comte." *Journal of Women's History* 8, no. 2 (1996): 10–40.

Pickering, Mary. *Auguste Comte: An Intellectual Biography.* 3 vols. Cambridge, UK: Cambridge University Press, 1993–2009.
Pickering, Mary. "Auguste Comte and the Curious Case of English Women." In *The Anthem Companion to Auguste Comte,* edited by Andrew Wernick, 175–204. London: Anthem, 2017.
Pickering, Mary. "Auguste Comte and the Saint-Simonians." *French Historical Studies* 18, no. 1 (1993): 211–36.
Pickering, Mary. "A New Look at Auguste Comte." In *Reclaiming the Sociological Classics: The State of the Scholarship,* edited by Charles Camic, 11–44. Oxford: Blackwell, 1997.
Picon, Antoine. "A propos du rôle social de la science: Auguste Comte et les Saint-Simoniens." In Petit, *Auguste Comte: Trajectoires positivistes,* 241–52.
Picon, Antoine. *Les Saint-Simoniens: raison, imaginaire, et utopie.* Paris: Belin, 2002.
Pilbeam, Pamela. *French Socialists before Marx.* Montreal: McGill-Queen's University Press, 2000.
Plé, Bernard. *Die "Welt" aus den Wissenschaften: Der Positivismus in Frankreich, England und Italien von 1848 bis ins zweite Jahrzehnt des 20. Jahrhunderts: Eine wissensoziologische Studie.* Stuttgart: Klett-Cotta, 1994.
Poincaré, Henri. *The Foundations of Science.* Translated by G. B. Halsted. New York: The Science Press, 1913.
Poincaré, Henri. *La valeur de la science.* 1905. Reprint, Paris: Flammarion, 1970.
Popper, Karl R. *The Open Society and Its Enemies.* 2 vols. London: Routledge & Kegan Paul, 1945.
Popper, Karl R. *The Poverty of Historicism.* 1957. Reprint, New York: Harper Torchbooks, 1964.
Pozzi, Regina. "Comte devant son siècle." In Bourdeau, Braunstein, and Petit, *Auguste Comte aujourd'hui,* 135–50.
Pyenson, Lewis. *The Passion of George Sarton: A Modern Marriage and Its Discipline.* Philadelphia: American Philosophical Society, 2007.
Quinn, Susan. *Marie Curie: A Life.* Cambridge, MA: Perseus, 1995.
Quirk, Robert E. *The Mexican Revolution and the Catholic Church, 1910–1929.* Bloomington: Indiana University Press, 1973.
Raat, William Dirk. "The Antipositivist Movement in Prerevolutionary Mexico, 1892–1911." *Journal of Interamerican Studies and World Affairs* 19, no. 1 (1977): 83–98.

Raat, William Dirk. "Augustín Aragón and Mexico's Religion of Humanity." In *Mexico: From Independence to Revolution 1810–1910*, edited by William D. Raat, 241–59. Lincoln: University of Nebraska Press, 1982.

Rabossi, Eduardo. "Latin American Philosophy." In *The Cambridge History of Philosophy, 1870–1945*, edited by Thomas Baldwin, 507–17. Cambridge, UK: Cambridge University Press, 2003.

Ramsaur, Ernest Edmondson. *The Young Turks: Prelude to the Revolution of 1908*. Princeton, NJ: Princeton University Press, 1957.

Reardon, Bernard M. G. *Religion in the Age of Romanticism: Studies in Early Nineteenth-Century Thought*. Cambridge, UK: Cambridge University Press, 1985.

Reisch, George A. *How the Cold War Transformed Philosophy of Science: To the Icy Slopes of Logic*. New York: Cambridge University Press, 2005.

Renouvier, Charles Bernard. "Contradictions positivistes au sujet de la loi des Trois États." *La Critique philosophique* 4, no. 1 (1875): 129–32.

Renouvier, Charles Bernard. "Le *Cours de philosophie positive* est-il au courant de la science?" *La Critique philosophique* 6, no. 1 (1877): 291–99, 327–36; no. 2 (1877): 1–7, 97–106, 113–20.

Renouvier, Charles Bernard. "De la Méthode scientifique." *La Critique philosophique* 4, no. 2 (1876): 401–4.

Renouvier, Charles Bernard. "De la philosophie du XIXe siècle en France." *L'Année philosophique* 1 (1867): 1–108.

Renouvier, Charles Bernard. "La loi des trois états." *La Critique philosophique* 4, no. 1 (1875): 81–86.

Renouvier, Charles Bernard. *Philosophie analytique de l'histoire*. Vol. 4. Paris, 1897.

Renouvier, Charles Bernard. "Les Prétensions de la Science." *La Critique philosophique* 2, no. 1 (1873): 227–35.

Renouvier, Charles Bernard. "Les Positivismes." *La Critique philosophique, politique, scientifique, littéraire* 4, no. 1 (1875): 33–37.

Renouvier, Charles Bernard. "Le sens de la méthode phénoméniste. Les réalités et les postulats." *La Critique philosophique* 13, no. 2 (1884): 161–68.

Renouvier, Charles Bernard. "Y a-t-il une loi du progrès?" *La Critique philosophique* 4, no. 1 (1875): 65–68.

Reynié, D. "L'opinion publique organique, Auguste Comte et la vraie théorie de l'opinion publique." *Archives de philosophie* 70 (2007): 95–114.

Richard, Nathalie. *Hippolyte Taine: Histoire, psychologie, littérature*. Paris: Classiques Garnier, 2013.

Richardson, Sarah S. "The Left Vienna Circle, Part 1. Carnap, Neurath, and the Left Vienna Circle Thesis." *Studies in History and Philosophy of Science* 40, no. 1 (2009): 14–24.

Richardson, Sarah S. "The Left Vienna Circle, Part 2. The Left Vienna Circle, Disciplinary History, and Feminist Philosophy of Science." *Studies in History and Philosophy of Science* 40, no. 2 (2009): 167–74.

Riot-Sarcey, Michèle. *La démocratie à l'épreuve des femmes: trois figures critiques du pouvoir, 1830–1848*. Paris: Albin Michel, 1994.

Ripoll, Roger. "Zola et le modèle positiviste." *Romantisme* 8, no. 21 (1978): 125–35.

Ritzer, George. *Classical Sociological Theory*. 3rd ed. Boston: McGraw-Hill, 2000.

Robin, Charles. "Sur la direction que se sont proposée en se réunissant les membres fondateurs de la Société de Biologie pour répondre au titre qu'ils ont choisi." In *Comptes rendus des séances et mémoires de la Société de Biologie*, 1st year (1849), i–xi. Paris, 1850.

Robinet, Jean-François Eugène. *Notice sur l'oeuvre et la vie d'Auguste Comte*. Paris, 1860.

[Rodrigues, Olinde]. "A Digression on the Work Entitled *The Third Book of the Industrialists' Catechism* by Auguste Comte, a Student of Saint Simon." In *The Doctrine of Saint-Simon: An Exposition, First Year, 1828–1829*, edited by Georg G. Iggers, 231–43. Boston: Beacon Press, 1958.

Romanell, Patrick. *Making of the Mexican Mind: A Study in Recent Mexican Thought*. Freeport, NY: Books for Libraries, 1969.

Rose, R. S. "Brazil's Military Positivists: Another Myth in Need of Explosion?" In Gilson and Levinson, *Latin American Positivism*, 133–52.

Sáenz, Mario. *The Identity of Liberation in Latin American Thought: Latin American Historicism and the Phenomenology of Leopoldo Zea*. New York: Lexington Books, 1999.

Saint-Simon, Henri. "Avant-propos au troisième cahier du *Catéchisme des industriels*." In *Henri Saint-Simon: Œuvres complètes*, edited by Juliette Grange, Pierre Musso, Philippe Régnier, and Franck Yonnet, vol. 4, 2976–77. Paris: Presses Universitaires de France, 2012.

Saint-Simon, Henri. *Œuvres de Claude-Henri de Saint-Simon*. 6 vols. Paris: Anthropos, 1966.

Saisselin, Rémy G. *The Rule of Reason and the Ruses of the Heart: A Philosophical Dictionary of Classical French Criticism, Critics, and Aesthetic Issues*. Cleveland, OH: Press of Case Western Reserve University, 1970.

Sánchez, Gonzalo J. *Pity in Fin-de siècle French Culture: "Liberté, Egalité, Pitié."* Westport, CT: Greenwood, 2004.

Sarti, Roland. *Italy: A Reference Guide from the Renaissance to the Present.* New York: Infobase, 2009.

Sartori, Eric. "Michel Houellebecq, romancier positiviste." In *Michel Houellebecq*, edited by Sabine van Wesemael, 143–51. Amsterdam: Rodopi, 2004.

Savary, Félix. "Sur la détermination des orbites que décrivent autour de leur centre de gravité deux étoiles très rapprochées l'une de l'autre." In *Connaissance des temps ou des mouvements célestes à l'usage des astronomes et des navigateurs, pour l'an 1830*; additions, 56–69, 163–71. Paris, 1827.

Scharff, Robert C. *Comte after Positivism*. Cambridge, UK: Cambridge University Press, 1995.

Scharff, Robert C. "Mill's Misreading of Comte on 'Interior Observation.'" *Journal of the History of Philosophy* 27, no. 4 (1989): 559–72.

Scharff, Robert C. "Positivism, Philosophy of Science, and Self-Understanding in Comte and Mill." *American Philosophical Quarterly* 26, no. 4 (1989): 253–68.

Schickore, Jutta. "More Thoughts on HPS: Another 20 Years Later." *Perspectives on Science* 19, no. 4 (2011): 453–81.

Schkolnyk, Claude. *Victoire Tinayre, 1831–1895: Du socialisme utopique au positivisme prolétaire*. Paris: L'Harmattan, 1997.

Schmaus, Warren. "The Concept of Analysis in Comte's Philosophy of Mathematics." *Philosophy Research Archives* 8 (1982): 205–22.

Schmaus, Warren. *Durkheim's Philosophy of Science and the Sociology of Knowledge: Creating an Intellectual Niche*. Chicago: University of Chicago Press, 1994.

Schmaus, Warren. "Hypotheses and Historical Analysis in Durkheim's Sociological Methodology: A Comtean Tradition." *Studies in History and Philosophy of Science* 16, no. 1 (1985): 1–30.

Schmaus, Warren. "Kant's Reception in France: Theories of the Categories in Academic Philosophy, Psychology, and Social Science." *Perspectives on Science* 11, no. 1 (2003): 3–34.

Schmaus, Warren. "Lévy-Bruhl, Durkheim, and the Positivist Roots of the Sociology of Knowledge." *Journal of the History of the Behavioral Sciences* 32, no. 4 (1996): 424–40.

Schmaus, Warren. "A Reappraisal of Comte's Three-State Law." *History and Theory* 21, no. 2 (1982): 248–66.

Schmaus, Warren. "Renouvier and the Method of Hypothesis." *Studies in History and Philosophy of Science* 38, no. 1 (2007): 132–48.

Schmaus, Warren. "Rescuing Auguste Comte from the Philosophy of History." *History and Theory* 47, no. 2 (2008): 291–301.

Schmaus, Warren. *Rethinking Durkheim and His Tradition*. Cambridge, UK: Cambridge University Press, 2004.

Schöttler, Peter. "From Comte to Carnap. Marcel Boll and the Introduction of the Vienna Circle." *Revue de synthèse* 136, no. 1–2 (2015): 207–36.

Scott, John. *Social Theory: Central Issues in Sociology*. London: Sage, 2006.

Sebestik, Jan. "Thomas Garrigue Masaryk ou le positivisme détourné." *Revues d'histoire des sciences humaines* 8, no. 1 (2003): 103–23.

Segond, Louis A. "Examen historique de la méthode suivie jusqu'à ce jour dans l'étude de l'organisation des animaux et exposition d'un plan définitif d'anatomie humaine." *Comptes rendus des séances et mémoires de la Société de Biologie, 1ère année, Mémoires*, 12–32. Paris, 1850.

Seys, Pascale. *Hippolyte Taine et l'avènement du naturalisme: Un intellectuel sous le Second Empire*. Paris: L'Harmattan, 1999.

Sherman, John W. *The Mexican Right: The End of Revolutionary Reform, 1929–1940*. Westport, CT: Praeger, 1997.

Shapin, Steven. *The Scientific Revolution*. Chicago: University of Chicago Press, 1996.

Shiner, Larry. *The Invention of Art: A Cultural History*. Chicago: University of Chicago Press, 2001.

Shorris, Earl. *The Life and Times of Mexico*. New York: W. W. Norton, 2004.

Shroder, Maurice Z. *Icarus: The Image of the Artist in French Romanticism*. Cambridge, MA: Harvard University Press, 1961.

Simon, Walter M. *European Positivism in the Nineteenth Century: An Essay in Intellectual History*. Ithaca, NY: Cornell University Press, 1963.

Singer, Michael. *The Legacy of Positivism*. New York: Palgrave Macmillan, 2005.

Smith, Barry. *Austrian Philosophy: The Legacy of Franz Brentano*. Chicago: Open Court, 1996.

Smith, Pamela H. *The Body of the Artisan: Art and Experience in the Scientific Revolution*. Chicago: University of Chicago Press, 2004.

Soltau, Roger. *French Political Thought in the Nineteenth Century*. New Haven, CT: Yale University Press, 1931.

Spektorowski, Alberto. *The Origins of Argentina's Revolution of the Right*. Notre Dame, IN: University of Notre Dame Press, 2003.

Spencer, Herbert. "The Genesis of Science." 1854. In *Essays, Scientific, Political, and Speculative*, 158–227. London, 1858.

Sperber, Jonathan. *Karl Marx: A Nineteenth-Century Life*. New York: Liveright, 2013.

Spurzheim, Johann Gaspard. *Observations sur la phraenologie*. Paris, 1818.

Spurzheim, Johann Gaspard. *Outlines of Phrenology*. London, 1829.

Stanguennec, André. *Ernest Renan: De l'idéalisme au scepticisme*. Paris: Honoré Champion, 2015.

Staten, Clifford L. *History of Nicaragua*. Westport, CT: Greenwood, 2010.

Steinmetz, George. "Introduction: Positivism and Its Others in the Social Sciences." In *The Politics of Method in the Human Sciences: Positivism and Its Epistemological Others*, edited by George Steinmetz, 1–56. Durham, NC: Duke University Press, 2005.

Stengers, Isabelle. "Intervention." In *Positivismes: Philosophie, Sociologie, Histoire, Sciences*, edited by Andrée Despy-Meyeter and Didier Devriese, 13–34. Turnhout: Brepols, 1999.

Sullivan, Cath. "Method and Theory in Qualitative Research." In *Doing Qualitative Research in Psychology: A Practical Guide*, edited by Michael A. Forrester, 15–38. London: Sage, 2010.

Sutton, Michael. *Nationalism, Positivism and Catholicism: The Politics of Charles Maurras and French Catholics, 1890–1914*. Cambridge, UK: Cambridge University Press, 1982.

Szporluk, Roman. *The Political Thought of Thomas G. Masaryk*. Boulder, CO: East European Monographs, 1981.

Tannery, Paul. *Recherches sur l'histoire de l'astronomie ancienne*. 1892. Reprint, Paris: Gabay, 1995.

Tatarkiewicz, Wladysław. *Outline of the History of Philosophy in Poland to World War II*. Translated by Christopher Kasparek. San Francisco: Polish Arts and Culture Foundation, 1948.

Taylor, Keith, ed. *Henri Saint-Simon (1760–1825): Selected Writings on Science, Industry, and Social Organization*. New York: Holmes and Meier, 1975.

Taylor, Michael W. *The Philosophy of Herbert Spencer*. London: Continuum, 2007.

Thibert, Marguerite. *Le rôle social de l'art d'après les Saint-Simoniens*. Paris: Librairie des sciences économiques et sociales, Marcel Rivière, n.d.

Thompson, Ken. *Emile Durkheim*. Rev. ed. London: Routledge, 2002.

Tocqueville, Alexis de. *L'Ancien Régime et la Révolution*. 1856. Vol. 2, bks.1–2, of *Œuvres complètes d'Alexis de Tocqueville*. Paris: Gallimard, 1952.

Tocqueville, Alexis de. *De la démocratie en Amérique*. 1835–1840. Vol. 1, bks.1–2, of *Œuvres complètes d'Alexis de Tocqueville*. Paris: Gallimard, 1951.

Tocqueville, Alexis de. *Democracy in America*. Translated by Henry Reeve. Cambridge, MA: Sever and Francis, 1863.

Tocqueville, Alexis de. *Sur l'Algérie*. Paris: Garnier Flammarion, 2003.

Tosh, John. *The Pursuit of History*. 3rd ed. London: Longman, 2002.

Tresch, John. "The Order of the Prophets: Series in Early French Social Science and Socialism." *History of Science* 48 (2010): 315–42.

Tresch, John. *The Romantic Machine: Utopian Science and Technology after Napoleon*. Chicago: University of Chicago Press, 2012.

Turgot, Anne-Robert-Jacques. *Turgot on Progress, Sociology, and Economics*. Translated and edited by Ronald L. Meek. Cambridge, UK: Cambridge University Press, 1973.

Turner, Jonathan. "Positivism." In *Encyclopedia of Sociology*, edited by Edgar F. Borgatta and Marie L. Borgatta, vol. 3, 2192–95. New York: Macmillan, 1992.

Turner, Jonathan. "Positivism: Sociological." In *International Encyclopedia of the Social and Behavioral Sciences*, edited by Neil Smelser and Paul Bates, vol. 17, 11827–31. New York: Elsevier, 2001.

Turner, Jonathan, Leonard Beeghley, and Charles H. Powers. *The Emergence of Sociological Theory*. 7th ed. Los Angeles: Sage, 2012.

Uebel, Thomas. "Political Philosophy of Science in Logical Empiricism: The Left Vienna Circle." *Studies in History and Philosophy of Science* 36, no. 4 (2005): 754–73.

Varley, Karine. *Under the Shadow of Defeat: The War of 1870–71 in French Memory*. New York: Palgrave Macmillan, 2008.

Vernon, Richard. "Auguste Comte and the Withering-Away of the State." *Journal of the History of Ideas* 45, no. 4 (1984): 549–66.

Walicki, Andrzej. *A History of Russian Thought: From the Enlightenment to Marxism*. Translated by Hilda Andrews-Rusiecka. Stanford, CA: Stanford University Press, 1979.

Wartelle, Jean-Claude. *L'Héritage d'Auguste Comte: Histoire de "l'église" positiviste, 1849–1946*. Paris: L'Harmattan, 2001.

Watson, Peter. *Ideas: A History of Thought and Invention, from Fire to Freud*. New York: HarperCollins, 2005.

Weinberg, Adelaide. *The Influence of Auguste Comte on the Economics of John Stuart Mill*. London: E. G. Weinberg, 1982.

Weinmann, Heinz. "Galileo Galilei: de la précision à l'exactitude." *Études françaises* 19, no. 2 (1983): 9–26.

Wernick, Andrew. "Auguste Comte (1798–1857)." In *The Cambridge Dictionary of Sociology*, edited by Bryan S. Turner, 85–86. Cambridge, UK: Cambridge University Press, 2006.

Wernick, Andrew. *Auguste Comte and the Religion of Humanity: The Post-Theistic Program of French Social Theory*. Cambridge, UK: Cambridge University Press, 2001.

Wernick, Andrew. "Comte, Auguste." In *Encyclopedia of Social Theory*, edited by G. Ritzer, vol. 1, 128–34. London: Sage Publications, 2005.

Whewell, William. *History of the Inductive Sciences, from the Earliest to the Present Time*. London, 1837.

Whewell, William. *On the Philosophy of Discovery*. 1860. Reprint, New York: Burt Franklin, 1971.

Whewell, William. *Philosophy of the Inductive Sciences, Founded upon Their History*. London, 1840.

Wiarda, Howard J. *The Soul of Latin America: The Cultural and Political Tradition*. New Haven, CT: Yale University Press, 2001.

Wils, Kaat. *De omweg van de wetenschap: het positivisme en de Belgische en Nederlandse intellectuele cultuur, 1845–1914*. Amsterdam: Amsterdam University Press, 2005.

Wils, Kaat. "Les sympathisants de Comte et la diffusion du positivisme aux Pays-Bas (1845–1880)." In Petit, *Auguste Comte: Trajectoires positivistes*, 333–50.

Wright, Terence R. *The Religion of Humanity: The Impact of Comtean Positivism on Victorian Britain*. Cambridge, UK: Cambridge University Press, 1986.

Wunderlich, Roger. *Low Living and High Thinking at Modern Times*. Syracuse, NY: Syracuse University Press, 1992.

Zammito, John H. *A Nice Derangement of Epistemes: Post-Positivism in the Study of Science from Quine to Latour*. Chicago: University of Chicago Press, 2004.

Young, Robert M. *Mind, Brain, and Adaptation in the Nineteenth Century: Cerebral Localization and Its Biological Context from Gall to Ferrier*. Oxford: Clarendon, 1970.

Zea, Leopoldo. "Positivism." In *Major Trends in Mexican Philosophy*, edited by Mario de La Cueva and Miguel Leon-Portilla, translated by A. R. Caponigri, 220–45. Notre Dame, IN: University of Notre Dame Press, 1966.

Zea, Leopoldo. "Positivism and Porfirism in Mexico." In *Latin American Philosophy: An Introduction with Readings*, edited by Susana Nuccetelli and Gary Seay, 198–218. Upper Saddle River, NJ: Prentice Hall, 2004.

Zea, Leopoldo. *Positivism in Mexico*. Translated by Josephine H. Schulte. Austin: University of Texas Press, 1974.

Zea, Leopoldo. *The Latin-American Mind*. Translated by James H. Abbott and Lowell Dunham. Norman: University of Oklahoma Press, 1963.

List of Contributors

Michel Blay, emeritus senior researcher at the Centre National de la Recherche Scientifique (CNRS), is a philosopher and historian of science. His main research areas are the mathematization of physics during the seventeenth and eighteenth centuries as well as the history of optics. Since 2010 he has headed the Committee for the History of the CNRS. He has published numerous works, including *Reasoning with the Infinite: From the Closed World to the Mathematical Universe* (University of Chicago Press, 1998); *Penser avec l'infini de Giordano Bruno aux Lumières* (Vuibert, 2010); and *Les ordres du Chef—Culte de l'autorité et ambitions technocratiques: Le CNRS sous Vichy* (Armand Colin, 2012).

Michel Bourdeau is currently emeritus director of research at the Institute for the History and Philosophy of Science and Technology. Before joining the Centre National de la Recherche Scientifique, in 1990, he taught at colleges and universities in France and abroad. He is author of *Les trois états: science, théologie et métaphysique chez Comte* (Editions du Cerf, 2006) and *Auguste Comte, Science et société* (SCEREN, 2013). Along with Laurent Clauzade and Frédéric Dupin, he also edited Comte's *Cours de philosophie positive: leçons 46–51* (Hermann, 2012).

Anastasios Brenner is professor of philosophy at the Université Paul-Valéry Montpellier and member of the Center of Interdisciplinary Research in Human and Social Sciences (CRISES). He is a specialist of French philosophy of science, and has explored the origins and development of this tradition as well as its interaction with logical empiricism. He has also examined the issue of rational values from the perspective of historical epistemology. Among his main publications are *Duhem: Science, réalité et apparence* (Vrin, 1990), *Les origines françaises de la philosophie des* *sciences* (PUF, 2003), *Raison scientifique et valeurs humaines* (PUF, 2011), and with Jean Gayon, *French Studies in the Philosophy of Science* (Springer, 2009).

Laurent Clauzade is associate professor of philosophy at the University of Caen-Normandy and associate researcher at the Institute for the History and Philosophy of Science and Technology, Paris. He is also treasurer of the international association Maison d'Auguste Comte. He is the author of *L'organe de la pensée: biologie et philosophie chez Auguste Comte* (2009). In collaboration with Michel Bourdeau and Frédéric Dupin, he has published an edition of the fourth volume of Comte's *Cours de philosophie positive*, and he is currently working on an edition of Comte's *System of Positive Polity*.

Vincent Guillin is currently associate professor at the University of Quebec, where he teaches the history and philosophy of the human sciences, social philosophy, and modern philosophy. His research focuses on nineteenth-century philosophy, and most notably Auguste Comte and John Stuart Mill, the use of comparison and experimentation in social science, and the democratic governance of science. He is the author of *Auguste Comte and John Stuart Mill on Sexual Equality* (Brill, 2009).

Jean Elisabeth Pedersen is associate professor of history at the University of Rochester and the author of *Legislating the French Family: Feminism, Theater, and Republican Politics, 1870–1920* (Rutgers University Press, 2003). Her articles and essays on Auguste Comte, Émile Durkheim, and the history of French sociology have appeared in the *Journal of the History of the Behavioral Sciences* (1998); *SIGNS: Journal of Women in Culture and Society* (2001); *Teaching Durkheim*, edited by Terry Godlove (Oxford University Press, 2005); and *Le pouvoir du genre: Laïcités et religions 1905–2005*, edited by Florence Rochefort (Presses universitaires du Mirail, 2007).

Mary Pickering is professor of modern European history at San Jose State University, specializing in cultural/intellectual history, social history, and women's history. She has written a three-volume work titled *Auguste Comte: An Intellectual Biography* (Cambridge University Press, 1993–2009). Her articles have appeared in the *Journal of the History of Ideas*, *French Historical Studies*, *Journal of Women's History*, *Historical Reflections*, *Revue philosophique*, *Revue internationale de philosophie*, and *Revue interdisciplinaire d'études juridiques*.

Warren Schmaus is professor of philosophy at the Illinois Institute of Technology in Chicago. He conducts research on the history and philosophy of science in nineteenth- and twentieth-century France, and has published two monographs on Émile Durkheim—*Durkheim's Philosophy of Science and the Sociology of Knowledge* (Chicago, 1994) and *Rethinking Durkheim and His Tradition* (Cambridge, 2004)—and articles on Comte, Victor Cousin, Maine de Biran, Charles Renouvier, and Lucien Lévy-Bruhl. He is a past president of HOPOS: The International Society for the History of Philosophy of Science, and has served as the book review editor of its journal, *HOPOS*.

Andrew Wernick is emeritus professor of cultural studies and sociology at Trent University in Canada and a life member of Clare Hall, Cambridge. A social theorist, intellectual historian, sociologist of culture, and jazz musician, he is the author of more than seventy essays on contemporary culture and cultural/social theory. His writings include *Promotional Culture: Advertising, Ideology and Symbolic Expression*; *Auguste Comte and the Religion of Humanity: the Post-theistic Project of French Social Theory*; and the coedited anthologies *Shadow of Spirit: Religion and Postmodernism* and *Images of Aging: Cultural Representations of Later Life*. His current work is on nihilism and the sacred, national imaginaries, and the political theory of the gift.

Index

abolitionism, 286–88, 297–98
academic philosophy, French, 6
Académie des Sciences, 13, 158
Action Française, 9
Adams, Henry, 302
Adams, Herbert Baxter, 302
affect. *See* emotions
Alberdi, Juan Bautista, 293–94
Allen, Joseph Henry, 279
altruism, 16, 116, 222, 232; egoism *vs.*, 170, 219; evolution toward, 105–6, 153; faculties for, 123–24; strengthening, 223–24
Amane, Nishi, 298
analysis, in mathematics, 45–46
analytical mechanics, 56
Ancien Régime, 164–65, 242; fall of, 128, 132, 221
Angiulli, Andrea, 270
animals, 110, 227; humans' relations with, 167, 185–86
anthropology, 261–62
Appel aux conservateurs, 16
Aragón, Augustin, 290, 292, 298
Arbousse-Bastide, Paul, 288
Ardigò, Roberto, 270
Argentina, positivism's influence in, 293–94
Arguedas, Alcides, 295
aristocracy, 286
Aristotle, 139, 168
Arlès, Philippe, 265
arts, 13, 190, 198, 203; Comte's multiple uses of term, 194–95; in education, 199–200; gender and, 191–94, 215; observation and imagination in, 196–97; relations among, 195–96; relation to affective life and gender, 191–94; relation to science, 192, 194, 196, 200–202, 213–14
Asia, positivism in, 298–300
Asociación Metodófila, 290
Aspiazu, Agustín, 295

astronomy, 34, 89; criticisms of Comte's, 83–84; history of, 8, 76–77; hypotheses and observation in, 46–48, 82–84; importance to Comte, 72–73, 75; positive state of, 39–40; precision of, 85–86; promoting positivism through, 79–80; religion *vs.*, 78–79
astronomy lectures: workers at, 77–78
astronomy lectures, Comte's, 81; goals of, 78, 81; popularity of, 14, 72; publication of, 14 (see also *Cours de philosophie positive*; *Traité philosophique d'astronomie populaire*); workers at, 73, 80–81
Atatürk, Mustafa Kemal, 9, 278, 304
atomic theory, 49–50, 259
Audiffrent, Georges, 345n16
Austin, John, 303
Austria, 272
authoritarianism, 294; Comte criticized for, 10, 266, 278; Latin Americans not put off by, 285, 291–93, 297; positivism and, 266, 293

Bachelard, Gaston, 9, 28, 87–88, 91, 261
Bacon, Francis, 238, 243–44
Báez, Cecilio, 295, 304
Bain, Alexander, 267
Bakunin, Mikhail, 276
Balmaceda, José Manuel, 293
Barreda, Gabino, 288–90
Barrès, Maurice, 193, 263
Barrios, Justo Rufino, 295–96, 304
Barthez, Paul-Joseph, 116
Batlle y Ordóñez, José, 295, 304
Bazard, Claire, 206–7
Bazard, Saint-Amand, 205–7, 243
Beesly, Edward Spencer, 267–68, 271
Beiser, Frederick, 271
Belgium, 269, 304
Bellamy, Edward, 281
Bensaude-Vincent, Bernadette, 212, 260

393

Bentham, Jeremy, 303
Berkeley, George, 7
Bernard, Claude, 88, 108, 259, 262, 265
Bernoulli, Jacob, 157
Berthelot, Marcelin, 259–60
Bertrand, Joseph, 83
Besant, Annie, 268, 300
Bhattacharyya, Ram Kamal and Krishna Kamal, 298
Bichat, Marie François Xavier, 98–103, 108, 111–12, 120
Bilbao, Francisco, 292
Binet, Jacques, 210
biocracy, 120, 185–86
biographies, of Comte, 11
biology, 34, 95, 209; architectonics of, 93, 96, 103–9; *Cours* on, 93, 104–17, 126; definition of, 99–100; development of terminology in, 94–95; hierarchy in, 109–11; influence of tissular anatomy on, 100–103; physiology and, 95, 121; sociology compared to, 50–51, 101, 147, 150; sociology's relation to, 124–26, 144, 151, 153–55, 325n142; *Système de politique positive* on, 104–5, 117–20, 126–27; unity in, 103–4, 116–17
biotaxy, 95–96, 107–11
Blainville, Henri Marie Ducrotay de, 98, 116, 147, 210; biology terminology by, 94–95; influence on Comte, 94–97, 105; modifying Bichat's theories, 101–3
Blanchard, Calvin, 279
Blanckaert, Claude, 262
Bliot, Sophie, 210
Bois-Reymond, Emil du, 270
Bolivia, 295
Boll, Marcel, 273
Bolsheviks, similarity to Russian positivists, 277
Bonald, Louis Gabriel Ambroise de, 246
Boutroux, Émile, 8, 43, 258–59
Boyer, Rosalie, 210
brains, 31–32, 120–24, 223
Brandão, Francisco Antonio, 284
Branting, Karl Hjalmar, 269, 304
Brasileira, Nísia Floresta, 284
Braunstein, Jean-François, 31
Brazil, 10, 284–88
Brenner, Anastasios, 257, 261, 346n44
Brentano, Franz, 272
Bridges, J. H., 267–68
Britain, 266–69, 304
Broca, Paul, 261–62
Broglie, Louis de, 71
Broussais, François, 33, 107, 143, 210
Brownson, Orestes, 280

Brunschvieg, Léon, 28, 87
Buchholz, Friedrich, 270
Buckle, Henry Thomas, 267, 302
Bulnes, Francisco, 291
Bulwer-Lytton, Edward, 266

Cabanis, Pierre Jean Georges, 114–15, 120, 143
calculus, 56
Canguilhem, Georges, 9, 28, 261
Carnap, Rudolf, 3, 272, 349n115
Cartesian rationalism, 6
Casanova, Jean-Claude, 188–89
Castilhos, Júlio de, 287, 304
Catéchisme positiviste, 15, 170
The Catechism of Positive Religion, 218
Catholic church, 242; blamed for backwardness of Latin America, 283, 292; efforts to reduce influence of, 242, 272, 284; influence on education, 264, 296; positivism used against, 19–21, 286, 288–89, 291–92, 295, 304; Religion of Humanity to replace, 236–37; virtues of, 237–38
Cattaneo, Carlo, 269
causation, positivism rejecting hypotheses of, 22, 30, 48, 60–61
centralization, 166–67, 182
Cercle des Prolétaires Positivistes, 263
Channing, William Henry, 280
Chapel of Humanity, 287
Charlton, D. G., 259
Chartier, Émile ("Alain"), 265
Chatterjee, Bankim Chandra, 299
Chatterjee, Guru Das, 299
chemistry, 34, 41
Chernyshevsky, Nikolai, 275
Child, Lydia Maria, 280
Chile, 292–93
Chimisso, Cristina, 28
Chmielowski, Piotr, 274
Christianity, 242, 244. *See also* Catholic church; religion
Church of Humanity, 267–68, 299
Científicos, 291–92
class conflict, 221
classification, 178–80; in biology, 96, 109; Comte's system of, 301; in philosophy, 96; of phrenological faculties, 123; of sciences, 13, 21–22, 34, 39, 43, 135, 168, 172, 252, 269, 290, 301; social, 173, 177–78; of species, 22
Clayton, Lawrence, 297
Clemenceau, Georges, 265, 304
Coen, Deborah, 272
Cointet, Jean-Paul, 256
Collège de France, 13, 254

Index

colonialism, 283; effects of, 294, 298; positivism used to overcome, 286, 289, 291, 296
colonization, 167, 183–84
Comité Positivista Argentino, 294
commemoration, in maintaining society, 20, 228–29
commonsense philosophy, Scottish, 6
communism, 168, 176–77
comparative method, 108–11, 151–52
Condillac, Étienne Bonnot de, 114–16
Condorcet, Nicolas de, 12, 140–41, 157, 237
Congreve, Richard, 267, 285, 298
Conniff, Michael, 297
"Considerations on the Spiritual Power," 238
Considérations philosophiques sur les sciences et les savants, 37
Considérations sur le pouvoir spirituel, 164, 168
Constant Botelho de Magalhães, Benjamin, 285–86
continuity/discontinuity, 58–59, 63–64, 110
Contreras Elizalde, Pedro, 288
Corra, Émile, 255
correspondence, Comte's, 14–15, 17
Cosmes, Francisco G., 291
Cotton, Henry, 298–99, 300
Coulanges, Fustel de, 302
Cournot, Antoine Augustin, 8, 40
Cours de philosophie positive, 5, 13, 19, 37, 60, 88, 134, 202; on arts, 198–99, 201; on astronomy, 73–75, 78–79; on biology, 93, 104–17, 126; compared to Whewell's *Philosophy of the Inductive Sciences*, 6; on gender, 190–91, 208–9, 215; law of three stages in, 135; on mathematical analysis, 45–46; on philosophy, 156, 163; reception of, 13, 251; on science, 64, 67, 300; on scientific methods, 84–85; scientific positivism in, 73; on sociology, 129, 139, 147, 158–59, 325n122; *Système de politique positive* and, 104–5, 110, 153, 194, 198, 213; *Système de politique positive* compared to, 217–18, 230, 240–41; translations of, 14, 267, 279, 298
Cousin, Victor, 6, 29–30, 32–33, 200, 258
Croly, David, 281
Croly, Herbert, 10, 281
Crosland, Maurice, 259–60
Courbertin, Pierre de, 265
Cult of Humanity, 16
Curie, Marie, 274
Cuvier, Georges, 106, 110
Czechs, Comte's influence on, 9–10

d'Alembert, Jean le Rond, 64–65, 85
Darwin, Charles, 44
Darwin, Erasmus, 94
Daston, Lorraine, 88–89

David, Menno, 269
death, Comte's, 17
d'Eichthal, Gustave, 90
Delambre, Jean-Baptiste, 76
Deroin, Jeanne, 207
Deroisin, Hippolyte Philémon, 264
Descartes, René, 29, 67, 88, 156
Destutt de Tracy, Antoine, 116
Díaz, Porfirio, 290–92, 304
Diaz Covarrbias, Francisco and José, 289
Diderot, Denis, 199
Dilthey, Wilhelm, 270
Discours sur l'ensemble du positivisme, 16, 168, 191; on arts, 195, 198–99, 201; astronomy lectures published as, 14, 72, 81; on intellect *vs.* affect, 202–4
Discours sur l'esprit positif, 5, 35–36, 81, 86–87
division of labor, 169, 186, 221
Dobroliubov, Nikolai, 275–76
Draper, John, 302
Duhem, Pierre, 69–71, 260, 261
Dühring, Eugen, 272
Durkheim, Émile, 9, 20, 193, 262, 265

Earth, Humanity's dependence on, 180–81
Echeverría, Esteban, 293
eclectic spiritualism, 6, 29–33, 144, 307n23
École Polytechnique, 5; Comte as teaching assistant at, 13, 32n152, 238; Comte attending, 11, 238; Comte dismissed by, 14, 238
economics, 171. *See also* political economy
economy, industrial, 245
Edger, Henry, 263–64, 280–81
education, 285, 300; arts in, 199–200; Catholic influence on, 264, 284, 296; Comte on teaching mathematics, 13–14; Comte's, 6, 11; for elite leaders, 277, 290; expansion of, 295–96; in functions of spiritual power, 176–80; goals of, 178–79; as key to progress, 289–90; leadership of, 12, 235–36; moral, 221–22, 242; reform of, 238, 293, 294, 295; in science, 10–11, 79, 171; universal, 177, 264, 274, 277; by women, 211–12, 232
egoism, 153, 170, 219, 223–24, 340n22
Einstein, Albert, 71
Eliot, George, 268
elites, 282, 284–85, 297, 299; education for, 277, 290; leadership by, 132–33, 266, 273–74, 277, 291
emotions, 143, 202; Comte's focus on, 18, 201; gender differences in, 205–13; power of, 17–18; reason *vs.*, 202–4; relation to gender and arts, 191–94; women linked to, 204–5, 215; women's superiority in, 191, 209

empiricism, 47, 85, 144; positivism and, 250, 259–60, 273, 302–3
Enfantin, Prosper, 205–7, 243–45, 336n77
English Positivist Committee, 267–68
Enlightenment, as metaphysical system, 131
epistemology, historical, 28
equality: gender, 192, 206–7, 209; racial, 287–88; *vs.* subordination, 177–78
Espinas, Alfred, 262, 265
Esquirold, Jean-Étienne, 210
ethics, 116, 126, 327n13, 331n90
Euler, Leonhard, 80, 85
Europe: centrality of, 182–83, 226, 294; positivism in, 269–78
evolution, 105–6, 147, 152–53; historical, 149–50; intellectual, 154–55
Ewerbeck, August Hermann, 270
experimentation, 50, 107, 151

family, 191, 205; gender relations and, 208–11; as unit of society, 247, 265. *See also* marriage and family
Ferguson, Adam, 144, 171, 178
Ferrari, Giuseppe, 270
Ferri, Enrico, 270
Ferry, Jules, 264, 265, 304
fetishism, 20, 239
Feuerbach, Ludwig, 241–42
finances, 14–15, 253, 266
Fiske, John, 279, 302
Fitzhugh, George, 279
Flammarion, Camille, 75
Flint, Robert, 40
Flórez, José Segundo, 283
Fontenelle, Bernard le Bovier de, 74
Forbes, Geraldine, 304
force, 205, 219; government resting on, 10, 22, 166, 171, 174–76; manifestations of, 175–76
Foucault, Michel, 9, 92, 346n51
Fourier, Joseph, 210
Fournel, Henri, 206
France, 5, 236, 264, 283; Comte's influence in, 9, 253–66; Comte's proposal for intendancies in, 167, 182
France, Anatole, 255
Francelle, Auguste, 79
Frank, Philipp, 3, 272
Frankfurt School, 273
French Idéologues, 12, 144
French Revolution, 11, 237; effects of, 128, 165, 221, 248; search for stability after, 131–32, 190, 221, 242, 245
Fresnel, Jean, 70

Galileo, 86
Galison, Peter, 88–89
Gall, Franz-Joseph, 120–23, 143, 202, 222
Gambetta, Léon, 264, 304
Gane, Mike, 192
Garofalo, Raffaele, 270
Gauthier, Théophile, 200
Geddes, James, 298–99
Geddes, Patrick, 268
gender: Comte's dedications and acknowledgments, 210; in *Cours*, 190–91; equality/inequality of, 192, 206–7, 209; relation to affective life and arts, 191–94; women subordinated to men, 202, 209, 211–12
gender differences, 215, 234; in capacities, 204–5, 209; in instinctual apparatus, 231–32; reason *vs.* emotions, 205–13
gender relations: complementarity of, 207–8; Comte's continuity of thought on, 214–15
gender roles, 232; in marriage and family, 209–12; in Religion of Humanity, 215–16
geometers, 156–58
Geometrical Analysis (Leslie), 46
Germany, 241–42, 251, 270–72
Ghosh, Girish Chunder, 299
Gillespie, William Mitchell, 279
Gissing, George, 268
Gökalp, Ziya, 278
Gomperz, Theodor, 272
Gouhier, Henri, 11
government, 10, 179; as an art, 170–71, 187; limits of, 171–73; in positive politics, 166–70; relation to society, 167–68, 186; roles of, 186–87
Great Being, 184, 225, 341n34; Humanity as, 225–33, 248
Guardia, Tomás, 296, 304
Guatemala, 295
Guindorf, Reine, 207
Guizot, François, 264
Gumplowicz, Ludwig, 274
Guzmán Blanco, Antonio, 295, 304

Haarbleicher, André, 255
Hacking, Ian, 7, 40–41, 301
Hahn, Hans, 3
Halbwachs, Maurice, 20
Hamilton, William, 32
Hardy, Thomas, 268
Harrison, Frederic, 267–68
Harvey, Joy, 261
Hayek, Friedrich, 36, 189, 240, 328n19
Hegel, Georg, 241–42
Heidegger, Martin, 247
Helmholtz, Hermann von, 88, 270

Index

Helvétius, Claude Adrien, 105
Hennequin, Émile, 256
Herschel, John, 4, 5, 27, 266
historicism, 134, 137
history, 131, 137, 152, 182; memorialization of the past, 228–29; positive, 270, 302
history of science, 33, 254, 264; astronomy and, 76–77; Comte on, 8–9, 27–28, 300; as new discipline, 261, 300; relation to philosophy of science, 28, 54
Hobbes, Thomas, 164
Hoff, Jacobus H. van't, 260
Holmes, George Frederick, 278–79
Houellebecq, Michel, 188, 263
Howard, Don, 3
Howe, Julia Ward, 280
Hughes, Henry, 279
humanism, 241–42, 246, 251–52
humanity: cult of, 248, 267–68, 299; fragility of, 246–47; as Great Being, 225–33, 248; love for, 239, 264
human nature, 152, 155, 257; errors in depictions of, 143, 146; theory of, 119–21, 124–26
Humboldt, Alexander von, 210
Hume, David, 7, 30–31, 33, 144
Hurd, Madeleine, 269
Huxley, Thomas, 88, 165, 236
Huygens, Christiaan, 61–62, 64–65, 67, 69–70
hypotheses, 54; observation and, 36, 46–50, 85; research to be guided by, 257–58; role of, 7–8, 49, 82–84; two types of, 67–68

imperialism, 266, 277
India, 298
individualism, 247; of, 221; fear of rising, 129–30; opposition to, 168, 208, 278; psychology omitted from sociology for, 141–45, 149–50; support for, 253, 292–93
industrialism, 81, 221, 245
industrialism school, 144
industrialization, 92, 128, 167, 292
Ingenieros, José, 294
instrument, astronomical, 75–76, 86
intellectualism, of sociology, 133–34, 137–38
International Positivist Society, 255
introspection, 32–33, 36
Italy, 269–70, 304

Jacobi, Friedrich, 30, 242
Jacobins, 166–67, 182, 242–43
Janet, Paul, 258
Jeannolle, Charles, 254
Jowett, Benjamin, 279

Juárez, Benito, 289, 304
Justo, Juan Bautista, 294

Kant, Immanuel, 33, 241–42; influence of, 29, 258; on objectivity, 88–90
Kareev, Nikolai, 276
Kavelin, Konstantin, 276
Keufer, Auguste, 263
kinematics, 60–67
Kitcher, Philip, 3
knowledge, 55, 239, 257; advances in, 38–39; Comte's vagueness about, 250–51; theory of, 33, 134; three-state law of, 34, 224–25; unity of, 117–18
Kovalevsky, Maksim, 276
Koyré, Alexandre, 28
Krupiński, Franciszek, 274
Kuhn, Thomas, 28, 145

Lafargue, Paul, 271
Laffitte, Pierre, 9, 254–55, 260, 264, 267–68, 277, 345n16
Lagarrigue, Juan Enrique, Jorge, and Luis, 293, 298
Lagrange, Joseph-Louis, 85
laïcité, 264, 277
Lamarck, Jean-Baptiste, 94, 105–6, 110
Langlois, Charles-Victor, 302
Laplace, Pierre-Simon, 74, 78, 85, 157
Lastarria, José Victorino, 292–93
Latin America, 10; explanations for backwardness of, 282–83, 292; positivism in, 283–98
Laudan, Larry, 35–36, 47–49
Lavisse, Ernest, 265, 302
Lavrov, Petr, 276
law of three stages, 7, 275, 308n40; criticisms of, 39–40; defense of, 36–41; in development of scientific methods, 34, 301; as historical evolution, 147, 149–50; of knowledge, 224–25; in sociology, 135, 138–39, 146
lectures, Comte's, 12, 236, 265. See also astronomy lectures; Cours
LeGouis, Catherine, 265
Leibniz, Gottfried Wilhelm, 29, 58–59, 342n48
Lemos, Miguel, 285–87, 298
Leroy, Charles Georges, 116
Le Roy, Édouard, 83, 260
Lesevich, Vladimir, 276
Leslie, John, 46
Letelier, Valentín, 292–93
Lévi-Strauss, Claude, 9
Lévy-Bruhl, Lucien, 3, 8, 28, 43, 261, 265; Comte's influence on, 9, 259

Lewes, George Henry, 267
liberalism, 188–89, 294; influence of positivism on, 10, 173, 282
life, 96–99, 118
Limantour, José, 291
Limburg Stirum, Menno David van, Count, 269
Lindenfeld, David, 272
literary criticism, proposed as science, 256, 275
Littré, Émile, 11, 276; break with Comte, 14, 187–88, 193, 240, 253; influence of, 264, 285; promoting positivism, 8, 253–55
Lobb, Samuel, 298
Locke, John, 31, 242
logical positivists (logical empiricists), 73, 272–73, 314n2, 349n115
Lombroso, Cesare, 270
London Positivist Committee, 267
London Positivist Society, 267
Longino, Helen, 3
Longueville, Henry, 32
love, 19, 228, 248

Macaulay, Thomas Babington, 5
Maćedo, Miguel, 291
Macedo, Pablo, 291
Mach, Ernst, 251, 272–73
Machado Dias, Antônio, 284
Macherey, Pierre, 140
Magendie, François, 108
Magnin, Fabien, 79–80, 263
Mahrburg, Adam, 274
Maikov, Valerian, 275
Maine de Biran, Pierre, 6, 29, 31–32
Maistre, Joseph de, 12, 165, 171, 178, 237, 246
Mantegazza, Paolo, 270
Marcuse, Herbert, 240
marriage and family, 253; Comte's, 12, 15, 208; Comte's proposals for, 205, 212; gender roles in, 208–9, 212; Saint-Simonians on, 211–12
Marselli, Niccola, 270
Martineau, Harriet, 14, 267
Marx, Karl, 187, 241–42, 271, 294
Masaryk, Tomás, 9–10, 273–74, 304
Massin, Caroline (wife), 12–13, 15, 208, 231
materialism/antimaterialism, 99, 113–14, 258
mathematics, 8; analysis in, 45–46, 157; Comte teaching, 13–14, 32n152; philosophy and, 156–57; physics and, 57–58, 71; positive state of, 39–40; relation to other sciences, 34, 156; sociology and, 157–58
Maurras, Charles, 193, 266
Mauss, Marcel, 9
McClintock, John, 278–79
McCosh, James, 279

Meinong, Alexius, 272
memorialization, 233, 249, 263–65
memory, collective, 20, 229
men: superior capacities of, 205, 215; women's role of perfecting, 231. *See also under* gender
mental illness, Comte's, 12–13, 250
metaphysical method, 34, 37, 55
metaphysical psychology, in eclectic spiritualism, 29–33
metaphysical realism, 43–45
metaphysical stage, 13, 145; collapse of, 131, 246; in law of three stages, 7, 39–40, 197, 283
metaphysical systems, 134
metaphysics, 117, 273; Comte rejecting, 73, 80, 87, 188; supporters of, 272–73
Metzger, Hélène, 28
Mexico, 288–92
Meyerson, Émile, 261
Michel, Louis, 263
Mikhailovskii, Nikolai, 276
Milhaud, Gaston, 83–84, 91
milieu, 97–100, 104–6, 255–56
military, positivism used against, 288, 291
Mill, James, 5
Mill, John Stuart, 5, 32, 50, 165, 253; on Comte, 17, 256, 265, 266; on Comte's books, 52, 79; Comte's disagreements with, 105, 192; positivism version of, 130, 266–67; split with Comte, 14, 187–88, 240, 266
Milosz, Czeslaw, 274
Milyutin, Vladimir, 275
mind, 169, 175; development of, 133–35, 137, 143–44, 153, 158–59; heart and, 201, 214; theory of, 30–32, 33
Mitter, Dwarkanath, 299
modernity, rejecting spiritual power, 173
modernization, 296; positivism seen as key to, 10, 278, 282–83, 287–88, 295, 304
Molenaar, Heinrich, 271
Monod, Gabriel, 265, 302
Montesquieu, 139–41, 172
Montucla, Jean-Étienne, 8
morality, 16, 191, 208, 260; education in, 221–22, 235, 242; gender and, 211, 232–33; positivism as doctrine of, 294–95; proposed as science, 20, 52, 259, 263; strengthening, 222–23
Morris, William, 268
Moses, Claire Goldberg, 205
motion, science of, 57–62, 312n9, 313n38, 314n50; Comte's laws of, 64–67
Mukherjee, Bhudev Chandra, 299
Mukherjee, Satish Chandra, 299
Murray, Gilbert, 268

Index

Napoleon III, 11, 16, 253, 290–91
nationalism, 247; positivism used to strengthen, 265–66, 296, 300; rise of, 142, 272
nature, 37–38, 44
Navier, Henri, 210
Netherlands, 269, 304
Neurath, Otto, 3, 272, 349n115
Newton, Isaac, 62–65, 67
New York Positivist Society, 280–81
Niboyet, Eugénie, 207
Nietzsche, Friedrich, 240, 246–47, 270
nihilism, 246–48
normativity, 52–54
novels, positivism in, 262–63, 268, 275
Nystrom, Anton, 269

objectivity, 74, 88–90, 302
observation, 60, 90, 151, 196; hypotheses and, 36, 46–50, 85; internal vs. external, 31–32
Ochorowicz, Julian, 274
O'Connell, James, 279
Oken, Lorenz, 94
optics, as positive science, 67–71
Opuscules de jeunesse, 129
order, 172; progress and, 164, 230; theory of, 168–70, 177
Orzeszkowa, Eliza, 274
Ostwald, Wilhelm, 271–72, 301
Ottoman Committee of Union and Progress (CUP), 277–78
Ottoman Empire, 277

Pardo, Manuel González, 295, 304
Pareto, Vilfredo, 270
Paraguay, 295
Parker, Theodore, 280
Parra, Porfirio, 290
Pater, Walter, 268
Paz, María de, 260
Peabody, Elizabeth, 280
Pearson, Karl, 301
Peirce, Charles, 301
Pereira Barreto, Luis, 284–86
personality, Comte's, 15
Peru, 295
Petit, Annie, 199, 212, 259–60
Petrashevsky Circle, 275
philosophy, 27, 156, 203, 251, 262; Comte's, 11, 193; gender roles in, 215–16; introspection in, 28–29; mathematics and, 156–57; science and, 86–87, 91, 253, 256–57. *See also* positive philosophy
philosophy of biology, 93
philosophy of history, 342n48

philosophy of science, 29, 163; Comte's influence on, 72–73, 261, 300–301; as descriptive vs. normative, 52–54; history of science's relation to, 28, 54; political philosophy's relation to, 4, 170–71, 187; as separate subdiscipline, 4–5, 27
phrenology, 153–54, 155, 223
physics, 136; hypotheses in, 82–83; mathematics and, 57–58, 71; as positive science, 34, 67–68
physiological phrenology, 120–21
physiology: animal and vegetal, 111–17; biology vs., 94, 95; cerebral, 111, 120–24
Pickering, Mary, 11
Pineda, Rosendo, 291
Pisarev, Dmitrii, 275
Plan des travaux scientifiques nécessaires pour réorganiser la société (Comte), 131, 134–35, 190, 201, 215; appended to *Système de politique positive*, 163–64; Saint-Simon accused of taking credit for, 12, 192
Poe, Edgar Allan, 278
Poincaré, Henri, 83, 261, 260, 265
Poinsot, Louis, 210
Poland, 274, 282–83, 304
political economy, 141–42, 144–46
political philosophy, 4, 164–65, 170–71, 187
politics, 133, 327n13; Comte's proposals for, 16–17; positivism and, 131, 163, 251–52, 263–64. *See also* positive politics
Pommier, Louis Edmond, 276
Popper, Karl, 38, 240
positive history, 270, 302
positive method, 34–35, 52, 157; definitions of, 86–87; supplanting others, 37, 53
positive philosophy, 95, 131; gender in, 190–91; observation and imagination in, 196–97; poetry and, 197–99
positive politics, 163; cosmic character of, 180–89; evaluation of, 165, 188; government in, 166–68; need for social maturation in, 134–35; observation and imagination in, 196–97; sociology and, 155–56, 160; spiritual power in, 164–66, 173–80
positive religion, 225–33, 236. *See also* Religion of Humanity
positive science, 67–71
positive social science, 139
positive sociology, 129, 147, 158; need for, 136–37; positive politics depending on, 155–56; three fundamental laws of, 118–19
positive stage, 7, 13, 39–41, 283
positivism, 195, 210, 214, 250; aesthetics in, 199–200; in Argentina, 293–94; in Asia, 298–300; characteristics of, 5–6; in Chile, 292–93;

positivism (*cont.*): components of, 73, 163, 251, 267; criticisms of, 73, 91, 257; education and, 176–77, 296; empiricism and, 302–3; evaluation of legacy of, 303–4; influence of, 250–51, 261–62, 263–64, 296; influence on liberalism, 10, 173, 282; in Latin America, 283–98; in Mexico, 288–92; multiple versions of, 130, 266–67, 300, 345n12; political stance of, 251–52; promotion of, 253–54, 254–55; as religious movement, 251; seen as key to modernization, 10, 278, 282–83, 287–88, 295, 304; spread of, 78, 298; support for, 259, 304; synonyms for, 86–87; used against Catholic church, 20–21, 264, 269, 286, 295; used to press for reforms, 282, 287; values of, 247, 298, 302; workers' acceptance of, 79–80
positivist calendar, 214–15, 218, 233–34
Positivist School of Criminology, 270
Positivist Society: Brazilian, 285, 287, 351n143; French, 16, 17, 79, 81, 250, 345n16; Indian, 298
"Positivist Subsidy," 14
Positivist Church of Brazil, 285
progress, 81, 169, 195, 295, 299; direction of, 230–31; education as key to, 289–90; order and, 164, 230
Progressive Movement, 281
progressivism, positivism supporting, 281, 286–87
Protestant church, 279, 304
Prus, Boleslaw, 274
psychology, 29–30, 34, 141–46, 262, 308n37
public, science education for, 10–11
public opinion, theory of, 176

race, 286–88, 294–95, 297–98
rationalism, 85
realism, Comte's, 43–44, 49
reality, and scientific theory, 260
reason, 18; emotions *vs.*, 202–4; gender differences in, 205–13; men's superior capacity for, 191, 205
Reid, Thomas, 6
Reisch, George, 3
relativity, of knowledge, 55
religion, 180, 183, 221, 223, 246, 264, 279, 286; astronomy *vs.*, 78–79; Comte and, 11, 18, 20, 337n88, 340n15; Comte on, 186, 257; efforts to reconcile with science, 279–80; as least popular part of positivism, 218–19, 240, 267; positivism used against, 269, 304; Religion of Humanity and, 79, 239–40; role of fictive beings in, 224–25; sciences used against, 275, 277–78

Religion of Humanity, 21, 184, 202, 219, 263, 327n13; Comte's writings on, 16, 19–20, 191; gender roles in, 215–16; influence of, 303–4; influences on, 15, 217–18, 237, 239–40; lack of support for, 278, 298; popularity of, 252, 298; positivists and, 183–84, 253, 254, 285; as posttheistic teleological humanism, 241–42; priesthood of, 19, 192, 234–36, 238; to replace traditional religion, 79, 236–37; similarity to Saint-Simon's "new Christianity," 244–46; supporters of, 267, 281, 287–88, 290; tenets of, 16–17; women's roles in, 192, 211–12, 232–33; worship in, 233–34
Renan, Ernest, 255–57, 260, 302, 345n2
Renouvier, Charles, 8, 40, 52–53, 251, 345n12
republicanism, 11, 286
Revolution of 1848, 21
Rey, Abel, 87, 261, 265
Ribbentrop, Adolphe von, 270
Ribbentrop, Marie de, 284
Ribeiro de Mendonça, Joaquim Alberto, 284
Ribeiro, Demétrio, 286
Ribot, Théodule, 262
Richard, Nathalie, 256
Richardson, Sarah, 3
Riza, Ahmed, 183, 277–78
Roberty, Eugène de, 276
Robin, Charles, 93, 256, 261–62, 264–65
Robinet, Jean-François Eugène, 11
Rodrigues, Olinde, 192
Rondon, Cândido Mariano da Silva, 287–88
Rosas, Juan Manuel de, 293–94
Ross, Edward Alsworth, 281–82
Rousseau, 242
Royer, Clémence, 254, 261–62
Royer-Collard, Pierre, 6
Russia, 183, 274–77, 282–83, 304

Saint-Simon, Henri de, 192, 206; Comte's split with, 144, 192, 208, 244; Comte writing for, 11–12, 20; "new Christianity" of, 243–46
Saint-Simonianism, 5, 183, 336n77; Comte's criticism of, 205–6; on marriage and family, 205–7, 211–12
Sarmiento, Domingo, 293–94, 304
Sarton, George, 261
Say, Jean-Baptiste, 171, 243
Scalabrini, Pedro, 294
Scharff, Robert, 32–33, 38, 52, 130
Schelling, Friedrich, 30
Scherer, Wilhelm, 272
Schlick, Moritz, 272
science of the soul, 222–25, 248

sciences, 168, 215, 312n12, 331n90; applicability of, 53–54, 83–85; arts' relation to, 192, 194, 196, 200–202, 213–14; characterizations of, 5, 35–36, 56–57, 238, 257; Comte on, 43, 90–92, 253, 257; Comte's focus on, 19, 53–54; development of, 7, 39–41, 80, 91, 135; different methods in, 50–51; dogmatic vs. historic mode of exposition in, 35–36, 41; education in, 10, 79, 171, 235; efforts to reconcile religion with, 279–80; hierarchy of, 34, 41–43, 135–36, 138–39, 141, 301; literary criticism proposed as, 256, 275; morality proposed as, 20, 52, 259, 263; philosophy and, 86–87, 253, 256–57; precision in, 86–88; relations among, 42–43, 156, 300–301; six- vs. seven-stage model of, 203–4; social character of, 7, 27, 29, 39, 55; sociology and, 129, 136, 153–55; support for, 253, 274–77, 288–90, 296–97; thinking in mode of, 5, 132–33, 300; tools in, 46–47, 60, 82–83, 178; uses of, 164, 264, 277–78, 283–84
scientificity, 85–90
scientific methods, 265, 300; in astronomy, 77, 78, 81; in biology, 107–11; Comte's influence through, 261, 266; differing among sciences, 50–51; influence of, 260, 303–4; in sociology, 146, 151–52; three-state law in development of, 34, 301; unity of, 43, 45–46
scientific socialism, 5
scientism, 276; positivism equated to, 144, 241, 250, 259–60
Scottish school, 143–44, 171
Segond, Louis Auguste, 105
Seignobos, Charles, 302
self, 116–17, 144
selfishness, 97, 116, 123, 221–22
Sémérie, Eugène, 254, 345n16
sentiments, in science of the soul, 222–25
Serres, Michel, 261
Sierra, Justo, 291–92
Simon, Jonathan, 260
Small, Albion, 281–82
Smith, Adam, 8, 144, 145, 171
Smith, Barry, 272
social dynamics, 149–50, 152, 301–2; intellectual evolution and, 154–55, 158–59; social statics vs., 4, 123
socialism, 294
social regeneration, 19–20
social sciences, Comte's influence on, 9
social statics, 125, 148–49, 166, 199, 202, 301–2; social dynamics vs., 4, 123; theory of order and, 168–70
Sociedad Positivista de Centro-América, 295
Sociedad Positivista of Mexico, 295
Société d'Anthropologie, 261–62
Société de Biologie, 93, 261
Société Française de Philosophie, 259
Société Positiviste, 182
society, 15, 190, 230; crisis following French Revolution, 131–32; emotional needs of, 18, 20; government's relation to, 167–68, 186; independence and interdependence in, 168–69, 220–21; laws of, 301–2; need for religion in, 186, 219; need for reorganization of, 132, 164–65, 187, 243; as organism, 245, 301–2; reform of, 297–98, 304; rising individualism in, 129–30; role of science in, 7, 81, 164, 297; three stages of, 13, 134
Society for Positivist Philosophy, 271
Society of Humanity, 281
sociocracy, 185–86, 285
sociology, 121, 134, 146, 160, 276, 322n57; biology compared to, 101, 147, 150–51; biology's relation to, 50–51, 124–26, 144, 325n142; development of, 139, 262, 301–3; goals for, 4, 28, 128–29, 218; on historical evolution, 149–50; history of, 138–41; intellectualism of, 133–34, 137–38; law of three stages in, 135, 138–39, 146; mathematics submitting to, 157–58; methodology in, 9, 51, 110–11, 151–52, 157, 178; psychology and political economy omitted from, 141–46; relation to other sciences, 34, 129, 153–56; as science, 41, 136, 159; in United States, 279, 281–82. See also positive sociology
Solovyov, Vladimir, 276
Solvay, Ernest, 269
Sorbonne, 259, 265, 343n82
Soto, Marco Aurelio, 296, 304
Spencer, Herbert, 40, 268–69, 294–96, 301
Spinoza, Baruch, 164
spiritualism. See eclectic spiritualism
spiritual power, 182, 187, 238, 330n65; functions of, 186, 237–38; in positive politics, 164, 165–66, 173–80; religion and, 165, 180; temporal power vs., 173–75, 186
Spurzheim, Johann, 123, 143, 202
static/dynamic mechanics, 95
statues: of Comte, 265, 290; popularity of, 265
subjective method, 51
Subjective Synthesis, 228
subjectivity, 117–19
Sumner, William Graham, 279
Supiński, Józef, 274
Sweden, 269, 304
Świętochowski, Aleksander, 274

Synthèse subjective ou Système universel des conceptions propres à l'état normal de l'Humanité, 17
Système de politique positive, 90, 112, 177, 191, 215, 298; on arts, 195, 198, 201; on astronomy, 73–75, 81; on biology, 104–5, 117–20, 126–27; *Cours* and, 104–5, 110, 153, 194, 198, 213; *Cours* compared to, 217–18, 230, 240–41; influences on, 105, 159; on intellect *vs.* affect, 202–4; on politics, 163, 188; positivist calendar in, 214–15; publication of, 14, 16, 163–64; religion in, 73, 186; on Religion of Humanity, 19–20, 217, 239

Taine, Hippolyte, 255–56, 302
Tannery, Paul, 76, 258, 261
Teixeira Mendes, Raimundo, 285–87, 298
Templo da Humanidade, 285, 287
Testament, 17
theological method, 34–35, 36–37, 55
theological stage, 13, 36, 135, 325n122; collapse of, 131, 140, 246; in law of three stages, 7, 197
theological system, 134
Third Republic: Comte's influence on, 9, 254–55; memorialization in, 264–65; positivism's influence on, 263–64
three stage-law. *See* law of three stages
Tinayre, Victoire, 263–64
tissular anatomy, 100–103, 108
Tommasi, Salvatore, 270
Tocqueville, Alexis de, 167, 171, 177, 183, 187
Traité de l'éducation universelle, 176
Traité élémentaire de géométrie analytique, 13–14, 45–46
Traité philosophique d'astronomie populaire, 35, 79, 84, 90; astronomy lectures published as, 14, 72–75, 78, 81; on scientific method, 78, 85
transcendentalists, 279–80
Treviranus, Gottfried Reinhold, 94
Turgenev, Ivan, 275
Turgot, Anne-Robert-Jacques, 308n40
Turkey, 183, 277–78

Uebel, Thomas, 3
United States, 278–82, 304
Uruguay, 294–95
utopianism, 181

Varela, José Pedro, 295
Vargas, Getúlio, 287, 304
Varignon, Pierre, 64–65
Vaux, Clotilde de, 210; Comte's relationship with, 15, 17, 193, 231; influence on Comte, 17, 159, 217–18
Venezuela, 295
Veret, Désirée, 207
Vienna Circle, 3, 130, 273, 314n2, 349n115
Villari, Pasquale, 270
Villavicencio, Rafael, 295
vitalism, 98–99, 112, 120

Wakeman, Thaddeus, 281
Ward, Lester, 281
Webb, Beatrice, 268
Webb, Sidney, 268
Weinmann, Heinz, 86
Whewell, William, 4, 6, 27, 266; criticism of Comte, 8, 40, 53–54
White, Andrew Dickson, 279
Wiarda, Howard, 297
will, 31–32, 36, 247
Williamson, Alexander, 260
women, 192, 210; Comte and, 15, 266–67; education of children by, 211–12, 232; linked to emotions, 204–5, 215; in Religion of Humanity, 211–12; rights of, 206–8; roles in society, 15, 213, 231–33; in Saint-Simonianism, 206–7, 336n77; subordinated to men, 211–12. *See also under* gender
workers, 16, 81, 221, 329n43; at astronomy lectures, 73, 77–78, 80–81; positivism and, 79–80, 269
workers' movement, 263
Wyrouboff, Grégoire, 9, 254, 276

Young, John Henry, 278
Young, Thomas, 70
Young Turks, 9, 183, 277–78, 282–83

Zaldivar, Rafael, 296, 304
Zelaya, José Santos, 296, 304
Zola, Émile, 262–63